AF830715

CARBON NANOTUBES FOR ENERGY AND ENVIRONMENTAL APPLICATIONS

CARBON NANOTUBES FOR ENERGY AND ENVIRONMENTAL APPLICATIONS

Edited by

Soney C. George, PhD

Jacob Philip, PhD

Ann Rose Abraham, PhD

A. K. Haghi, PhD

AAP | APPLE ACADEMIC PRESS

First edition published 2023

Apple Academic Press Inc.
1265 Goldenrod Circle, NE,
Palm Bay, FL 32905 USA

760 Laurentian Drive, Unit 19,
Burlington, ON L7N 0A4, CANADA

CRC Press
6000 Broken Sound Parkway NW,
Suite 300, Boca Raton, FL 33487-2742 USA

4 Park Square, Milton Park,
Abingdon, Oxon, OX14 4RN UK

© 2023 by Apple Academic Press, Inc.

Apple Academic Press exclusively co-publishes with CRC Press, an imprint of Taylor & Francis Group, LLC

Reasonable efforts have been made to publish reliable data and information, but the authors, editors, and publisher cannot assume responsibility for the validity of all materials or the consequences of their use. The authors, editors, and publishers have attempted to trace the copyright holders of all material reproduced in this publication and apologize to copyright holders if permission to publish in this form has not been obtained. If any copyright material has not been acknowledged, please write and let us know so we may rectify in any future reprint.

Except as permitted under U.S. Copyright Law, no part of this book may be reprinted, reproduced, transmitted, or utilized in any form by any electronic, mechanical, or other means, now known or hereafter invented, including photocopying, microfilming, and recording, or in any information storage or retrieval system, without written permission from the publishers.

For permission to photocopy or use material electronically from this work, access www.copyright.com or contact the Copyright Clearance Center, Inc. (CCC), 222 Rosewood Drive, Danvers, MA 01923, 978-750-8400. For works that are not available on CCC please contact mpkbookspermissions@tandf.co.uk

Trademark notice: Product or corporate names may be trademarks or registered trademarks and are used only for identification and explanation without intent to infringe.

Library and Archives Canada Cataloguing in Publication

Title: Carbon nanotubes for energy and environmental applications / edited by Soney C. George, PhD, Jacob Philip, PhD, Ann Rose Abraham, PhD, A.K. Haghi, PhD.
Names: George, Soney C., editor. | Philip, Jacob, editor. | Abraham, Ann Rose, editor. | Haghi, A. K., editor.
Description: First edition. | Includes bibliographical references and index.
Identifiers: Canadiana (print) 20220205639 | Canadiana (ebook) 20220206155 | ISBN 9781774637173 (hardcover) | ISBN 9781774638002 (softcover) | ISBN 9781003189619 (ebook)
Subjects: LCSH: Carbon nanotubes—Environmental aspects. | LCSH: Carbon nanotubes—Industrial applications.
Classification: LCC TA455.C3 C37 2023 | DDC 620.1/93—dc23

Library of Congress Cataloging-in-Publication Data

CIP data on file with US Library of Congress

ISBN: 978-1-77463-717-3 (hbk)
ISBN: 978-1-77463-800-2 (pbk)
ISBN: 978-1-00318-961-9 (ebk)

About the Editors

Soney C. George, PhD

Dean of Research and Director,
Amal Jyothi Centre for Nanoscience and Technology, Kerala, India

Soney C George, PhD, is Dean of Research and Director of the Amal Jyothi Centre for Nanoscience and Technology, Kerala, India. He is a Fellow of the Royal Society of Chemistry, London, and a recipient of "best researcher of the year" award in 2018 from APJ Abdul Kalam Technological University, Thiruvananthapuram, India. He has also received awards such as best faculty award from the Indian Society for Technical Education, best citation award from the *International Journal of Hydrogen Energy,* a fast-track award of young scientists by the Department of Science & Technology, India, and an Indian young scientist award instituted by the Indian Science Congress Association. He did his postdoctoral studies at the University of Blaise Pascal, France, and Inha University, South Korea. He has published and presented almost 200 publications in journals and at conferences. His major research fields are polymer nanocomposites, polymer membranes, polymer tribology, pervaporation, and supercapacitors. He has guided eight PhD scholars and 95 student projects.

Jacob Philip, PhD

Dean (Academic), Amal Jyothi College of Engineering, Kanjirappally,
Kottayam 686518, Kerala, India
Former Professor, Department of Instrumentation,
Cochin University of Science and Technology, Cochin 682022, India

Jacob Philip, PhD, is the Dean (Academic) at Amal Jyothi College of Engineering, Kanjirappally, Kottayam, Kerala, India, and a former Professor in the Department of Instrumentation at Cochin University of Science and Technology, India. His research has been on diverse areas of condensed matter physics, focused to elastic, thermal, piezoelectric, and optical properties of multifunctional materials. His contributions include techniques to study thermal and elastic properties of solids, such as photoacoustic and photothermal techniques, physical ultrasonics, laser-induced surface acoustic waves, sensors, etc. He was a fellow of the Alexander von Humboldt Foundation at the University of Heidelberg and a visiting scientist at several institutions in India and abroad. He has authored over 200 journal publications, two books, one edited volume, and more than 100 papers in conference proceedings. He has delivered invited lectures and presented papers at several international and national conferences. He has successfully guided 27 researchers to secure their doctoral degrees. He is a fellow/member of a number of science societies, including the Institute of Physics (UK). Dr. Philip received MSc and PhD degrees in Physics from the University of Kerala and had postdoctoral training at the University of Tennessee, Knoxville, USA.

Ann Rose Abraham, PhD

Assistant Professor, Sacred Heart College (autonomous), Thevara

Ann Rose Abraham, PhD, is currently Assistant Professor of Physics at Sacred Heart College, Thevara, Cochin, India. She is the author of ten book chapters and editor of two books. She has a number of publications to her credit in many peer-reviewed high impact journals of international repute, such as the Journal of Physical Chemistry C, Physical Chemistry Chemical Physics, New Journal of Chemistry, Philosophical Magazine, etc. She has research experience at various national institutes, including Bose Institute; SAHA Institute of Nuclear Physics; UGC-DAE CSR Centre, Kolkata; and has collaborated with various international laboratories, such as the University of Johannesburg, South Africa; Institute of Physics Belgrade; etc. She is the recipient of young researcher award in the area of physics, a prestigious forum to showcase intellectual capability. Dr. Abraham received her MSc, MPhil, and PhD degrees in Physics from the School of Pure and Applied Physics, Mahatma Gandhi University, Kerala, India. She has expertise in the field of materials science, nanomagnetic materials, multiferroics, and polymeric nanocomposites, etc.

A. K. Haghi, PhD

Professor Emeritus of Engineering Sciences, Former Editor-in-Chief, International Journal of Chemoinformatics and Chemical Engineering and Polymers Research Journal; Member, Canadian Research and Development Center of Sciences and Culture.

A. K. Haghi, PhD, is the author and editor of 200 books, as well as over 1000 published papers in various journals and conference proceedings. Dr. Haghi has received several grants, consulted for a number of major corporations, and is a frequent speaker to national and international audiences. Since 1983, he served as professor at several universities. He is the former Editor-in-Chief of the *International Journal of Chemoinformatics and Chemical Engineering* and *Polymers Research Journal* and is on the editorial boards of many international journals. He is also a member of the Canadian Research and Development Center of Sciences and Cultures (CRDCSC), Montreal, Quebec, Canada. He holds a BSc in urban and environmental engineering from the University of North Carolina (USA), an MSc in mechanical engineering from North Carolina A&T State University (USA), a DEA in applied mechanics, acoustics and materials from the Université de Technologie de Compiègne (France), and a PhD in engineering sciences from Université de Franche-Comté (France).

Contents

Contributors

Jiji Abraham
Vimala College (Autonomous) Thrissur, Kerala, India

Avinash V. Borgaonkar
Department of Mechanical Engineering, National Institute of Technology Warangal, Warangal, Telangana, India; E-mail: avi.borgaonkar@gmail.com

Anu Rose Chacko
School of Chemical Sciences, Mahatma Gandhi University, Kottayam, Kerala, India

Sijo Francis
St. Joseph's College, Moolamattom, Kerala, India; E-mail: srsijofrancis@gmail.com

Cincy George
Newman College, Thodupuzha, Kerala, India

Soney C. George
Centre for Nanoscience and Nanotechnology, Amal Jyothi College of Engineering, Kottayam, Kerala, India; E-mail: jijiabraham02@gmail.com

John C. Jimtha
Department of Forest pathology, Kerala Forest Research Institute, Peechi, Thrissur, Kerala, India; E-mail: jimthajoseph@gmail.com

Sona John
Newman College, Thodupuzha, Kerala, India

Cintil Jose
Newman College, Thodupuzha, Kerala, India

Nimitha K. C.
Vimala College (Autonomous) Thrissur, Kerala, India

Binila K. Korah
School of Chemical Sciences, Mahatma Gandhi University, Kottayam, Kerala, India

Ebey P. Koshy
St. Joseph's College, Moolamattom, Kerala, India

Beena Mathew
School of Chemical Sciences, Mahatma Gandhi University, Kottayam, India
E-mail: beenamscs@gmail.com

Sneha Mathew
School of Chemical Sciences, Mahatma Gandhi University, Kottayam, Kerala, India

Narvadeshwar
Department of Chemistry, MMV, Banaras Hindu University, Varanasi, Uttar Pradesh, India

Shital B. Potdar
Department of Chemical Engineering, National Institute of Technology Warangal, Warangal, Telangana, India

Mamatha Susan Punnoose
School of Chemical Sciences, Mahatma Gandhi University, Kottayam, India

Aparna Raj
School of Pure and Applied Physics, Mahatma Gandhi University, Kottayam, Kerala, India

Arunima Reghunadhan
International and Inter University Centre for Nanoscience and Nanotechnology, Mahatma Gandhi University, P. D. Hills, Kottayam, Kerala, India
Department of Chemistry, TKM College of Engineering, Karicode, Kollam, India

Piyush Kumar Sonkar
Department of Chemistry, MMV, Banaras Hindu University, Varanasi, Uttar Pradesh, India; E-mail:piyush.sonkar37@gmail.com, piyushchem.mmv@bhu.ac.in

Bino Thomas
St. Josephs College of Engineering and Technology, Palai, Kerala, India

Elvin Thomas
School of Environmental Sciences, Mahatma Gandhi University Priyadarsini Hills P.O. Kottayam, Kerala, India; E-mail: elvinthomas1992@gmail.com

Riju K. Thomas
Bharata Mata College, Thrikkakara, Ernakulam, Kerala, India; E-mail: rijukthomas@gmail.com

Sabu Thomas
School of Chemical Sciences, International and Inter University Centre for Nanoscience and Nanotechnology, Mahatma Gandhi University, P.D. Hills, Kottayam, Kerala, India

Remya Vijayan
School of Chemical Sciences, Mahatma Gandhi University, Kottayam, Kerala, India

Abbreviations

AAS	atomic absorption spectrometry
AC	activated carbon
AChE	acetylcholinesterase
ACNTs	aligned CNTs
AES	atomic emission spectrometry
AW	antiwear
BAP	6-benzylaminopurine
BDD	boron-doped diamond
CCVD	catalytic chemical vapor deposition
CD	cyclodextrin
CE	counter electrode
CIGS	copper indium gallium selenide
CNFs	carbon nanofibers
CNMs	carbon nanomaterials
CNPs	carbon nanopearls
CNPE	carbon nanotube paste electrode
CNSC	carbon nanotube solar cell
CNTs	carbon nanotubes
CT	cholera toxin
CTH	core–shell structures
CVD	chemical vapor decomposition
DBPs	disinfection by-products
DMC	dimethyl carbonate
DMF	*N-N*-dimethylformamide
DND	detonation nanodiamonds
DOE	Department of Energy
DPASV	differential pulse anodic stripping voltammetry
DPV	differential pulse voltammetry
DSSC	dye-sensitized solar cells
DWCNT	double-walled carbon nanotube
EC	ethylene carbonate
EDLC	electric double layer capacitor
EELS	electron energy-loss spectroscopy

EFCs	enzymatic biofuel cells
EP	extreme pressure
FET	field-effect transistors
FG	fluorographene
FHB	fusarium head blight
FSCs	flexible supercapacitors
FTIR	Fourier transform infrared spectroscopy
GAC	granular activated carbon
GaAs	gallium arsenide
GC	gas chromatography
GC	glassy carbon
GC/MS	gas chromatography with mass spectrometry
GCE	glassy carbon electrode
GO	graphene oxide
HPLC	high-performance liquid chromatography
HRP	horseradish peroxidase
HR-TEM	high resolution transmission electroscopy
HTC	hydrothermal carbonization
HTMs	hole-transporting materials
ICP-MS	inductively coupled plasma mass spectrometry
IF	inorganic fullerene-like
ITO	indium-tin oxide
LCA	life cycle assessment
LC/MS	liquid chromatography with mass spectrometry
LIBs	lithium-ion batteries
LOD	limit of detection
MCF	microbial fuel cell
MCTs	mesoporous carbon nanotubes
MECs	microbial electrolysis cells
MF	microfiltration
MFCs	microbial fuel cells
MGP	modified graphene platelets
MIP	molecularly imprinted polymers
MLG-Cu	multilayer graphene impregnated with copper
MO	methyl orange
MoDDP	molybdenum dialkyldithiophosphate
MPE-CVD	microwave plasma-enhanced chemical vapor deposition
MPI	microwave plasma irradiation

MWCNTs	multi-walled carbon nanotubes
ND	nanodiamonds
NF	nanofiltration
NIR	near-infrared
NOx	nitrogen oxides
NPs	nanoparticles
NTs	nanotubes
OLC	onion-like carbon
PAN	polyacrylonitrile
PAN-ND	polyaniline embedded nanodiamonds
PAO	polyalphaolefin
PBSU	polybutylene succinate
PC	propene carbonate
PCE	power conversion efficiency
PCBM	[6,6]-phenyl-C61-butyric acid methyl ester
Pd	palladium
PECVD	plasma-enhanced CVD
PEG	polyethylene-glycol
PEI	polyethylenemine
PET	photo-induced electron transfer
PFPE	perfluoropolyether
PHT	photo-induced hole transfer
PLD	pulsed laser deposition
PMMA	poly(methyl methacrylate)
PMT	pyrimethanil
POU	point of use
PPF	poly propylene fumarate
PPS	polyphenylene sulfide
PPy	polypyrrole
PV	photovoltaic
PVDF	poly(vinylidene fluoride)
QD	quantum dot
RBM	radial breathing mode
RFCVD	radiofrequency plasma-assisted chemical vapor deposition process
RFID	radiofrequency identification
rGO	reduced graphene oxide
RO	reverse osmosis

RP	random pyramid
SDBS	sodium dodecyl benzene sulfonate
SDS	sodium dodecyl sulfate
SEB	Staphylococcal enterotoxin B
SEM	scanning electron microscope
siRNA	small interfering RNA
SPR	surface plasmon resonance
SWCNTs	single-walled carbon nanotubes
SWNT	single-walled nanotube
TBA	tetrabutylammonium acetate
TCO	transparent conducting oxide
TEM	transmission electron microscope
THF	tetrahydrofuran
TiO_2	titanium oxide
TPD	temperature programmed desorption
UF	ultrafiltration
UF-MWNTs	urea functionalized MWCNTs
UNCD	ultra-nanocrystalline diamond
US	ultrasounds
WE	working electrode
WHO	World Health Organization
WSD	wear scar diameter
XANES	X-ray absorption near-edge structure
XPS	X-ray photoelectron spectroscopy
XRD	X-ray diffraction

Preface

This book covers the many aspects of the green applications of carbon nanotubes. Carbon nanotubes for remediating *environmental* contaminants and their utility in *green* nanocomposites are the main focus of the book. Carbon nanotubes for sensing of environmentally hazardous chemicals, water purification, environment protection, and energy are also discussed. CNTs serve as sorbents, nanofilters, sensors in food and agriculture, nanomedicinal and antimicrobial agents to remove organic and inorganic contaminants, as well as pathogenic microorganisms, in wastewater treatment. Development of highly sensitive CNT-based gas sensors for air pollution monitoring, green synthesis of carbon nanotubes, and green energy applications are addressed as well. Impacts of carbon nanotubes in environment and catalysis are also topics of discussion in the book. The book covers many aspects on carbon nanotubes for green environment. The book on green applications of carbon nanotubes is very timely and fills a need of the hour to create awareness among people to combat the damage to our environment.

CHAPTER 1

Nanodiamonds for Energy Storage

MAMATHA SUSAN PUNNOOSE and BEENA MATHEW*

School of Chemical Sciences, Mahatma Gandhi University, Kottayam, India

**Corresponding author. E-mail: beenamscs@gmail.com*

ABSTRACT

All allotropes of carbon nanomaterials are always fascinating, especially nanodiamonds (ND), due to their potential applications in the biomedical- and energy-related fields, including drug delivery, protein mimics, catalysis, photovoltaics, microelectronics, optoelectronics, and biosensing. The large surface area, hardness, chemical inertness, excellent mechanical properties, and good electrical conductance of carbon NDs have greatly contributed toward the new emerging research field of energy storage. This chapter emphasizes briefly on the structure, properties, characterizations, and synthetic strategies of diamond nanoparticles. It also focuses on the recent advancement of NDs in the energy storage devices, such as lithium-ion batteries (LIBs) and supercapacitors.

1.1 INTRODUCTION

Among the different allotropes of nanoscale carbon such as graphene, carbon nanotubes (CNTs), fullerenes and carbon dots, nanodiamonds (NDs) are of much interest. Diamond nanoparticles commonly called as NDs are diamond particles with a size less than 1 μm. NDs have become a key platform in the field of nanotechnology due to their wide applications in the fields of medicine, bioimaging, microelectronics, optoelectronics, and sensing.[1,2] These exceptional applications are attributed to their morphology and easy surface functionalization. Their electronic, optical, and chemical properties are also of much importance. The first reported synthesis of NDs is by three scientists K.V. Volkov, V.V. Danilenko, and V.I. Elin in 1963 by shock compression of nondiamond carbon modifications in blast chambers.[3] Astronomical observations suggest the presence of 1500 ppm sized NDs in primitive meteorites, along with isotopically anomalous noble gases like nitrogen, hydrogen, and other elements. Some interstellar dusts, proto-planetary disk, and nebulae also show the presence of NDs, though the origin of these cosmic sources is still under investigation.[4,5]

"Nanodiamond" is a broad term used for a variety of nanoscaled diamond-based materials including single diamond clusters to bulk nanocrystalline films.[6] Based on the molecular arrangement, diamonds are widely classified into diamondoids of particles in the range of 1–2 nm, ultra-nanocrystalline diamond (UNCD) of particle size less than 10 nm and nanocrystalline (NUNCD) particles of less than 100 nm in size. Differences in morphology arise from the growth process.[7]

1.2 NANODIAMONDS: A GENERAL OUTLOOK

1.2.1 SYNTHESIS

Different general approaches for the synthesis of diamond nanoparticles include detonation technique, ion irradiation of graphite, laser ablation, energy-assisted chemical vapor deposition, ultrasound cavitation, electrochemical, autoclave synthesis using super critical fluids, ball milling at high conditions of temperature and pressure, and other methods.[8–12] Certain natural sources of carbon are also used for the production of NDs, such as coal, carbon black, and soot.[13,14]

1.2.2 STRUCTURE AND PROPERTIES

The crystalline ND core consists of sp^3 hybridized carbon atoms wrapped in a layer of carbon atoms in their sp^2 hybridized state. The intermediate portion between the core and carbon cover comprises of carbon atoms in both sp^2 and sp^3 hybridization states.[15] The ND core is much different from all other nanomaterials due to its unique properties of high stability and chemical inertness. The sp^3 hybridized carbon atoms, that is, tetragonal arrangement of cubic configuration in the core structure of diamond, is responsible for the high refractive index, band gap of 5.49 eV and good transparency in infrared and ultraviolet regions (Figure 1.1).[16,17]

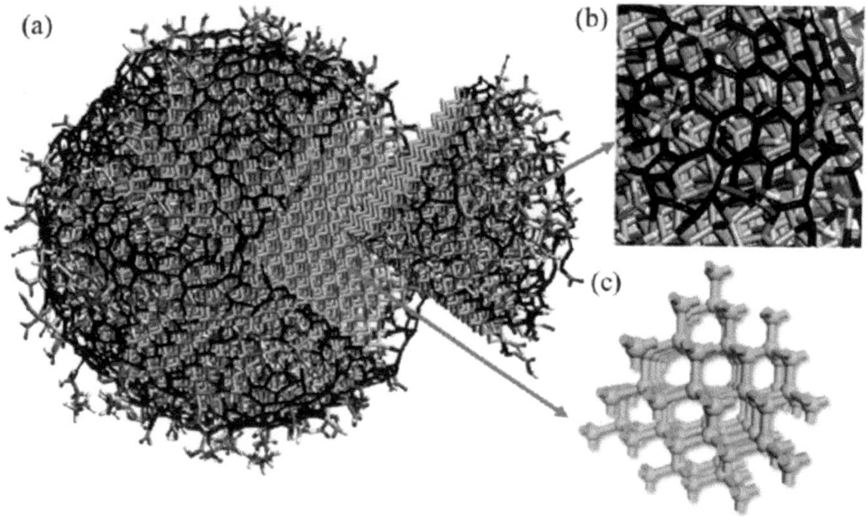

FIGURE 1.1 (a) A model structure of detonation carbon nanodiamond. (b) Closer view of nanodiamond surface covered by a layer of surface functionalization groups and sp^2 hybridized carbon atoms. (c) Highly ordered diamond core of sp^3 hybridized carbon. (Reproduced from Duan, X. et al.[18] with permission. © 2019 American Chemical Society)

Detonation techniques result in the production of spherical ND particles of 4–5 nm in dimension.[19] Its definite shape and size results in huge surface area of about 300–400 m2/g and are well suited as a lubricant.[20] Comparing with all other nanocarbon materials, NDs can easily undergo sophisticated surface modifications via different functional groups or active biomolecules without compromising the unique characteristics of the diamond core.[21] ND inherits most of its properties from their bulk and transmits them into the nanometer range. These include hardness, chemical stability, high dispersibility, electrical conductivity, biocompatibility, fluorescence, and optical properties.[22,23]

1.2.3 CHARACTERIZATIONS

The crystallinity, elemental composition, chemical states, and surface chemistry of NDs were examined using different characterization techniques like Fourier transform infrared spectroscopy (FTIR), temperature programmed desorption (TPD), and X-ray photoelectron spectroscopy (XPS).[24–26] Depending upon the sp^2/sp^3 ratio, composition, and homogeneity of NDs the intensity of D and G bands varies in Raman spectrum.[27] Electron energy-loss spectroscopy (EELS),[28] X-ray absorption near-edge structure (XANES) spectroscopy,[29.30] and solid-state ^{13}C NMR[31] techniques were utilized for the quantitative and qualitative estimation of graphitic and diamond carbon proportions in the sp^2/sp^3 nanohybrids. X-ray diffraction (XRD) patterns at (002) and (111) planes corresponding to the graphite and diamond crystals accordingly can be visually distinguished from the lattice spacing of 0.34 and 0.21 nm respectively in high resolution transmission electroscopy (HR-TEM).[32,33]

1.3 APPLICATIONS OF NANODIAMONDS IN ENERGY STORAGE

Advancement of nanotechnology leads to development of energy management by means of increasing the energy storage capacity in devices. The high surface area, good chemical stability, excellent electrical, and mechanical properties facilitated the application of NDs in energy devices such as supercapacitors and LIBs. The easy surface modifications of ND by funtionalization or doping with suitable metals or metal oxides also contributed their applications in the energy-related fields.

1.3.1 LITHIUM-ION BATTERIES (LIBS)

LIBs find commercial uses in the fields of electronics, transmission, energy storage devices for electric vehicles, and medical devices.[34,35] They have gained an unrivaled vivacity in the last decades as the energy storage of many portable devices such as cell phones, laptops, and computers. The negative electrochemical potential and the high theoretical specific capacity make lithium ideal to be used as anode in rechargeable batteries.[36] The high surface area and chemical inertness of NDs make it suitable anodic material in LIBs. They are rarely utilized as cathode moreover.[37]

The use of detonation nanodiamonds (DND) as an important component in the anode showed improved circling performances in lithium-ion based devices. The high energy storage and lithium-ion transition rates were attributed toward the hardness, immense surface area, and chemical inertness of DND. The covalent bonding of DND with the graphene nanosheets increased the Coulombic efficiency, reversible, and discharge capacities of anode than those without DND.[34]

The stability of the interface between the lithium metal and the electrolyte plays a critical role in the efficient and secure operation of LIBs. ND thin films show good electrochemical stability, high mechanical modulus of >200 GPa for dendrite suppression, and improved cycling efficiency. The introduction of unique double layer of NDs for lithium anodes maximized the defect tolerance and uniformity of anode ion flux. The interfacial layer improved the Coulombic efficiency by up to 99.4% and increased the capacity of retention for at least 400 cycles. The defects in one layer can be easily shielded by the other layer. This ND interface shows stability of Li half-cell and Li-S full cell configurations (Figure 1.2).[38]

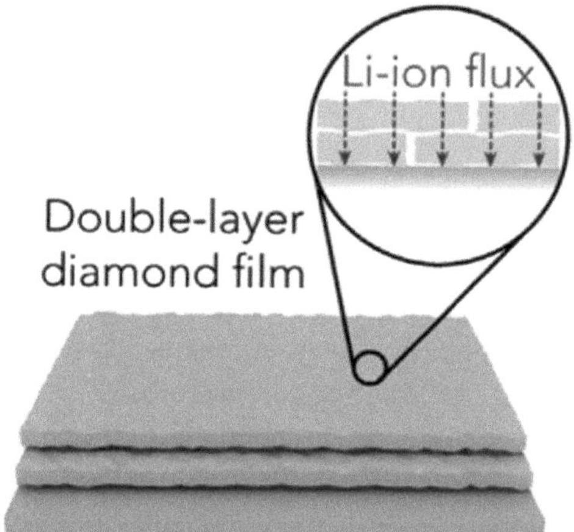

FIGURE 1.2 Double-layer nanodimaond film. (Reproduced from Liu, Y. et al.[38] with permission. © 2018 Elsevier)

UNCD films incorporated with nitrogen grain boundaries (N-UNCD) exhibits high performance, hence used as encapsulating coating on natural graphite-copper (NG/Cu) composite anodes in LIBs. N-UNCD films are composed of sp^2 carbon atoms in the grain boundaries. These grain boundaries act as effective conductive channels for lithium ions between graphite and the electrolyte. The co-intercalation of electrolytes in the anode is greatly suppressed by the electrochemically inert N-UNCD. Discharge-charge cycles became flat after 50 and 10 cycles for the N-UNCD coated and uncoated electrodes, respectively. Thus, enhanced capacity retentions of LIBs were exhibited by the N-UNCD coated on NG/Cu electrode than on uncoated NG/Cu electrodes.[39]

NDs supported by $Li_4Ti_5O_{12}$-Fe_2O_3 (LTO-Fe-ND) composite, synthesized by means of sol-gel method, are used as an efficient anodic material. High conductivity, low cost, good mechanical properties, and large surface area of NDs repress the low electrical conductivity of spinel $Li_4Ti_5O_{12}$ (LTO) material. The support of Fe and ND enhances the capacity, conductivity, and cyclic ability of LTO. Smaller values of the composite in EIS spectra suggest the improvement in electrical conductivity. LTO-Fe-ND composite demonstrated high retention capacity of 230.1 mg^{-1} after 50 cycles.[40]

Metal oxide cathode of traditional LIB is replaced by detonated NDs. These carbon ND particles are derived from soot by means of acid treatment and annealing process. The electrically conductive ND soot particles of average mesopore size of ~9 nm consists of sp^3 carbon core, coated with sp^2 graphitic or amorphous carbon. Their low cost, large surface area, mechanical properties, and good conductivity yield high energy storage capacities, which are evident from high specific capacitance provided by the pseudocapacitive reaction in cyclic voltammetric and charge-discharge process.[37]

The intercalation of lithium ions onto the multi-walled CNTs embedded into a nanohoneycomb diamond of diameter 400 nm was studied using cyclic voltammetry, electrochemical impedance and galvanostatic measurements. Studies emphasize on the decrease of CNT density and increase in the double layer capacitance of diamonds with the improvement in energy density and specific power of LIBs.[37]

The co-deposition of NDs as an electrolyte additive in LIBs suppresses the lithium dendrite growth by acting as heterogeneous seeds for Li plating. The columnar structure of lithium shows a decrease in average diameter upon from ~0.7-0.8 to ~0.3-0.4 μM upon ND particles deposition onto the lithium surface. The dendrite free morphology of uniform lithium deposition leads to 96% Coulombic efficiency, enhanced electrochemical process and stable cycling up to 200 and 150 h at 1.0 and 2.0 mA/cm^2,[36] respectively.

1.3.2 SUPERCAPACITORS

Electrochemical supercapacitors act as energy storage devices by means of accumulation of oppositely charged ions on a double layer at an electrode of high surface area, electrochemical stability, and good electrical conductance.[41,42] Supercapacitors also known as electrochemical capacitors or ultracapacitors have higher power densities, wide range of operational temperature, longer life span, and safer and faster charging/discharging rates.[43–45] These properties are attributed to the wide applications of supercapacitors in electronic equipment, electric vehicles, and charge storage devices.

The use of onion-like carbon particles synthesized by the annealing of NDs at 1800°C followed by electrophoretic deposition technique as electrode material resulted in an enhancement of energy and power densities of the electrochemical capacitors.[41] They showed a specific capacitance of 0.9 mF/cm^2 at 100 V/s. The resultant onion-like carbon species from the vacuum annealing of DND at 1170–2170 K exhibited a specific electrochemical capacitance of 70-100 F/g in acidic electrolyte and 20–50 F/g in alkaline electrolyte. These values of capacitance were much close to the other reported carbon nanomaterials.[42] The charge accumulation on the electrical double layer was estimated using voltammetric technique.

Symmetric supercapacitors containing polyaniline embedded nanodiamonds (PAN-ND) possess a specific capacitance of 640 F/g and up to 10,000 galvanostatic cycles in an acidic electrolyte. ND powder greatly improves the retention capacity, cycle stability and specific capacitance of the electrochemically active polyaniline polymer. The large capacitance value is almost 15 times higher than that of NDs and onion-like carbon.[46]

Electrochemical behavior of reduced graphene oxide (rGO) modified with ND composites (rGO/ND) of specific capacitance 241 F/g is studied using cyclic voltammetry and galvanostatic tests.[47] The enhancement in electrochemical properties and large surface area offers high potential for energy storage of rGO/ND[47] and diamond-nickel hydroxide[48] composites.

The fabrication of three dimensional diamond network with water-soluble redox couple electrolyte exhibits a high capacitance of 73.42 mF/cm^2 at a charging/discharging current density of 1 mA/cm^2. This value is four orders of magnitude higher than the magnitude of diamond electric double layer capacitors in the same potential range and capacitance remains constant even after 12,000 cycles. The power and energy densities of the novel network of diamond-based symmetric supercapacitor device were estimated to be 13.7 kW/kg and 56.50 Wh/k, respectively. This combinational approach toward the construction of supercapacitors bridges normal dielectric capacitors and rechargeable batteries.[49]

Boron-doped diamond (BDD) with p-type conductivity is extensively used as electrode material in supercapacitors. Electrochemical studies on the effect of morphology and size of the crystals, surface modification, effect of doping concentration, redox behavior of diamond electrode, and electrochemical pretreatment effects on BDD were examined.[50–54] BDD functionalized with different sp^2 carbon materials and pseudo capacitive species show high performance leading to the improvements in nanostructure-based capacitor electrodes of supercapacitors.[48,55,56]

Incorporation of BDD with vertically aligned multi-walled CNTs[57] and carbon nanofibers (CNFs)[56] resulted in a CNT/BDD composite of honeycomb like nanostructure and CNFs/BDD hybrid films, respectively. The CNT/BDD composite electrode showed a capacitance that is ~450 times higher than the conventional flat electrodes, which suggests its application in supercapacitor.[57] These electrochemical capacitors resulted in high capacitance retention, high power, and energy densities (Figure 1.3).[56]

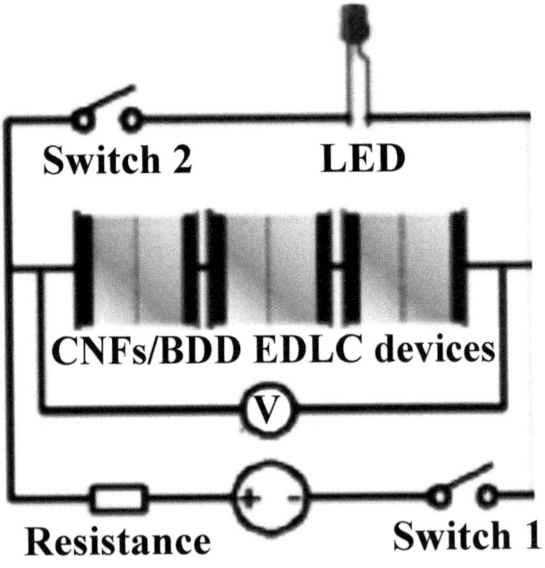

FIGURE 1.3 Electrochemical circuit of a supercapacitor. (Reproduced from Yu, S. et al.[56] with permission. © 2018 John Wiley and Sons).

Electrochemical and physicochemical studies of novel nanostructure composite of TiO_2 nanotubes (NTs) on BDD produced by the technique of microwave plasma enhanced chemical vapor deposition (PE MWCVD) showed enhanced capacitive effect.[58] A porous titanium carbide (TiC)/BDD composite film synthesized by the chemical vapor deposition of BDD on a thick TiO_2 layer during a plasma electrolytic oxidation is used as a high performance capacitor electrode in both aqueous and organic solutions. Combining this composite electrode with suitable aqueous redox electrolyte resulted in a high capacitance of 46.3 mF/cm^2 at a current density of 1 mA/cm^2, and 92% of the initial capacitance remains even after 10,000 charge/discharge cycles.[59] Deposition of hydrous ruthenium oxide on conductive powder surface of BDD resulted in the enhancement of both specific capacitance and specific charge. The good stability of response, excellent robustness at high positive potentials, and extremely low substrate effects of BDD contribute toward their advancement as an excellent electrochemical capacitor.[60] The combination of manganese oxide with BDD resulted in a capacitance of 384 μF, and it was reduced by 5% after 1000 cycles.[55]

Only a few studies on the electrochemical properties of the nitrogen-[61] or phosphorus-doped[62,63] diamond with n-type conductivity were reported due to their insufficient doping level, difficulty in incorporating large atoms such as phosphorus on the diamond lattice and problems associated with their preparation.

Phosphorus-doped conducting nanocrystalline diamond films were used as n-type semiconductor electrodes in capacitors having capacitance of 11.40 $\mu F/cm^2$ in 1.0 M Na_2SO_4.[62] Phosphorus-doped UNCD (P-UNCD)[63] and N-UNCD were produced via MWCVD method. After 3 h of exposure to oxygen plasma, electrochemical capacitance of N-UNCD increased from 17 $\mu F/cm^2$ to more than 1 mF/cm^2 due to enhancement of surface area and disorder of sp^2/sp^3 carbonaceous phases.[64]

1.4 CONCLUSION

The surface modulations and doping effects on the ND structure and properties are properly addressed. Recent research developments in NDs for the energy storage in LIBs and supercapacitors are discussed in detail. In addition to energy storage, they are used in electrochemical catalysis, optoelectronic, thermoelectric energy conversion, and other related applications in sustainable energy. High stability, hardness, good conductance, and easy tunable nanosurface make ND an excellent candidate for diverse applications including nanofluids and water treatment process. A brief outlook on the structure, properties, characterization techniques, and various methods for the synthesis of diamond nanoparticles are also included in this chapter.

KEYWORDS

- **nanodiamonds**
- **onion-like carbon**
- **detonation**
- **energy storage**
- **lithium-ion batteries**
- **supercapacitors**
- **capacitance**

REFERENCES

1. Chen, M.; Pierstorff, E. D.; Lam R.; Li S. Y.; Huang H.; Osawa E., et al. Nanodiamond Mediated Delivery of Water-Insoluble Therapeutics. *ACS Nano.* **2009,** 3, 2016–2022.
2. Deák, P.; Aradi, B.; Kaviani, M.; Frauenheim, T.; Gali, A. Formation of NV Centers in Diamond: A Theoretical Study Based on Calculated Transitions and Migration of Nitrogen and Vacancy Related Defects. *Phys. Rev. B.* **2014**, *89*, 075203.
3. Danilenko, V. V. On the History of the Discovery of Nanodiamond Synthesis. *Phys. Solid State* **2004,** *46*, 595–599.
4. Huss, G. R. Meteoritic Nanodiamonds: Messengers from the Stars. *Elements* **2005,** *1*, 97–100.
5. Guillois, O.; Ledoux, G.; Reynaud, C. Diamond Infrared Emission Bands in Circumstellar Media. *Astrophys. J.* **1999,** *521*, L133–L136.
6. Shenderova, O. A.; Zhirnov, V. V.; Brenner, D. W. Carbon Nanostructures. *Crit. Rev. Solid State Mater. Sci.* **2002,** *27*, 227–356.
7. Williams, O. A.; Nesladek, M.; Daenen, M.; Michaelson, S.; Hoffman, A.; Osawa, E.; Haenen, K.; Jackman, R. B. Growth, Electronic Properties and Applications of Nanodiamond. *Diamond Relat. Mater.* **2008,** *17*, 1080–1088.
8. Yang, G.-W.; Wang, J.-B.; Liu, Q.-X. Preparation of Nano-crystalline Diamonds Using Pulsed Laser Induced Reactive Quenching. *J. Phys. Condensed Matter*. **1998,** *10*, 7923–7927.

9. Boudou, J.-P.; Curmi, P. A.; Jelezko, F.; Wrachtrup, J.; Aubert, P.; Sennour, M.; Balasubramanian, G.; Reuter, R.; Thorel, A.; Gaffet, E. High Yield Fabrication of Fluorescent Nanodiamonds. *Nanotechnology* **2009,** *20*, 235602.
10. Frenklach, M.; Howard, W.; Huang, D.; Yuan, J.; Spear, K. E.; Koba, R. Induced Nucleation of Diamond Powder. *Appl. Phys. Lett.* **1991,** *59*, 546–548.
11. Gogotsi, Y. G.; Nickel, K. G.; Bahloul-Hourlier, D.; Merle-Mejean, T.; Khomenko, G. E.; Skjerlie, K. P. Structure of Carbon Produced by Hydrothermal Treatment of β-SiC Powder. *J. Mater. Chem.* **1996,** *6*, 595–604.
12. Galimov, É. M.; Kudin, A. M.; Skorobogatskii, V. N.; Plotnichenko, V. G.; Bondarev, O. L.; Zarubin, B. G.; Strazdovskii, V. V.; Aronin, A. S.; Fisenko, A. V.; Bykov, I. V.; Barinov, A. Y. Experimental Corroboration of the Synthesis of Diamond in the Cavitation Process. *Doklady Phys.* **2004,** *49*, 150–153.
13. Das, T.; Saikia, B. K. Nanodiamonds Produced from Low-Grade Indian Coals. *ACS Sustain. Chem. Eng.* **2017,** *5*, 9619–9624.
14. Hu, S.; Tian, F.; Bai, P.; Cao, S.; Sun, J.; Yang, J. Synthesis and Luminescence of Nanodiamonds from Carbon Black. *Mater. Sci. Eng B.* **2009,** *157*, 11–14.
15. Boudou, J.-P.; Tisler, J.; Reuter, R.; Thorel, A.; Curmi, P. A.; Jelezko, F.; Wrachtrup, J. Fluorescent Nanodiamonds Derived from HPHT with a Size of Less Than 10 nm. *Diamond Relat. Mater.* **2013,** *37*, 80–86.
16. Schuelke. T.; Grotjohn, T. A. Diamond Polishing. *Diamond Relat. Mater.* **2013,** *32*, 17–26.
17. Alkahtani, M. H.; Alghannam, F.; Jiang, L.; Rampersaud, A. A.; Brick, R.; Gomes, C. L.; Scully, M. O.; Hemmer, P. R. Fluorescent Nanodiamonds for Luminescent Thermometry in the Biological Transparency Window. *Optics Lett.* **2018,** *43*, 3317.
18. Duan, X.; Tian, W.; Zhang, H.; Sun, H.; Ao, Z.; Shao, Z.; Wang, S. sp^2/sp^3 Framework from Diamond Nanocrystals: A Key Bridge of Carbonaceous Structure to Carbocatalysis. *ACS Cataly.* **2019,** *9*, 7494–7519.
19. Turcheniuk, K.; Mochalin, V. N. Biomedical Applications of Nanodiamond (Review). *Nanotechnology* **2017,** *28*, 252001.
20. Rondin, L.; Tetienne, J.-P.; Hingant, T.; Roch, J.-F.; Maletinsky, P.; Jacques, V. Magnetometry with Nitrogen-Vacancy Defects in Diamond. *Reports Progr. Phys.* **2014,** *77*, 056503.
21. Meinhardt, T.; Lang, D.; Dill, H.; Krueger, A. Pushing the Functionality of Diamond Nanoparticles to New Horizons: Orthogonally Functionalized Nanodiamond Using Click Chemistry. *Adv. Funct. Mater.* **2010,** *21*, 494–500.
22. Jensen, K.; Kehayias, P.; Budker, D. Magnetometry with Nitrogen-Vacancy Centers in Diamond. In *High Sensitivity Magnetometers*. Springer, Cham. 2017, 553–576.
23. Fang, X.; Mao, J.; Levin, E. M.; Schmidt-Rohr, K. Nonaromatic Core−Shell Structure of Nanodiamond from Solid-State NMR Spectroscopy. *J. Am. Chem. Soc.* **2009,** *131*, 1426–1435.
24. Petit, T.; Puskar, L. FTIR Spectroscopy of Nanodiamonds: Methods and Interpretation. *Diamond and Relat. Mater.* **2018,** *89*, 52–66.
25. Roldán, L.; Benito, A. M.; García-Bordejé, E. Self-assembled Graphene Aerogel and Nanodiamond Hybrids as High Performance Catalysts in Oxidative Propane Dehydrogenation. *J. Mater. Chem A.* **2015,** *3*, 24379–24388.

26. Shenderova, O.; Panich, A. M.; Moseenkov, S.; Hens, S. C.; Kuznetsov, V.; Vieth, H.-M. Hydroxylated Detonation Nanodiamond: FTIR, XPS, and NMR Studies. *J. Phys. Chem. C.* **2011,** *115*, 19005–19011.
27. Ferrari, A. C.; Robertson, J. Raman Spectroscopy of Amorphous, Nanostructured, Diamond–like Carbon, and Nanodiamond. *Philos. Trans. R. Soc. London. Ser. A: Math. Phys. Eng. Sciences*. **2004,** *362*, 2477–2512.
28. Arenal, R. EELS Studies on Nanodiamonds and Amorphous Diamond-like Carbon Materials. *Microsc. Microanaly*. **2017,** *23*, 2274–2275.
29. Duan, X.; Sun, H.; Wang, S. Comment on Activation of Persulfate by Graphitized Nanodiamonds for Removal of Organic Compounds. *Environ. Sci. Technol.* **2017,** *51*, 5351–5352.
30. Takata, Y.; Edamatsu, K.; Yokoyama, T.; Seki, K.; Tohnan, M.; Okada, T.; Ohta, T. XANES and Raman Spectroscopic Studies of Diamond Films Synthesized by Hot Filament CVD. *Jpn J. Appl. Phys*. **1989,** *28*, L1282–L1285.
31. Cui, J.-F.; Fang, X.-W.; Schmidt-Rohr, K. Quantification of C=C and C=O Surface Carbons in Detonation Nanodiamond by NMR. *J. Phys. Chem. C.* **2014,** *118*, 9621–9627.
32. Pang, H.; Wang, X.; Zhang, G.; Chen, H.; Lv, G.; Yang, S. Characterization of Diamond-like Carbon Films by SEM, XRD and Raman Spectroscopy. *Appl. Surf. Sci.* **2010,** *256*, 6403–6407.
33. Vlasov, I.; Lebedev, O. I.; Ralchenko, V. G.; Goovaerts, E.; Bertoni, G.; Van Tendeloo, G.; Konov, V. I. Hybrid Diamond-Graphite Nanowires Produced by Microwave Plasma Chemical Vapor Deposition. *Adv. Mater*. **2007,** *19*, 4058–4062.
34. Song, Y.; Li, H.; Wang, L.; Qiu, D.; Ma, Y.; Pei, K.; Zou, G.; Yu, K. Nanodiamonds: A Critical Component of Anodes for High Performance Lithium-ion Batteries. *Chem. Commun.* **2016,** *52*, 10497–10500.
35. Honda, K.; Yoshimura, M.; Kawakita, K.; Fujishima, A.; Sakamoto, Y.; Yasui, K.; Nishio, N.; Masuda, H. Electrochemical Characterization of Carbon Nanotube/Nanohoneycomb Diamond Composite Electrodes for a Hybrid Anode of Li-Ion Battery and Super Capacitor. *J. Electrochem. Soc*. **2004,** *151*, A532.
36. Cheng, X.-B.; Zhao, M.-Q.; Chen, C.; Pentecost, A.; Maleski, K.; Mathis, T.; Zhang, X.-Q.; Zhang, Q.; Jiang, J.; Gogotsi, Y. Nanodiamonds Suppress the Growth of Lithium Dendrites. *Nat. Commun.* **2017**, *8*, 1–9.
37. Gu, W.; Peters, N.; Yushin, G. Functionalized Carbon Onions, Detonation Nanodiamond and Mesoporous Carbon as Cathodes in Li-ion Electrochemical Energy Storage Devices. *Carbon* **2013,** *53*, 292–301.
38. Liu, Y.; Tzeng, Y.-K.;Lin, D.; Pei, A.;Lu, H.; Melosh, N. A.; Shen, Z.-X.; Chu, S.; Cui, Y. An Ultrastrong Double-Layer Nanodiamond Interface for Stable Lithium Metal Anodes. *Joule* **2018,** *2*, 1595–1609.
39. Cheng, Y.-W.; Lin, C.-K.; Chu, Y.-C.; Abouimrane, A.; Chen, Z.; Ren, Y.; Liu, C.-P.; Tzeng, Y.; Auciello, O. Electrically Conductive Ultrananocrystalline Diamond-Coated Natural Graphite-Copper Anode for New Long Life Lithium-Ion Battery. *Adv. Mater*. **2014,** *26*, 3724–3729.
40. Wang, Q.; Lu, M.; Miao, J.; Hao, Y. Synthesis and Electrochemical Performance of Nanodiamond Supported Li4Ti5O12–Fe2O3 Composites as Anode Materials for Lithium Ion Batteries. *Mater. Lett.* **2016,** *164*, 1–4.

41. Pech, D.; Brunet, M.; Durou, H.; Huang, P.; Mochalin, V.; Gogotsi, Y.; Taberna, P.-L.; Simon, P. Ultrahigh-Power Micrometre-Sized Supercapacitors Based on Onion-Like Carbon. *Nat. Nanotechnol.* **2010,** *5*, 651–654.
42. Bushueva, E. G.; Galkin, P. S.; Okotrub, A. V.; Bulusheva, L. G.; Gavrilov, N. N.; Kuznetsov, V. L.; Moiseekov, S. I. Double Layer Supercapacitor Properties of Onion-Like Carbon Materials. *Phys. Status Solidi (B)* **2008,** *245*, 2296–2299.
43. Simon, P.; Gogotsi, Y.; Dunn, B. Where Do Batteries End and Supercapacitors Begin. *Science* **2014,** *343*, 1210–1211.
44. Wang, Y.; Xia, Y. Recent Progress in Supercapacitors: From Materials Design to System Construction. *Adv. Mater.* **2013,** *25*, 5336–5342.
45. Shao, Y.; El-Kady, M. F.; Sun, J.; Li, Y.; Zhang, Q.; Zhu, M.; Wang, H.; Dunn, B.; Kaner, R. B. Design and Mechanisms of Asymmetric Supercapacitors. *Chem. Rev.* **2018,** *118*, 9233–9280.
46. Kovalenko, I.; Bucknall, D. G.; Yushin, G. Detonation Nanodiamond and Onion-Like-Carbon-Embedded Polyaniline for Supercapacitors. *Adv. Funct. Mater.* **2010,** *20*, 3979–3986.
47. Wang, Q.; Plylahan, N.; Shelke, M. V.; Devarapalli, R. R.; Li, M.; Subramanian, P.; Djenizian, T.; Boukherroub, R.; Szunerits, S. Nanodiamond Particles/Reduced Graphene Oxide Composites as Efficient Supercapacitor Electrodes. *Carbon* **2014,** *68*,175–184.
48. Gao, F.; Nebel, C. E. Diamond Nanowire Forest Decorated with Nickel Hydroxide as a Pseudocapacitive Material for Fast Charging-Discharging. *Phys. Status Solidi (A).* **2015,** *212*, 2533–2538.
49. Yu, S.; Yang, N.; Zhuang, H.; Mandal, S.; Williams, O. A.; Yang, B.; Huang, N.; Jiang, X. Battery-like Supercapacitors from Diamond Networks and Water-Soluble Redox Electrolytes. *J. Mater. Chem A* **2017,** *5*, 1778–1785.
50. Yang, N.; Foord, J. S.; Jiang, X. Diamond Electrochemistry at the Nanoscale: A Review. *Carbon* **2016,** *99*, 90–110.
51. Liu, F. B.; Wang, J. D.; Liu, B.; Li, X. M.; Chen, D. R. Effect of Electronic Structures on Electrochemical Behaviors of Surface-Terminated Boron-Doped Diamond Film Electrodes. *Diamond and Relat. Mater.* **2007,** *16*, 454–460.
52. Azevedo, A. F.; Baldan, M. R.; Ferreira, N. G. Nanodiamond Films for Applications in Electrochemical Systems. *Int. J. Electrochem. 2012*, **2012,** 1–16.
53. Yamamoto, T.; Tryk, D. A.; Hashimoto, K.; Fujishima, A.; Okawa, M. Electrochemical Reduction of CO [sub 2] in the Micropores of Activated Carbon Fibers. *J. Electrochem. Soc.* **2000,** *147*, 3393.
54. Kulandainathan, M. A.; Hall, C.; Wolverson, D.; Foord, J. S.; MacDonald, S. M.; Marken, F. Boron-doped Diamond Electrodes in Organic Media: Electrochemical Activation and Selectivity Effects. *J. Electroanaly. Chem.* **2007,** *606*, 150–158.
55. Yu, S.; Yang, N.; Zhuang, H.; Meyer, J.; Mandal, S.; Williams, O. A.; Lilge, I.; Schönherr, H.; Jiang, X. Electrochemical Supercapacitors from Diamond. *J. Phys. Chem C.* **2015,** *119*, 18918–18926.
56. Yu, S.; Yang, N.; Vogel, M.; Mandal, S.; Williams, O. A.; Jiang, S.; Schönherr, H.; Yang, B.; Jiang, X. Battery-like Supercapacitors from Vertically Aligned Carbon Nanofiber Coated Diamond: Design and Demonstrator. *Adv. Energ Mater.* **2018,** *8*, 1702947.

57. Zanin, H.; May, P. W.; Fermin, D. J.; Plana, D.; Vieira, S. M. C.; Milne, W. I.; Corat, E. J. Porous Boron-Doped Diamond/Carbon Nanotube Electrodes. *ACS Appl. Mater. Interf.* **2014,** *6*, 990–995.
58. Sawczak, M.; Sobaszek, M.; Siuzdak, K.; Ryl, J.; Bogdanowicz, R.; Darowicki, K.; Gazda, M.; Cenian, A. Formation of Highly Conductive Boron-Doped Diamond on TiO_2 Nanotubes Composite for Supercapacitor or Energy Storage Devices. *J. Electrochem. Soc.* **2015,** *162*, A2085–A2092.
59. Xu, J.; Yang, N.; Heuser, S.; Yu, S.; Schulte, A.; Schönherr, H.; Jiang, X. Achieving Ultrahigh Energy Densities of Supercapacitors with Porous Titanium Carbide/Boron-Doped Diamond Composite Electrodes. *Adv. Energ. Mater.* **2019,** *9*, 1803623.
60. Spătaru, N.; Zhang, X.; Spătaru, T.; Tryk, D. A.; Fujishima, A. Anodic Deposition of RuO[sub x]·nH[sub 2]O at Conductive Diamond Films and Conductive Diamond Powder for Electrochemical Capacitors. *J. Electrochem. Soc.* **2008,** *155*, D73.
61. Raina, S.; Kang, W. P.; Davidson, J. L. Nitrogen Incorporated Nanodiamond Film with "Ridge" Surface Morphology for Detection of Bio-analyte. *Diamond and Relat. Mater.* **2009,** *18*, 574–577.
62. Yu, S.; Xu, J.; Kato, H.; Yang, N.; Schulte, A.; Schönherr, H.; Jiang, X. Phosphorus-Doped Nanocrystalline Diamond for Supercapacitor Application. *ChemElectroChem.* **2018,** *6*, 1088–1093.
63. Vlčková Živcová, Z.; Frank, O.; Drijkoningen, S.; Haenen, K.; Mortet, V.; Kavan, L. n-Type Phosphorus-Doped Nanocrystalline Diamond: Electrochemical and In Situ Raman Spectroelectrochemical Study. *RSC Adv.* **2016,** *6*, 51387–51393.
64. Tong, W.; Fox, K.; Zamani, A.; Turnley, A. M.; Ganesan, K.; Ahnood, A.; Cicione, R.; Meffin, H.; Prawer, S.; Stacey, A.; Garrett, D. J. Optimizing Growth and Post Treatment of Diamond for High Capacitance Neural Interfaces. *Biomaterials* **2016,** *104*, 32–42.

CHAPTER 2

Carbon Nanotubes for Energy Applications

CINTIL JOSE[1], SONA JOHN[1], BINO THOMAS[2], CINCY GEORGE[1], and SABU THOMAS[3*]

[1]*Newman College, Thodupuzha, Kerala, India*

[2]*St. Josephs College of Engineering and Technology, Palai, Kerala, India*

[3]*School of Chemical Sciences, Mahatma Gandhi University, Kottayam, Kerala, India*

[*]*Corresponding author. E-mail: sabuthomas@mgu.ac.in*

ABSTRACT

Use of energy and realizing a sustainable production are the greatest challenges of this century. All human activities require and presume the availability of energy, and it is the center of societal development. Nanotechnology plays an essential role in device applications for energy conversion and storage, green engineering of environmental friendly materials, and in environmental monitoring. Due to their unique structural, electronic, and mechanical properties, carbon nanotubes (CNTs) and their hybrid nanocomposites received immense research attention for their applications in different fields. Here the applications of CNTs in different energy conversion and storage devices are reviewed. Development of CNTs in integrated energy conversion technologies is a promising progress toward the efforts to solve the energy challenge for future.

2.1 INTRODUCTION

Great attention has been paid in recent years to nano-structured materials of different chemical composition, produced as nanoparticles, nanowires, or nanotubes.[1] A new era of interest within the field of nanotechnology began with the identification of carbon nanotubes (CNTs) in 1991. The report of this new type of carbon fiber triggered intense research focused on harnessing the unique properties of these nanostructures. Recent studies have involved the possible applications of SWNTs as nano test tubes, H_2 storage media, batteries, field-emission materials, tips for scanning probe microscopy, chromatographic stationary phases, transistors, diodes, and sensors.[2,3] SWNTs have interesting physical, morphological, and electronic properties and find applications in a variety of areas.

Realizing a sustainable production and use of energy is one of the grand challenges of the 21st century.[4] Energy lies at the heart of societal development and all human activities require and presume the availability of energy. There is a roughly linear correlation between per-capita energy consumption and gross domestic product, but it is time to change this development model and realize the transition to a society that uses energy in a sustainable manner. The world population is increasing, and in addition, the fraction of the world population that accesses large amounts of energy is rising. On the other hand, the availability of fossil fuels is decreasing, and their production rate is close to reaching its maximum—after which this rate will enter terminal decline. The third element of this scenario is the impact of the current energy system, largely dominated (>85%) by fossil fuels, on the global environment, particularly the changes associated with greenhouse gas emissions.

These three synergetic elements create an explosive mixture, and solutions for these require a new approach to science and technology. Nanotechnology and nanomaterials are at the core of the effort toward sustainable energy,[2] which requires a multipronged approach: Energy saving. For example, reducing the consumption of energy and raw materials by introducing new processes, materials, and technologies. This can be achieved by: (i) new, more selective processes and/or processes that consume less energy; (ii) new materials that reduce energy dissipation, from cars to lighting (currently, more than half of the energy used is wasted); and (iii) new technologies to recover electrical energy from waste low-temperature streams, or to save energy.[5]

The issue of the sustainability of energy supply has attracted worldwide concern given the rapid depletion of fossil energy sources and the drive to develop alternative, environment-friendly, renewable energy sources. Therefore, the latest research progress and the current situation pertaining to hydrogen storage in CNTs are discussed in this chapter. Moreover, CNTs can have wide applications as alternative energy storage media, fully utilizing their unique structural characteristics.[3,6,7]

2.2 HYDROGEN STORAGE

The recent discovery of the high and reversible hydrogen storage of CNTs makes such a system very promising.[8] CNTs are considered as good adsorbents for hydrogen because of their high surface area and percentage of pore volume. The hydrogen adsorption is proportional to both parameters, and it is also known that gas adsorption depends on pressure and temperature for any adsorbent material.[9] Hydrogen is a gas, so the adsorption operates in the space unoccupied by CNTs.[10] Dillon et al.[11] published experimental data about hydrogen adsorption on CNTs. They found that SWCNTs could have high energy storage at temperatures and pressures near room conditions. Their results show that SWCNTs with diameter between 1.6 and 2 nm could reach the target H_2 adsorption density of 6.5 wt%.[23,24] As stated by the review of Ong et al.,[20] there is controversy over claims that CNTs display great performance as hydrogen storage since the high storage capacity could not be reproduced.[24]

Recent studies proved very low hydrogen storage capacities of CNTs at room temperature.[25] Tibbetts et al.[23] investigated the adsorption of hydrogen on various CNTs. They observed that none of the CNTs they investigated showed good hydrogen adsorption. The largest sorption is about 0.1 wt.%.[24]

2.2.1 *HYDROGEN STORAGE CAPACITIES OF CNT*

It depends on its structure, geometry, structural defects, pretreatments, operating pressure, etc. One advantage with CNTs is the carbon structure. It helps correlating observed data with theoretical predictions.[26] The study by Chen et al.[27] proved that CNT doped with alkali metals absorbed 20 wt.% of hydrogen. However, the presence of water influenced this result.

Yang[28] again studied these data by synthesizing the doped nanotubes and measured the adsorption/desorption with TGA. Later these results were independently confirmed by Pinkerton et al.[30] These results are again confirmed by studies of Hirscher et al.[29] SWNTs and MWNTs were studied for their hydrogen storage capacity.

Shaijumon and Ramaprabhu[31] prepared CNTs by pyrolysis of acetylene with quartz reactor with argon flow. The hydrogen adsorption on nanotubes was observed, increased with treatment of heat and acid. Maximum storage capacity was 2.1 wt.% at 298 K and 100 bar. Liu et al.[32] studied various SWCNTs and MWCNTS with (air-oxidized, CO_2-oxidized, and KOH-activated) and without post-treatment. The obtained storage capacities for the samples were less than 1.7 wt.%. Rzepka et al.[33] observed that hydrogen uptake has linear relation with tube diameter. It is due to hydrogen uptake being proportional to the surface area, which increases with increase in diameter of the tube. Increasing SWNT diameter to 1.85 nm through an arc discharge with FeS gives a hydrogen adsorption of 4.2 wt.%. Gayathri et al.[34] have reported the influence of defects in CNTs and reported that defects possess an important contribution to the adsorption mechanism of single-walled CNTs. Gayathri et al.[35,36] have also studied coiled CNTs as a potential material for hydrogen storage. The CCNTs have structural defects that influenced the hydrogen adsorption. CCNTs also showed good adsorption binding energies.

2.3 SOLAR CELLS

Researchers are working on solar cells to improve its efficiency and reduce its price. The main strategy to implement CNTs into solar cells consists of mixing CNTs with a solution of an electron-donating conjugated polymer.[8] This mixture is then spin-coated onto a thin transparent conductive electrode (60–120 nm).[37] These electrodes are usually glass and plastic, covered with indium tin oxide and a layer of poly(3,4-ethylenedioxythiophene)-poly(styrenesulfonate) (PEDOT-PSS).[38] Moreover, to improve the CNTs' solubility in solvents for incorporation into a matrix, the physical and electric properties of CNTs can be modified by functionalization, using, for example, nucleic acids. It was shown that polymers mixed with CNTs are an efficient and rapid way for exciton dissociation, due to their high electron affinity. The excitons are dissociated into free electrons with high yield. This results in obtaining higher energy conversion efficiency.[39] Kymakis and Amaratunga were the first to study CNTs in organic solar cell as electron acceptors.[40] The conjugation of CNTs and the polymer poly(3-octylthiophene) (P3OT) in solar cell was studied.[41,42] The PV performance of P3OT devices was highly increased with SWCNT doping, revealing an increase in photocurrent and in the open-circuit voltage. It was demonstrated[43] that the introduction of just a small amount of CNTs in the polymer matrix highly improves the PV properties. Kymakis and Amaratunga[41] and Geng and Zeng[42] studied the interaction between SWCNTs and poly(3-hexylthiophene) (P3HT). SWCNTs have good homogeneous dispersions in P3HT matrices due to their high solubility in P3HT, thus leading to a large interface between SWCNTs and polymers. This large interface guarantees efficient utilization of absorbed light for exciton generation. In their experiments, SWCNTs are used to increase the crystallinity of P3HT, which is important to increase the performance of the PV devices. They are also used as electron acceptors and electron transport tunnel, and in that way, the efficiency is increased due to the exciton dissociation and enhanced charge transport. Since organic solar cells are flexible and have low production costs in contrast to silicon solar cells, the organic ones have attracted much research interests and commercial interests.

Colin et al. investigated photoactive side of semiconducting nanotubes for the purpose of photoconversion.[43] They optimized that cells of lower concentration show better performance than cells with higher concentration of CNTs. This is contrary to the expectation that larger number of nanotubes will give more photoconversion and so more power generation. The reason is the presence of metallic nanotubes. They also investigated optimization strategies that alter cell efficiency by orders of magnitude; if it is possible to prepare semiconducting-only CNT films, that give the greatest efficiency improvement. Figure 2.1 shows the carbon nanotube solar cells and its comparison with dye-sensitized solar cells.

2.3.1 CARBON NANOTUBES AS ELECTRODE MATERIALS IN PHOTOVOLTAIC DEVICES

One great application of carbon nanotubes in photovoltaic devices is their use as electrode materials.[44–46] The methods for preparing carbon nanotube films are growing them directly (CVD), or by coating solutions of carbon nanotubes.[47] This type of CNT films are used both in the OTE (i.e., working electrode)[48–50] and in the counter-electrode in dye-sensitized and organic solar cells.[51–53] They have many advantages like structural strength, flexibility, and cost, allowing the possibility of flexible polymer or liquid-phase photovoltaic devices. As the vacuum deposition processes of conventional OTE materials are very expensive, CNTs have the ability to lower the costs in solar cell devices.[54] Some studies showed a density-gradient method of sorting metallic nanotubes into a film, which showed great promise and demonstrated high conductivity of films at a similar optical transmission.[53–55]

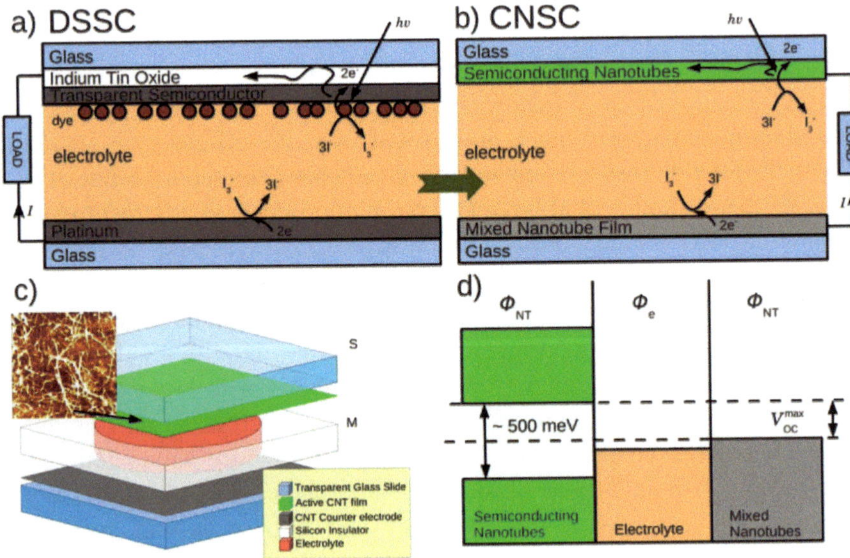

FIGURE 2.1 Carbon nanotube solar cells; comparison to Dye-Sensitized Solar Cells (DSSC), construction, and energetics. (a) DSSC. (b) Carbon Nanotube Solar Cell (CNSC). (c) Layout of a CNSC. (d) Band diagram of the CNSC.

Source: Reprinted from Ref. 42. © 2012 Klinger et al. https://creativecommons.org/licenses/by/4.0/

2.4 LITHIUM BATTERY

LIBs have many applications from portable electronics to electric vehicles, because of their superior energy density than other rechargeable batteries. Even then, the societal demands for lighter and higher capacity

LIBs facilitate ongoing research of new materials with better properties. It requires a concerted development of both electrodes and electrolyte to strengthen battery capacity, cycle life, and charge–discharge rates with high degree of safety. CNTs can also be used in LIBs because of their unique set of electrochemical and mechanical properties. The mixing of CNTs with conventional carbon gives a more effective and advanced strategy to produce an electrical percolation network. CNTs also have the ability to be assembled into free-standing as a physical support of ultra high capacity anode materials, say silicon or germanium. Efforts to use pre-lithiation methods and modification of the single-walled CNT bundling are going to increase the energy density of future CNT batteries.[56]

2.4.1 CARBON NANOTUBES FOR LITHIUM-ION BATTERIES

Lithium-ion batteries (LIBs) are popular modern materials of electrochemistry. Although already commercially available, the energy, power density per volume, and stability are not enough for many applications, such as electrical cars. In an intercalation-type LIB, Li ions are extracted from the $LiMO_2$ electrode during the charging process and simultaneously inserted into the graphitic carbon electrode, coupling this process with negatively charged electrons to keep overall charge neutrality.[4] During the discharging process, the reverse occurs. There are two critical problems: accelerating the kinetics of these processes and maintaining reversibility in long cycling.[57] Conventional LIBs employ a lithiated metal oxide ($LiCoO_2$ is the most common) for the positive electrode (cathode) and a graphitic carbon as the negative electrode (anode). The active materials are combined with a binder (e.g., polyvinylidene fluoride) and conductive additives (e.g., carbon black, graphite, or others) prior to being deposited onto metal foil current collectors. The electrodes are electrically insulated in the battery through a polymer separator that allows Li ion diffusion. The composite thickness can range from 75 to 200 mm for typical high-energy-density LIBs.

2.4.2 APPLICATION OF CNTs IN ANODE MATERIALS

CNTs has a hollow structure of multilayer and coaxial circular tube which involves a dozen of layers of carbon atoms arrayed in hexagonal pattern.[58]

The low density of carbon atom and its specific structure permit the imbedding of Li ion. Li ion can be imbedded on anywhere and also between layers.[6] Udomvech et al.[58] reported that it is easy for Li/Li+ to spread along the walls inside; thus, an effective opening on the end or adding more defects will help Li/Li+ to spread inside. Nishidate and Hasegawa[59] have reported that Li+ can be imbedded not only anywhere inside the tube but also in the gap. Studies on electron density have reported that complete charge transfer occurs between Li+ and CNTs after Li+ is imbedded in the tube. The imbedding makes slight structural deformations in nanotubes. Inside CNTs, the gap between layers allows Li+ to enter.[60] Wu et al.[61] found that the structures of the CNT decide both specific capacity and cycle life.

2.4.3 RAW CNTs AS ANODE MATERIALS

The special graphite structure of nanotubes allows CNTs to find application as the electrode material instead of graphite. Landi et al.[62] discovered that melting point of propene carbonate (PC) is very low; the graphitic layer can be peeled off when graphite is used as the anode material; and if SWCNT is used as the anode material, when PC is added to the system consisting of ethylene carbonate (EC) and dimethyl carbonate (DMC), the reversible specific capacity can be improved effectively, retaining 95% with the specific capacity of 520 mAh/g at the 10th cycle. A thin layer of PEDOT of good conductivity and electrochemical activity had been polymerized on the top of MWCNT array by Chen et al.[63] to enhance the conductivity between the nanotubes. Pushparaj et al.[64] prepared a multilayer nanocomposite in a way similar to that of building blocks. The specific capacity of 373 mAh/g and good cyclability has been realized by Zhang et al.[65] using the CNTs array as the electrode. This study also reported the lithium storage mechanism of carbon tube. During the process of charging decades, to improve the performance of CNT electrode, materials with high specific capacity such as metallic oxide and Si are combined with CNTs in many studies, which have the advantage of huge specific surface area and the role of supporting and conducting as the frame. Ren et al.[66] have produced flexible fibrous LIBs through spinning CNTs into fibers on which MnO_2 was deposited. As an N-type wide-band-gap semiconductor (300 K, Eg = 3.6 eV), SnO_2 is one kind of anode material of great application potential.[67] However large volume change during lithiation/delithiation process makes the electrochemical active particles crack and lose the electrical contact, which results in fast decline of reversible capacity. MWCNTs composite material coated by a layer of even SnO_2 is produced by Wang et al.[68] using thioglycolic acid assisted hydrothermal method; the thickness of SnO_2 can be controlled by reaction conditions. SnCl0.5 and Sb nanoparticles are produced on the external wall of CNTs through reduction reaction by Chen et al.[69] Zhang et al.[70] have produced the CNTs array with CVD before drawing the array into a CNT film on which nano SnO_2 is produced. The specific capacity is as high as 850 mAh/g, and it retains 100% after 65 cycles at voltage ranges of 0.01–3 V. The Co_3O_4 was reported to show a high theoretic reversible specific capacity of 890 mAh/g.[71] Park et al.[72] fabricated vertically aligned mesoporous carbon nanotubes (MCTs) by a dual template method: a hard template-assisted sol-gel process. And this kind of CNTs was filled with Co_3O_4 nanoparticles (NPs). As an anode material, it shows a high reversible capacity of about 627 mAh/g after the 50th discharge.

2.5 CARBON NANOTUBES FOR SUPERCAPACITORS

A supercapacitor, or electric double layer capacitor (EDLC), can be defined as a device that stores charge electrostatically by the reversible adsorption of ions from an electrolyte into active materials which are electrochemically very stable and expose a large surface area.[4] Electrochemical capacitors might be fully charged or discharged in few seconds. Their energy density is less than batteries, but a very high power ideally polarizable delivery or uptake is possible within seconds of time. EDLCs involve electrodes that are dipped in an electrolyte and a porous membrane, that physically separates it, through which its ions can diffuse.[73,74] The charges are stored in a very less distance in the electric double layer, which involves the interface between electrode and adjacent electrolyte. A supercapacitor possesses high operating cell voltage, large capacitance value, and minimum equivalent series resistance.

In these days, the most often used materials are carbon black or activated carbons, because of their moderate cost and high specific surface areas. Even then, it is necessary to enhance the performance and energy density for applications for portable electronics and hybrid electric vehicles, for keeping the long cycling life. One disadvantage of MWCNTs in supercapacitor applications is its very low specific surface area than activated carbon.[75–77] But, CNTs are highly entangled and form a well-developed network that promises the fast transport of ions from the electrolyte to electrode surface. As a result, the response time to an external potential change is very fast, which results in the capability for energy extraction at high frequencies. CNTs show voltammograms with a regular shape that is the peculiarity of an entirely electrostatic attraction.[78] Due to their theoretically higher surface area and higher conductivity, SWCNTs have been mostly used as carbon electrodes for EDLCs.

The name "supercapacitors" and "ultracapacitors" developed from their very high energy density than conventional capacitors. The word "EDLC" is sometimes used to indicate this system, in which the electric energy is stored in the electrochemical double layer, at the interface between polarizable electrodes and the solution of electrolyte.[79] The most commonly used electrode materials for supercapacitors are activated carbons or AC fibers with high specific surface areas. The specific capacitance of AC-based supercapacitors is very high as that expected from their specific surface area. That is why, the micropores contained in ACs are not accessible to the electrolyte ions. The preferable features of the electrodes of supercapacitors are large specific surface area and high mesopore volume. Commercially available supercapacitors of high specific surface area with AC as electrodes can reach a specific capacitance in excess of 7 F/g and a pulse power over 1 kW/kg.[80] There are high needs regarding the use of CNTs as the electrodes of supercapacitors, because of their unique pore structure, low electrical resistivity, good chemical stability, and good chemical stability. Graphical diagram and SEM image of PANI/SWNT composite for pseudocapacitors are shown in Figure 2.2.

2.5.1 CARBON NANOTUBES IN PSEUDOCAPACITORS

CNTs are used in pseudocapacitors in a functionalized form or as a composite material with other materials like conductive polymers and metal oxides. Through chemical or electrochemical methods, CNTs can be functionalized. One method to functionalize CNTs is acidic oxidation to introduce surface carboxyl groups.[82] By this method, the increased hydrophilicity of the electrodes and the introduction of pseuducapacitance increases the specific capacitance of CNTs by 3.2 times. If we treat CNTs with NaOH followed by ultrasonication in we can improve the specific capacitance from 28 F/g for the pristine CNTs to 85 F/g.[83] But, the oxidation treatments cause some defects to degrade the structure of the specific capacitance of

CNTs increased by 3.2 times and decreases the electrical conductivity. So, a balance between structure integrity and the electrode performance is important for high performance pseudocapacitors. Conducting polymers with good electrical conductivities and redox activities show high specific capacitances with CNTs. In the study of Bai et al., the energy density of a CNT-based PC was increased up to 11.3 Wh/kg, by compositing PEDOT onto the CNT electrode through in situ polymerization.[84] And also, polypyrrole (PPy)/CNT composite electrodes prepared yielded a specific capacitance of 165 F/g in 1 M KCl solution.[85] If we compare, PEDOT and PPy, PANI possesses very high theoretical specific capacitance,[86] as confirmed by high specific capacitance of 501.8 F/g reported for flexible PANI/SWNT composite films prepared through in situ electrochemical polymerization.[87] The electrode gradation further increased the specific capacitance to 706.7 F/g. During the charge–discharge process, PANI changes its color. PANI/CNT composites are used for high-performance smart supercapacitors.[88]

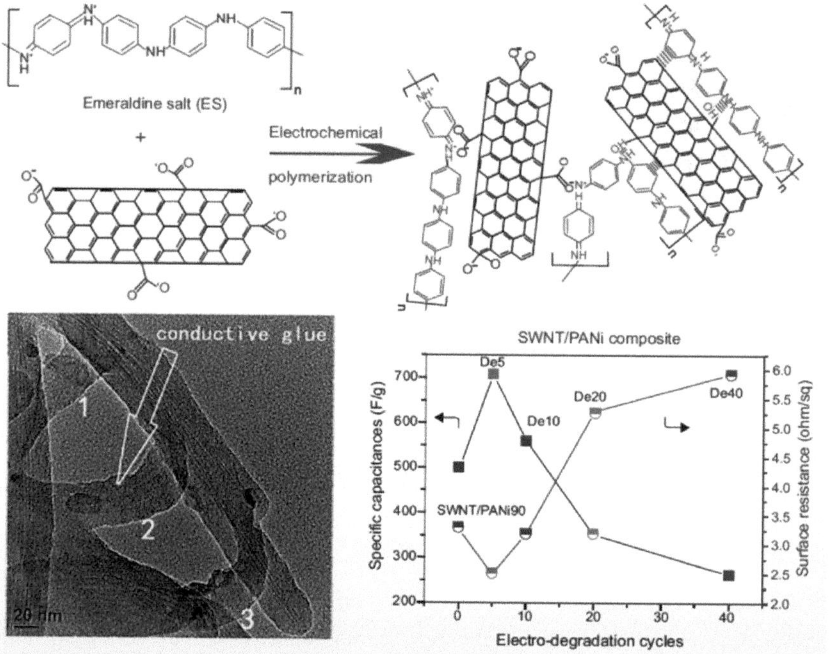

FIGURE 2.2 Graphical representation and SEM image of PANI/SWNT composite for pseudocapacitor.

Source: Reprinted with permission from Ref. 80. © 2010 American Chemical Society.

2.5.2 CARBON-BASED BENDABLE SUPERCAPACITORS

The advances in the development of flexible and wearable electronics have popularized attention as advanced power sources. Due to their large surface area, electrical properties, excellent mechanical, and high electrochemical stability, carbon nanomaterials are promising as electrode materials for flexible supercapacitors (FSCs). Chen et al. prepared flexible and transparent supercapacitors based on In_2O_3 nanowire/CNT heterogeneous films, and optimized an increase in specific capacitance up to 64 F/g with the numbers of In_2O_3 nanowires (up to 0.007 mg) dispersed on the CNT films.[90] In another investigation, a 2-mm-thick film-based FSC made of MnO_2 nanosheet-decorated carbon nanofiber electrodes showed a gravimetric capacitance of 142 F/g at a slow scan rate (10 Mv/s); in this case the electrode was interfaced with PVA-H4SiW12O40·*n*H2O.[91] TiO2/MWNT/PEDOT can also use other materials as the electrode in FSCs for composited carbon fibers[92] and various carbon papers of fiber aerogel and nanotubes.[93] Liu et al. prepared highly flexible porous films of carbon nanofibers (P-CNFs) with an electro-spinning technique combined with a Co ion-assistant acid corrosion process.[94] The resultant fibers showed high conductivity and excellent mechanical flexibility, with small change in resistance value under repeated bending. The P-CNF electrode exhibited a specific capacitance of 104.5 F/g in 0.5 M H_2SO_4 at 0.2 A/g current, which effectively improved cycling stability and 94% retention of specific capacitance, along with a retention rate of 89.4% capacitance after 500 bending cycles.[94] These performances are attributable to the high graphitization degree and the unique hierarchical pore structures of P-CNF.[94] A low processing cost for flexible electrode manufacturing is always advantageous.

2.6 CONCLUSIONS

The use of energy and the challenge of a sustainable production have focused research on the nanostructure of materials. It is of critical importance in improving the performance of advanced materials/electrodes to solve the demanding expectations. CNTs play a key role in the production of advanced solutions for sustainable energy applications. The research area of growing interest from their potential uses in future technologies are the applications of CNTs in the field of energy conversion and storage. CNTs showed environmental applications with their outstanding structural, electronic, and mechanical properties. The development of CNTs in integrated energy conversion technologies showed efficient energy storage system with promising progress to solve the energy challenge for sustainable and clean energy source. Even then their use is still largely based on phenomenological observations. There is a gap in the complexities of the nanostructure and characteristics of these materials. We need a better understanding of the major aspects of CNTs for their optimization in various applications, such as solar cells, fuel cells, LIBs, and supercapacitors.

KEYWORDS

- **carbon nanotubes**
- **energy**
- **solar cells**
- **hydrogen storage**
- **pseudocapacitors**

REFERENCES

1. Trojanowicz, M. Analytical Applications of Carbon Nanotubes: A Review. *Trends Analy. Chem.* **2006,** *25* (5).
2. Centi, G.; Perathoner, S. Facing the Energy Challenge through Chemistry in a Changing World. In *The Chemical Element. Chemistry's Contribution to Our Global Future*; Garcia-Martinez, J., Serrano-Torregrosa, E., Eds.; Wiley: Weinheim, 2011.
3. *Nanotechnology for the Energy Challenge*; Garcia-Martinez, J., Ed.; Wiley- VCH: Weinheim, 2010.
4. Centi, G.; Perathoner, S. Carbon Nanotubes for Sustainable Energy Applications. *ChemSusChem* **2011,** *4*, 913–925. DOI: 10.1002/cssc.201100084.
5. Vairavapandian, D.; Vichchulada, P.; Lay, M. D. Preparation and Modification of Carbon Nanotubes: Review of Recent Advances and Applications in Catalysis and Sensing. *Analytica Chimica Acta* **2008,** 626, 119–129.
6. Smalley, R. E. Grand Challenges in Energy, presented at BES Science in 2018, Arlington, May 29, 2003.
7. Liu, C.; Cheng, H-M. Carbon Nanotubes for Clean Energy Applications. *J. Phys. D: Appl. Phys.* **2005,** *38*, R231–R252. DOI:10.1088/0022-3727/38/14/R01

8. Zhang, Q.; Bai, Z.; Du, F.; Dai, L. Carbon Nanotube Energy Applications, Nanotube Superfiber Materials. https://doi.org/10.1016/B978-0-12-812667-7.00028-8
9. Cheng, H.; Yang, Q.; Liu, C. Hydrogen Storage in Carbon Nanotubes. *Carbon* **2001,** *39*, 1447–1454.
10. F.L. Darkrim, P. Malbrunot, G.P. Tartaglia, Review of Hydrogen Storage by Adsorption in Carbon Nanotubes. *Int. J. Hydrog. Energ.* **2002,** 27, 193–202.
11. S. Banerjee, I. Puri, Enhancement in Hydrogen Storage in Carbon Nanotubes under Modified Conditions. *Nanotechnology* **2008,** *19*,155702.
12. C. Chen, C. Huang, Enhancement of Hydrogen Spillover onto Carbon Nanotubes with Defect Deature. *Micropor. Mesopor. Mater.* **2008,** *109*, 549–559.
13. Gayathri, V.; Geetha, R. Hydrogen Adsorption in Defected Carbon Nanotubes. *Adsorption* **2007,** *13*, 53–59.
14. Dillon, A. C.; Jones, K. M.; Bekkedahl, T. A.; Kiang, C. H.; Bethune, D. S.; Heben, M. J. Storage of Hydrogen in Single Walled Carbon Nanotubes. *Nature* **1997,** *386*, 377–379.
15. Darkrim, F. High Adsorptive Property of Opened Carbon Nanotubes at 77 K. *J. Phys. Chem. B* **2000,** *104*, 6773–6776.
16. Wang, Q.; Johnson, J. K. Optimization of Carbon Nanotube Arrays for Hydrogen Adsorption. *J. Phys. Chem. B* **1999,** *103*, 4809–4813.
17. Yin, Y.; Mays, T.; McEnaney, B. Molecular Simulations of Hydrogen Storage in Carbon Nanotube Arrays. *Langmuir* **2000,** *16*, 10521–10527.
18. Dillon, A. C.; Whitney, E.; Engtrakul, C.; Curtis, C. J.; O'Neill, K. J.; Parilla, P. A.; Simpson, L. J.; Heben, M. J.; Zhao, Y.; Kim, Y. H.; Zhang, S. B. Novel Organometallic Fullerene Complexes for Vehicular Hydrogen Storage. *Phys. Status Solidi B* **2007,** *244* (11), 4319–4322.
19. Liu, C.; Fan, Y. Y.; Liu, M.; Cong, H. T.; Cheng, H. M.; Dresselhaus, M. S. Hydrogen Storage in Single-Walled Carbon Nanotubes at Room. *Science* **1999,** *286*, 1127–1129.
20. Ong, Y.; Ahmad, A.; Zein, S.; Tan, S. A Review on Carbon Nanotubes in an Environmental Protection and Green Engineering Perspective. *Braz. J. Chem. Eng.* **2010,** *27* (2), 227–242.
21. Hirscher, M.; Becher, M.; Haluska, M.; Quintel, A.; Skakalova, V.; Choi, Y.; Dettlaff-Weglikowska, U.; Roth, S.; Stepanek, I.; Bernier, P.; Leonhardt, A.; Fink, J. Hydrogen Storage in Carbon Nanostructures. *J. Alloys Compd.* **2002,** *330–332*, 654–658.
22. Tibbetts, G.; Meisner, G.; Olk, C. Hydrogen Storage Capacity of Carbon Nanotubes, Filaments, and Vapor Grown Fibers. *Carbon* **2001,** *39*, 2291–2301.
23. Kajiura, H.; Tsutsui, S.; Kadono, K.; Kakuta, M.; Ata, M. Hydrogen Storage Capacity of Commercially Available Carbon Materials at Room Temperature. *Appl. Phys. Lett.* **2003,** *82*, 1105–1107.
24. Liu, C.; Chen, Y.; Wu, C.; Xu, S.; Cheng, H. Hydrogen Storage in Carbon Nanotubes Revisited. *Carbon* **2010,** *48*, 452–455.
25. Mohan, M.; SharmaV. K., Kumar, E. A.; Gayathri, V. Hydrogen Storage in Carbon Materials—A Review. *Energ. Storage* **2019,***1*, e35. https://doi.org/10.1002/est2.35.
26. Chen, P.; Wu, X.; Lin, J.; Tan, K. L. High H2 Uptake by Alkali-doped Carbon Nanotubes under Ambient Pressure and Moderate Temperatures. *Science* **1999,** *285*, 91–93.
27. Yang, R. T. Hydrogen Storage by Alkali-doped Carbon Nanotubes–Revisited. *Carbon* **2000**, *38*, 623–641.

28. Hirscher, M.; Becher, M.; Haluska, M., et al. Hydrogen Storage in Sonicated Carbon Materials. *Appl. Phys. Mater. Sci. Process.* **2001,** *72,* 129–132.
29. Pinkerton, F. E.; Wicke, B. G.; Olk, C. H., et al. Thermogravimetric Measurement of Hydrogen Absorption in Alkali-Modified Carbon Materials. *J. Phy. Chem. B* **2000,** *104,* 9460–9467.
30. Shaijumon, M. M.; Ramaprabhu, S. Studies of Yield and Nature of Carbon Nanostructures Synthesized by Pyrolysis of Ferrocene and Hydrogen Adsorption Studies of Carbon Nanotubes. *Int. J. Hydrogen Energ.* **2005,** *30,* 311–317.
31. Liu, C.; Chen, Y.; Wu, C. Z.; Xu, S. T.; Cheng, H. M. Hydrogen Storage in Carbon Nanotubes Revisited. *Carbon* **2010,** *48,* 452–455.
32. Rzepka, M.; Lamp, P.; Casa Lillo, M. A. Physisorption of Hydrogen on Microporous Carbon and Carbon Nanotubes. *J. Phys. Chem. B* **1998,** *102,* 10894–10898.
33. Geetha, R.; Gayathri, V. Comparative Study on Gas Adsorption in Defected Carbon and Boron Nitrate Nanotubes. *Curr. Nanosci.* **2010,** *6,* 131–136.
34. Gayathri, V.; Geetha, R. Hydrogen Adsorption in Defected Carbon Nanotubes. *Adsorption* **2007,** *13,* 53–59.
35. Gayathri, V.; Devi, N. R.; Geetha, R. Hydrogen Storage in Coiled Nanotubes. *Int. J. Hydrogen Energ.* **2010,** *35,* 1313–1320.
36. Sgobba, V.; Guldi, D. Carbon Nanotubes as Integrative Materials for Organic Photovoltaic Devices. *J. Mater. Chem.* **2008,** *18* (2), 141–248.
37. Cataldo, S.; Salice, P.; Menna, E.; Pignataro, B. Carbon Nanotubes and Organic Solar Cells. *Energ. Environ. Sci.* **2012,** *5,* 5919–5940.
38. Scharber, M. C.; Muhlbacher, D.; Koppe, M.; Denk, P.; Waldauf, C.; Heeger, A. J.; Brabec, C. J. Design Rules for Donors in Bulk-Heterojunction Solar Cells-Towards 10% Energy-Conversion Efficiency. *Adv. Mater.* **2006,** *18* (6), 789–794.
39. Kymakis, K.; Amaratunga, G. A. J. Single-wall Carbon Nanotube/Conjugated Polymer Photovoltaic Devices. *Appl. Phys. Lett.* **2002,** *80* (1), 112–114.
40. Kymakis, E.; Amaratunga, G. A. J. Carbon Nanotubes as Electron Acceptors in Polymeric Photovoltaic. *Rev. Adv. Mater. Sci.* **2005,** *10,* 300–305.
41. Geng, J.; Zeng, T. Influence of Single-walled Carbon Nanotubes Induced Crystallinity Enhancement and Morphology Change on Polymer Photovoltaic Devices. *J. Am. Chem. Soc.* **2006,** *128,* 16827–16833.
42. Klinger, C., Patel, Y.; Postma H. W. Ch. Carbon Nanotube Solar Cells. *PLoS ONE* **2012,** *7* (5), e37806. DOI:10.1371/journal.pone.0037806.
43. Baker, B. A.; Zhang, H.; Cha, T.-G.; and Choi, J. H. *Carbon Nanotube Solar Cells*; Woodhead Publishing Limited, **2013**; Chapter 9. DO : 10.1533/9780857098627.3.241.
44. Hu, L. B.; Hecht, D. S.; Gruner, G. Carbon Nanotube Thin Films: Fabrication, Properties, and Applications. *Chem. Rev.* **2010,** *110* (10), 5790–5844.
45. Kumar, A.; Zhou, C. W. The Race to Replace Tin-doped Indium Oxide: Which Material Will Win? *ACS Nano* **2010,** *4* (1), 11–14.
46. Pasquier, A. D.; Unalan, H. E.; Kanwal, A.; Miller, S.; Chhowalla, M. Conducting and Transparent Single-wall Carbon Nanotube Electrodes for Polymer-fullerene Solar Cells. *Appl. Phys. Lett.* **2005,** *87* (20), 3.
47. van de Lagemaat, J.; Barnes, T. M.; Rumbles, G.; Shaheen, S. E.; Coutts, T. J.; Weeks, C.; Levitsky, I.; Peltola, J.; Glatkowski, P. Organic Solar Cells with Carbon Nanotubes Replacing In2O3: Sn as the Transparent Electrode. *Appl. Phys. Lett.* **2006,** *88* (23).

48. Rowell, M. W.; Topinka, M. A.; McGehee, M. D.; Prall, H. J.; Dennler, G.; Sariciftci, N. S.; Hu, L. B.; Gruner, G. Organic Solar Cells with Carbon Nanotube Network Electrodes. *Appl. Phys. Lett.* **2006,** *88* (23), 3.
49. Ramasamy, E.; Lee, W. J.; Lee, D. Y.; Song, J. S. Spray Coated Multi-wall Carbon Nanotube Counter Electrode for Tri-iodide (I-3(-)) Reduction in Dye-sensitized Solar Cells. *Electrochem. Commun.* **2008,** *10* (7), 1087–1089.
50. Lee, K. S.; Lee, W. J.; Park, N. G.; Kim, S. O.; Park, J. H., Transferred Vertically Aligned N-doped Carbon Nanotube Arrays: Use in Dye-Sensitized Solar Cells as Counter Electrodes. *Chem. Commun.* **2011,** *47* (14), 4264–4266.
51. Suzuki, K.; Yamaguchi, M.; Kumagai, M.; Yanagida, S. Application of Carbon Nanotubes to Counter Electrodes of Dye-sensitized Solar Cells. *Chem. Lett.* **2003,** *32* (1), 28–29.
52. McCreery, R. L. Advanced Carbon Electrode Materials for Molecular Electrochemistry. *Chem. Rev.* **2008,** *108* (7), 2646–2687.
53. Ulbricht, R.; Lee, S. B.; Jiang, X. M.; Inoue, K.; Zhang, M.; Fang, S. L.; Baughman, R. H. and Zakhidov, A. A., Transparent Carbon Nanotube Sheets as 3-D Charge Collectors in Organic Solar Cells. *Sol. Energ. Mater. Sol. Cells* **2007,** 91 (5), 416–419.
54. Green, A. A. and Hersam, M. C., Colored Semitransparent Conductive Coatings Consisting of Monodisperse Metallic Single-Walled Carbon Nanotubes. *Nano Lett.* **2008,** 8 (5), 1417–1422.
55. Landi, B. J.; Ganter, M. J.; Cress, C. D.; DiLeo, R. A.; Raffaelle, R. P. Carbon Nanotubes for Lithium Ion Batteries. *Energ. Environ. Sci.* **2009,** *2*, 638–654.
56. Li, L.; Yang, H.; Zhou, D.; Zhou, Y. Progress in Application of CNTs in Lithium-Ion Batteries. *J. Nanomater.* **2014,** Article ID 187891, 8 pages, http://dx.doi.org/10.1155/2014/187891.
57. Udomvech, A.; Kerdcharoen, T.; Osotchan, T. First Principles Study of Li and Li+ Adsorbed on Carbon Nanotube: Variation of Tubule Diameter and Length. *Chem. Phys. Lett.* **2005,** *406* (1–3), 161–166.
58. Nishidate, K.; Hasegawa, M. Energetics of Lithium Ion Adsorption on Defective Carbon Nanotubes. *Phys. Rev. B: Condensed Matter Mater. Phys.* **2005,** 71 (24), Article ID 245418.
59. Morris, R. S.; Dixon, B. G.; Gennett, T.; Raffaelle, R.; Heben, M. J. High-energy, Rechargeable Li-ion Battery Based on Carbon Nanotube Technology. *J. Power Sources* **2004,** 138 (1–2), 277–280.
60. Wu, G. T.; Wang, C. S.; Zhang, X. B. et al. Structure and Lithium Insertion Properties of Carbon Nanotubes. *J. Electrochem. Soc.* **1999,** 146 (5), 1696–1701.
61. Landi, B. J.; Ganter, M. J.; Schauerman, C. M.; Cress, C. D.; Raffaelle, R. P. Lithium Ion Capacity of Single Wall Carbon Nanotube Paper Electrodes. *J. Phys. Chem.* **2008,** *112* (19), 7509–7515.
62. Chen, J.; Liu, Y.; Minett, A. I.; Lynam, C.; Wang, J.; Wallace, G. G. Flexible, Aligned Carbon Nanotube/Conducting Polymer Electrodes for a Lithium-ion Battery. *Chem. Mater.* **2007,** *19* (15), 3595–3597.
63. Pushparaj, V. L.; Shaijumon, M. M.; Kumar, A., et al. Flexible Energy Storage Devices Based on Nanocomposite Paper. *Proc. Natl. Acad. Sci.USA* **2007,** *104* (3)4, 13574–13577.

64. Zhang, H.; Cao, G.; Wang, Z.; Yang, Y.; Shi, Z.; Gu, Z. Carbon Nanotube Array Anodes for High-rate Li-ion Batteries. *Electrochim. Acta* **2010,** *55* (8), 2873–2877.
65. Ren, J.; Li, L.; Chen, C., et al. Twisting Carbon Nanotube Fibers for Both Wire-shaped Micro-supercapacitor and Micro-battery. *Adv. Mater.* **2013,** *25* (8), 1155–1159.
66. Du, N.; Zhang, H.; Chen, B., et al. Synthesis of Polycrystalline SnO2 Nanotubes on Carbon Nanotube Template for Anode Material of Lithium-ion Battery. *Mater. Res. Bull.* **2009,** *44* (1), 211–215.
67. Wang, Z.; Chen, G.; Xia, D. Coating of Multi-walled Carbon Nanotube with SnO_2 Films of Controlled Thickness and Its Application for Li-ion Battery. *J. Power Sources* **2008,** *184* (2), 432–436.
68. Chen, W. X.; Lee, J. Y.; Liu, Z. The Nanocomposites of Carbon Nanotube with Sb and SnSb0.5 as Li-ion Battery Anodes. *Carbon* **2003,** *41* (5), 959–966.
69. Zhang, H.; Feng, C.; Zhai, Y.; Jiang, K.; Li, Q.; Fan, S. Crossstacked Carbon Nanotube Sheets Uniformly Loaded with SnO_2 Nanoparticles: A Novel Binder-free and High-capacity Anode Material for Lithium-ion Batteries. *Adv. Mater.* **2009,** *21* (22), 2299–2304.
70. Cabana, J.; Monconduit, L.; Larcher, D.; Palac´ın, M. R. Beyond Intercalation-based Li-ion Batteries: The State of the Art and Challenges of Electrode Materials Reacting through Conversion Reactions. *Adv. Mater.* **2010,** *22*, (35), E170–E192.
71. Park, J.; Moon, W. G.; Kim, G. P., et al. Three-dimensional Aligned Mesoporous Carbon Nanotubes Filled with Co_3O_4 Nanoparticles for Li-ion Battery Anode Applications. *Electrochim. Acta* **2013,** *105*, 110–114.
72. Pandolfo, A. G.; Hollenkamp, A. F. *J. Power Sources* **2006,** *157*, 11–27.
73. Zhang, L. L.; Zhao, X. S. *Chem. Soc. Rev*. **2009,** *38*, 2520–2531.
74. Niu, C.; Sichel, E. K.; Hoch, R.; Moyand, D.; Tennet, H. *Appl. Phys. Lett*. **1997,** *70*, 1480–1482.
75. Ma, R. Z.; Liang, J.; Wei, B. Q.; Zhang, B.; Xu, C. L.; Wu, D. H. *J. Power Sources* **1999,** *84*, 126–129.
76. Frackowiak, E.; Jurewicz, K.; Delpeux, S.; Beguin, F. *J. Power Sources* **2001,** *97*, 822–825.
77. (a) Frackowiak, E.; Beguin, F. *Carbon* **2001,** *39*, 937–950; (b) Frackowiak, E.; Delpcux, S.; Jurewice, K.; Szostak, K.; Cazorla-Amoros, D.; Beguin, F. *Chem. Phys. Lett*. **2002,** *361*, 35– 41.
78. Lee, Y. H.; An, K. H.; Lee, J. Y.; Lim, S. C. Carbon Nanotube-based Supercapacitors. In *Encyclopedia of Nanoscience and Nanotechnology*; Nalwa, H. S., Ed., Vol. 1; American Scientific Publishers: Stevenson Ranch, CA, 2004; p 625.
79. Chen, X.; Paul R.; Dai, L. Carbon-based Supercapacitors for Efficient, Energy Storage. *Natl. Sci. Rev.* **2017,** *0*, 1–37. DOI: 10.1093/nsr/nwx009
80. Liu, J.; Sun, J.; Gao, L. A Promising Way to Enhance the Electrochemical Behavior of Flexible Single-walled Carbon Nanotube/Polyaniline Composite Films. *J. Phys. Chem. C* **2010,** *114*, 19614–20.
81. Obreja, V. V. N. On the Performance of Supercapacitors with Electrodes Based on Carbon Nanotubes and Carbon Activated Material: A Review. *Phys. E: Lowdimen Sys Nanostruct* **2008,** *40*, 2596–605.
82. Wang, G.; Liang, R.; Liu, L., et al. Improving the Specific Capacitance of Carbon Nanotubes-Based Supercapacitors by Combining Introducing Functional Groups on

Carbon Nanotubes with Using Redox-Active Electrolyte. *Electrochim. Acta* **2014,** *115*, 183–188.

83. Bai, X.; Hu, X.; Zhou, S., et al. In situ Polymerization and Characterization of Grafted Poly (3,4-ethylenedioxythiophene)/Multiwalled Carbon Nanotubes Composite with High Electrochemical Performances. *Electrochim. Acta* **2013,** *87*, 394–400.
84. Paul, S.; Choi, K. S.; Lee, D. J., et al. Factors Affecting the Performance of Supercapacitors Assembled with Polypyrrole/multi-walled Carbon Nanotube Composite Electrodes. *Electrochim. Acta* **2012,** *78*, 649–655.
85. Li, H.; Wang, J.; Chu, Q., et al. Theoretical and Experimental Specific Capacitance of Polyaniline in Sulfuric Acid. *J. Power Sources* **2009,** *190*, 578–586.
86. Liu, J.; Sun, J.; Gao, L. A Promising Way to Enhance the Electrochemical Behavior of Flexible Single-walled Carbon Nanotube/Polyaniline Composite Films. *J. Phys. Chem. C* **2010,** *114*, 19614–19620.
87. Chen, X.; Lin, H.; Chen, P., et al. Smart, Stretchable Supercapacitors. *Adv. Mater.* **2014,** *26*, 4444–4449.
88. Du, L.; Yang, P.; Yu, X., et al. Flexible Supercapacitors Based on Carbon Nanotube/ MnO_2 Nanotube Hybrid Porous Films for Wearable Electronic Devices. *J. Mater. Chem. A* **2014,** *2*, 17561–17567.
89. Chen, P. C.; Shen, G.; Sukcharoenchoke, S., et al. Flexible and Transparent Supercapacitor Based on In_2O_3 Nanowire/Carbon Nanotube Heterogeneous Films. *Appl. Phys. Lett.* **2009,** *94*, 043113 1–3.
90. Nataraj, S. K.; Song, Q.; Al-Muhtaseb, S. A., et al. Thin, Flexible Supercapacitors Made from Carbon Nanofiber Electrodes Decorated at Room Temperature with Manganese Oxide Nanosheets. *J. Nanomater.* **2013,** *272093*, 1–6.
91. Chien, C. J.; Deora, S. S.; Chang, P., et al. Flexible Symmetric Supercapacitors Based on TiO_2 and Carbon Nanotubes. *IEEE Trans. Nanotechnol.* **2011,** *10*, 706–709.
92. Zopf, S. F.; Panzer, M. J. Integration of UV-cured Ionogel Electrolyte with Carbon Paper Electrodes. *AIMS Mater. Sci.* **2014,** *1*, 59–69.
93. Liu, Y.; Zhou, J.; Chen, L., et al. Flexible Hybrid Membranes with $Ni(OH)_2$ Nanoplatelets Vertically Grown on Electrospun Carbon Nanofibers for High Performance Supercapacitors. *ACS Appl. Mater. Interf.* **2015,** *7*, 23515–23520.

CHAPTER 3

Understanding the Impact of Carbon Nanotubes in Environmental Protection

JOHN C. JIMTHA*

Department of Forest pathology, Kerala Forest Research Institute, Peechi, Thrissur, Kerala, India

**Corresponding author. E-mail: jimthajoseph@gmail.com*

ABSTRACT

Recent developments in engineering have created carbon nanotubes (CNTs) one among the foremost studied nanomaterials. A good vary of applications in numerous engineering fields are recommended by taking advantage of the exceptional physical, chemical, and electronic properties of CNTs. The contribution of CNTs to sustainable environments and green technologies similar to wastewater treatment, air pollution control, biotechnology, renewable energy technologies, supercapacitors, and green nanocomposites is mentioned in this chapter. Thought of CNTs from cost perspective and doable risks for large-scale implementation is additionally addressed. CNTs show great potential as a promising material to be used in various environmental fields, supported the literature reviewed.

3.1 INTRODUCTION

Due to the invention of carbon nanotubes (CNTs), the sector of carbon nanotubes has bit by bit been revolutionized future nanotechnology. CNTs are macromolecules having cylindrical shape with radius of some nanometers and a length of up to a few micrometers.[6,26] The walls of those tubes are created of a polygon lattice of carbon atoms, capped by structures like fullerenes. The distinctive structure of CNTs may be divided mainly into multiwalled carbon nanotubes (MWCNTs) and single-walled carbon nanotubes (SWCNTs). MWCNTs contains two or additional focused cylindrical shells of coaxially organized graphene sheets round the central hollow space with layer spacing. By comparison, SWCNTs consist of one cylinder graphite sheet bound along by van der Waals.[4,54] Present synthesis methods, such as electrical arc-discharge, laser ablation, and chemical vapor decomposition (CVD) are used to produce large quantities of CNTs.[14,21,30] CNTs' mutable hybridization states and structural sensitivity to disturbances underneath synthesis conditions exploit their specific physical, chemical, and electronic properties that inspire creativity in new technologies and applications. In addition, these distinctive and tunable properties supply potential advances from proactive (prevention of environmental degradation, improvement of energy efficiency) to retroactive (reuse of wastewater, transformation of pollutants) in environmental systems.[42] Current applications of CNTs in effluent treatment, pollution control, biotechnology, renewable energy and supercapacitors are addressed during this short analysis and a proposal for a green nanocomposite system that follows the 3R principle has additionally been discussed. At the end of the day, the assessment of CNTs in large-scale applications is checked out from the angle of prices and potential hazards.

3.1.1 CARBON NANOTUBES IN THE EFFLUENT TREATMENT

Waste water discharge from domestic, industrial, or agricultural sources contain varieties of pollutants and has attracted international concern because it adversely affects the standard of water quality .[21,24,29,34,38,39,71,72,76,87] Contaminants in wastewater, such as 1,2-dichlorobenzene, dioxin, and heavy metal ions are nondegradable, extremely harmful and should cause accumulative toxicity, cancer, and harm to the nervous system .[18,33,37]

The expulsion of specific contaminations relies upon the sorbent's sorption conduct. CNTs have an uncommon sorption potential and high-sorption proficiency contrasted with conventional granular and powder-activated carbon, which have intrinsic constraints, for example, surface-active site and sorption-activation energy. Broad investigations have demonstrated that the limit of CNTs to adsorb is subject to both the functional surface groups and the type of the sorbate. For instance, the measure of surface acidity (carboxylic, lactonic, and phenolic gatherings) favors the adsorption of polar compounds.[71] Then again, it has been demonstrated that the nonfunctional surface of the CNT has a higher adsorption limit toward nonpolar compounds, for example, polycyclic aromatic hydrocarbons.[80,86] The sorption capacity of CNTs include chemical interactions with polar compounds and physical interactions of nonpolar compounds. Sorption of both polar and nonpolar compounds is normally furnished with Langmuir or Freundlich isotherms.[37,38,54,85,86]

The sorption capacity of CNTs is efficient over a wide range of pHs. Optimum performance in the pH range of 7–10 has been accounted (Rao et al., 2007). Ionization and rivalry between ion species could happen other than this pH range. Despite the fact that CNTs are more costly than conventional activated carbon, also their sorption, desorption cycles are more.[38] Sorption/desorption studies have indicated the accessibility of CNTs in various sorption and desorption cycles. The regeneration study reported a slight decrease in Ni_2+ adsorption and desorption in CNTs but a sharp decrease in granular activated carbon (GAC) after several cycles.[7,38]

This is explained by the fact that the porous structure of the GAC makes it more difficult to dissolve the Ni^2+ because the ions have to move from the inside to the outside. Apart from acting as a sorbent for organic and inorganic pollutants, the existing technology has used CNTs as nanofilters to minimize particulate matter concentrations in wastewater. This miracle could be explained by the way that the porous structure of the GAC makes it all hard to break down the Ni^2+ because the particles need to move from inside to the outside. Besides going probably as a sorbent for natural and inorganic toxins, the present development has used CNTs as nanofilters to restrict particulate issue fixations in wastewater.[74] Like the sorbents, the particular selectivity of the CNT filters can be controlled by attaching different functions to the pore entrances.[20]

Notwithstanding its hydrophobic qualities, CNTs have indicated outstanding execution in the transport of water. Contrasted with sorbents, it is conceivable to control the exact selectivity of CNT filters by including various functions at the pore entrances.[20] Another reason given by Hummer et al.[25] is that the frictionless progression of water is by nanoscale regulation, which diminishes the energy distribution of contact and limits contact with water.[25] Likewise, recent CNT filtration trials additionally uncovered the capacity of CNT nanofilters to wipe out pathogenic microorganisms, for example, protozoa and microscopic organisms in wastewater get retained on the CNT surface utilizing a depth filtration system.[49] There are many efficient means to remove *E.coli* bacteria, using SWCNT filters. As the inclusion of the microporous membrane based on poly(vinylidene fluoride) (PVDF) was completely captured, the nanotube bundles inside the SWCNTs were proposed to additionally improve the filter's robustness, reusability, and thermal resistance by immobilizing the SWCNTs on a ceramic filter. In a further report, they utilized a shower pyrolysis technique to synthesize a controllable nanoscale CNT-based porosity filter and watched greatest productivity in the expulsion of the MS2 virus at a pressure of 8–11 bar.[9]

CNTs are not restricted to filtration and sorbent application in wastewater treatment; a few scientists have recognized efficient antimicrobial properties of CNTs. This conduct permits CNTs to replace chemical disinfectants as another and proficient approach to oversee microbial pathogens.[13,31,32,51,63] Applying CNTs to the treatment of water purification forestalls the formation of harmful disinfection by-products (DBPs, for example, trihalomethane, haloacetic acids, and aldehydes since they are not strong oxidants and are generally inert in water. Surfactants or polymers, for example, sodium dodecyl benzenesulfonate, polyvinylpyrolidone, or Triton X are usually used to advance their dispersion. Exceptionally purified CNTs display strong antimicrobial activity toward both Gram-positive and Gram-negative bacteria and bacterial spores. Activities brought about by antimicrobial properties might be attributed to impaired pathogen cell function by destruction of major components, interference with pathogen cell metabolic processes, and inhibition of pathogen's growth. *E.coli* cells with SWCNTs cause critical harm to the membranes and inactivation of the cells. Because of their incredible mechanical properties, a few reports have additionally proposed the utilization of CNTs as scaffolding for antimicrobial agents, for example, Ag nanoparticles, and lysozyme antimicrobials.

3.1.2 CARBON NANOTUBES IN REMOVAL OF AIR POLLUTANTS

The remarkable electrical, electrochemical, and optical properties of CNTs have aroused the enthusiasm of researchers to investigate the potential applications of CNTs as sensing elements for detecting and monitoring the concentration of poisonous gases delivered to the climate.[8,15,40,57,83,90] CNTs have exceptional and tunable electronic properties whereby their semiconductivity is significantly influenced by their cylindrical shaped one-dimensional structure, for example, size and chirality. CNT-based gas sensors deliver a range of advantages over conventional semiconductor metal oxide gas sensors including less power consumption, less operating temperature, and more sensitivity.[17] The individual CNTs make a high electrical field close to their ultrafine tips and increase the overall field so as to quicken the process of gas ionization. This gas sensor identification depends on changes in the resistance or conductivity of CNTs because of direct gas contact.[54]

CNT gas sensors were used to detect nitrogen oxides (NOx) in several studies at room temperature.[11,47,52,73,77] Despite the results demonstrated a strong and prompt response from the gas sensors, tedious recuperation is an issue. A few strategies have been recommended to improve the desorption of gas from the sensor. For instance, warming the sensor utilizing bright (UV) light and expanding the pace of cleansing gas transition will support the sensor's gas desorption. Endeavors have additionally been made to improve the sensitivity and affinity of CNT-based gas sensors by polymer functionalization. The inclusion of CNTs in polymers, for example, polyaniline and polypyrrole increases the sensitivity of the sensor to selected gases or fumes.[40]

An investigation of polymer coating by Qi et al.[58] reported that the CNT-coated polyethylenemine (PEI) gas sensor indicated high affinity for NO_2 detection at concentrations below 1 ppb without NH_3 interference because of low binding affinity and NH_3 sticking coefficient for electron-rich CNTs.[58] A Nafion- coated CNT-based gas sensor (a polymeric perfluorinated sulfonic acid ionomer) permitted higher selectivity for NH_3 sensing because of NO_2 blockage on CNTs, in contrast to the PEI-coated sensor. A gas sensor based on a CNT/poly(methyl methacrylate) (PMMA) composite chemeristor has additionally demonstrated an improvement in sensitivity.[1] In spite of the fact that CNT-based gas sensors show intense competition for traditional metal oxide sensors, consistent advancements have been made to improve the conventional gas sensor. Integrating CNTs in metal oxide sensors defeat high operating temperature requirement.[41] Also, significant increases in sensitivity and response time were observed for the detection of pollutant gases, for example, NO_2 and NH_3 at room temperature in CNT/metal oxide sensors. The expanded yield was because of the efficient access of the gas to nanopasses and the fluctuation in conductivity following the adsorption of the gas created by the CNTs.[22]

3.1.3 BIOTECHNOLOGICAL ASPECTS OF CARBON NANOTUBES

Biotechnology has grown quickly in recent years because of an expanding demand for biotechnological products. Innovative, environmentally friendly technologies in which living organisms are used to produce products for specific uses. The rise of biotechnology furnishes CNTs with an opportunity to participate in biofuel cells specifically. As Palmore and Whites[55] have described, biofuel cells are fuel cells that depend on biocatalytic activity to produce electricity.[55] They can typically be classified as either microbial fuel cells (MFCs) or enzymatic biofuel cells (EFCs). MFCs exploit microbial catabolic activities to produce electricity. Potential choices for wastewater treatment have been considered as a range of materials that can be utilized as source materials, including complex organic waste and sustainable biomass for wastewater.[36,82] In principle, however, the system is not feasible, largely due to poor performance and lack of technological maturity.

Because of their high conductivity and huge surface region, intensive research has focused on the utilization and modification of CNTs as electrodes to increase the power output of MFCs.[75] Another sort of electrode architecture is prepared by setting CNTs on a carbon cloth to make an exceptionally conductive electrode with a profoundly specific surface area of the MFC. The presence of CNTs brought about a 250% expansion in power density contrasted with the non-CNT-coated electrode. It is Sharma et al.[67] who contrasted the performance of MFC, which is electrode dependent on the CNT electrodeagainst the plain graphite ; the CNT electrode demonstrated an increase in power density of around six times that of the graphite cathode. Additional optimization of the power density can be accomplished by utilizing the biocompatibility of CNTs with microorganisms in MFCs.[67]

In spite of the fact that CNTs have been claimed to be antimicrobials, the behavior of cytotoxicity diminishes after their modification and functionalization. Research completed by Morozan et al.[48] about the biocompatibility of *Staphylococcus aureus* with CNTs showed rapid growth of microorganisms in CNT-adjusted cell culture media.[48,64] The ability to reduce power generation losses when applied to the anodized architecture of MFCs, EFC uses enzyme or protein catalysis instead of microbial catalytic activity to convert chemical energy into electrical energy. To date, their short life span, poor enzyme stability, and low power density have hindered the use of EFC as a power source for low-power sensors and communication devices.[44] The use of CNTs as bioelectrodes is still a significant step forward in the EFC process. In addition to mediating expenditure, CNTs have shown great efficiency in the EFC by providing a strong forum for enzyme immobilization. CNTs allowed the covalent attachment of enzyme molecules to their surface and encouraged high enzyme loading by allowing cross-linked enzyme aggregate coating.[92,19] Furthermore, the nanoscale environment provided by CNTs could improve the stability of the enzyme. Enzymes covalently attached to CNTs have a high degree of stability due to their covalent relationship, which offered significant resistance to denaturation.[68] The stability of the enzyme in EFC is critical as it provides excellent operational stability, which is expected to pave the way for an increase in power density and a prolongation of EFC's lifetime.[3,19]

3.1.4 CARBON NANOTUBES IN RENEWABLE ENERGY INDUSTRY

The global marketable energy demand is projected to increase by 57% between 2004 and 2030. This trend envisages the need for innovative renewable energy technologies to meet the long-term energy demand challenge and maintain an environmental balance. The major breakthrough that CNTs have contributed to the solar energy sector is their use in photovoltaic devices. Photovoltaic devices produce electricity by absorbing photons that are absorbed by the light. To date, many disadvantages have been reported for commercially available silicone and semiconductor-based photovoltaic systems, including high cost and low illumination stability.[60] CNTs are being used as an alternative material in different solar cell architectures due to their affordability and remarkable energy conversion, especially in silicon-based solar cells, organic solar cells, and dye-sensitized solar cells.

The active nanoscale surface area of CNTs also allows for massive absorption of photons for solar energy harvesting, but due to the presence of a delocalized electron system it enhances the mobility of charge transfer.[65] Good alignment between CNTs will improve their photoconductivity after lighting.[35] Silicon-based solar cells use the simplest pn junction to detach electrons/holes and generate electricity after illumination. When embedded in silicone, CNTs serve as a heterojunction for charging separation, as a highly conductive percolated network for charging, and as a transparent electrode for light illumination and charging. Some heterojunctions observed modest cell efficiency with improved stability.[29] The creation of organic solar cells has drawn a great deal of interest from researchers as a consequence of being flexible and having low production costs compared with silicone-based solar cells. For light absorption and charge transfer, organic solar cells rely on conductive organic polymers, such as poly(3-octylthiophene)(P3OT), poly(3-hexylthiophene)(P3HT), or [6,6]-phenyl-C61-butyric acid methyl ester (PCBM).[28] New work indicates an increase in performance with the introduction of CNTs into the top electrode of the photoactive layer and the back electrode of the organic solar cells. CNTs function as photoactive material in the photoactive layer and improve cell efficiency by providing efficient transport of the hole or electron to the CNT/polymer interface.[59,78] The photoactive part designed with P3OT/CNT exhibited a higher open-circuit voltage using the high capacity of the CNTs for the transport of electrons. CNTs manage to have a wide surface area in the top and back of the electron for high optical transmittance and low sheet resistance to reduce power loss.

As a result, an increase in photocurrent was observed for the design of the top electrode consisting of CNT films and indium–tin oxide (ITO).[56] Dye-sensitized solar cells (DSSCs) have been praised for their low cost and easy preparation as a promising solar cell.'.[53] Electricity is generated in DSSC when semiconductor materials generate an electron hole pair and transfer the charge in an electrolyte solution.[43] Applying CNTs as a photoelectrode can help increase carrier mobility when exposed to visible light. However, due to the ultrafast recombination of photogenerated charge carriers, the direct use of CNTs as photoelectrodes result in better performance. Brown et al.[10] therefore proposed CNT/TiO_2 as a technique to boost the separation of charges and facilitate the flow of charges because TiO_2 particles on CNTs are capable of injecting electrons from their excited state.[10] In addition to photoelectrodes, CNTs are also a common option for DSSC counter-electrode manufacturing materials. The CNT counter-electrode cells are expected to offer several advantages, including nanoscale conducting channels, lightweight, and low cost, as well as improved mechanical properties and thermal stability, in addition to improving conversion efficiency.[93]

In addition to solar energy, recent advances in hydrogen storage media have centered on CNTs as one of the important research areas underway.

Hydrogen, a fairly clean fuel compared with traditional fuel, has been seen as an enticing solution for the production of green energy technologies. The U.S. Department of Energy (DOE) had proposed 6.5 wt.% minimum hydrogen storage to meet the need for commercial storage requirements (U.S. Department of Energy: Washington, 2007). CNTs have received exceptional attention as a possible storage medium from the investigation into techniques used to achieve hydrogen storage because of their affordability, recycling properties, low density, distribution of nanoscale pore size and acceptable chemical stability.[62] Hydrogen storage at CNTs requires primarily physiosorption. Theoretical studies performed by Dodziuk and Dolgonos[16] showed that the quantity of hydrogen adsorbed depends on the size of the CNT array and orientation, with hydrogen adsorption.[16]

For their large bulk density, SWCNTs outperform activated carbon which enhances volumetric storage. Hydrogen storage in CNTs was lower than 1 wt.% at ambient temperature and high pressure, but that acceptable hydrogen storage at higher pressure and lower temperature could be achieved.[5] In addition, physiosorption of hydrogen alone was insufficient to meet the DOE specification. Thus, it has centered on hybrid CNT/metal compounds to enhance chemisorption. Transition or alkali metal-doped CNTs, with s-p-d hybridization, helped to reinforce the significant increase in hydrogen storage.[88] A nearly 30% increase in hydrogen storage capacity was reported at 2 MPa under normal room temperature for palladium and vanadium-doped CNTs. Schaller et al.[66] reinforced magnesium–nickel (Mg-23.5 wt.% Ni) with CNT using a system of powder metallurgy and reported storage up to 6.1 wt.% of hydrogen.[66] Another work conducted by Iyakutti et al.[27] suggested that CNTs coated with aluminum hydride would bind up to four hydrogen molecules, resulting in an increase in hydrogen storage ability of 8.3 wt.%.[27] Compared with the metal doping processes, structural defects also appear as another possible solution to improve hydrogen chemisorptions in CNTs.[81] With an increase in surface area and pore volume the presence of structural defects in CNTs can be expected. Such structures certainly improve the interaction between CNTs and hydrogen, allowing an increase of up to 50% in adsorption binding strength.

Sadly, so far there is controversy both experimentally and theoretically around reports that CNTs have extraordinary performance as hydrogen storage because other researchers in the same field could not replicate the high storage capacity.[91] Until now, CNTs' storage capacity is still far from achieving the DOE goal.

3.1.5 CARBON NANOTUBE SUPERCAPACITORS

Due to their miniature scale, high power capacity, long life cycle and high energy efficiency, electrochemical condensers or supercapacitors have been considered as an alternative to replacing conventional batteries with potential to minimize waste disposal for the environment.[12]

Recent nanotechnology advances have proposed the use of CNTs as an electrode material for the condenser. The use of the large surface area of CNTs in the pair of electrodes with a thinner layer between the electrode and the electrolyte improves the capacitor's ability to store higher energy densities. In addition, the use of vertically aligned CNTs with multiple atomic diameters in width will increase the supercapacitor capacity and power density substantially as a result of the drastically increased electrode surface area. While they have high surface stability and good mechanical properties, because of their low size, CNTs are not preferred to be used alone as an electrode.[23] CNTs have also been proposed as a substrate for the transformation of high specific potential metal oxides, such as manganese oxides and ruthenium oxide.[69] Supercapacitors and energy densities up to 5Wh/kg, which is about 10 times higher than traditional capacitors, with just 0.5Wh/kg were recorded with capacitances of about 5000F.[23]

3.1.6 CARBON NANOTUBES IN DESIGN OF GREEN NANOCOMPOSITES

Waste generation is proportional to the economic growth across the world. Wastes, especially synthetic polymer wastes, cause adverse environmental impacts. Thus, the European Community has developed a waste management framework based on two complementary strategies as a solution to this problem: Avoiding waste by increasing the recycling and reuse of waste with a focus on life cycle assessment (LCA) in order to generate transparent and comprehensive environmental impact assessments.[45]

The potential outlook for addressing waste management problems has raised the task of synthesizing green nanocomposites by using biodegradable polymers as part of the next generation wave of materials.[46] The green nanocomposite movement using natural renewable resources follows the LCA principle promoting waste recycling and reuse. Biodegradable polymers are of great commercial potential and have attracted much interest from researchers as an alternative to replacing nonrenewable petroleum-based polymers due to their degradability. Most biodegradable polymers, however, possess weaker mechanical properties and higher heat distortion temperatures, limiting their use in large applications. CNTs may also serve as nanoreinforcements for the biodegradable polymers to provide improvement in mechanical properties, increased longevity and improved thermal stability for a suite of composite materials. CNT alignment, CNT-biodegradable polymer adhesion and CNT dispersion in the biodegradable polymer matrix typically determine the consistency of the biodegradable polymer/CNT nanocomposite.[79]

Song and Qiu (2009) reported that upon integration with CNTs, poly-butylene succinate (PBSU) thermal stability had improved by about 10°C. Sitharaman et al.[70] demonstrated that ultrashort CNTs can greatly enhance the use of polypropylene fumarate (PPF), the inferior mechanical properties of which often restrict its use in load bearings.[70] In addition, CNT/PPF nanocomposites exhibit a comparable biocompatibility *in vivo* that provides the potential for use as a prototype bone tissue scaffold.

Another benefit that green nanocomposites give is their potential to recycle the embedded CNTs due to the biodegradable polymer's degradability. Degradation of the biologically degradable polymer can be accomplished by either microbial degradation or enzymatic degradation under varying pH and temperature conditions. The recovered CNTs serve as a reinforcement filler for the production of new composites following degradation. CNT's reuse and recycle process could minimize waste disposal and at the same time be cost-effective for processed materials.

3.1.7 CARBON NANOTUBES IN ENVIRONMENT

CNTs have been hailed as promising materials to be applied in different areas of the environment. There are some problems that need to be addressed if large-scale implementation is to make them practically feasible.

The high cost of synthesis is still one of the key issues blocking larger scale application of CNTs. Nevertheless, recent experimental studies have shown the possibility of mass production of high quality CNTs with lower cost.[2] CNTs with a production rate of 595 kg/h were recorded by hydrocarbon decomposition using a catalytic CVD (CCVD) method in either a plug-flow reactor or a fluidized bed reactor with a solid catalyst.[2] Nanothinx, was able to reduce the cost of producing CNT using a low cost novel catalyst produced by them. Studies conducted by Zhang et al.[89] have shown that using liquefied petroleum gas as a carbon source material, and the ceramic sphere as a substratum, high purity CNTs can be easily developed at low cost on a large-scale.[89] The ability to mass-produce CNTs has catalyzed the dream to market and apply CNT technology in large-scale applications.

Despite the high cost of the synthesis, questions about the possible environmental and human health effects of CNTs have also attracted considerable interest. The widespread applications envisaged for CNTs that result in their disposal of air and water into the environmental compartments. Disposed CNTs can enter the human body via inhalation or through the food chain and cause damage.[84] The adverse effects emerge from their physical aspect,[84] physical condition, impurity presence, and chemical treatment. Studies of the *in vivo* CNT found that CNTs could induce pulmonary inflammation and lung cell proliferation[50] and inhibit heart muscle growth (CBC, 2007) in rats. The toxicity of CNTs that are well dispersed is less than that of agglomerated CNTs. Functionalized CNTs with hydrophilic groups therefore have less of a detrimental impact on the viability of mammalian cells as their dispersion is increased. This behavior is contrary to what was found in bacterial cells, where dispersal increased toxicity. Until now, there is still insufficient research on the environmental and human health effects of CNTs.[31] Source reduction is therefore the safest way to reduce human and environmental exposure to CNTs.[61]

3.2 CONCLUSIONS

Potential constructive and retroactive applications of CNTs in ambient systems were addressed in this short study. CNTs play a major role in wastewater treatment and air pollution monitoring because of their excellent mechanical, electrical, physical, and chemical properties. CNTs act as sorbents, nanofilters, and antimicrobials for the removal of organic and inorganic pollutants, as well as pathogenic microorganisms in wastewater treatment. Design of CNT-based gas sensors results in high sensitivity in air pollution control, with prompt sensor response to pollutant gases. Furthermore, CNTs are also identified as one of the main challenges in the development of "green" energy, involving clean combustion. Electricity generation by "natural" energy technologies focused on biomass catalytic reaction, renewable sources, such as solar and hydrogen (hydrogen fuel cell) avoid the release of harmful gases into the environment and reduce high demand for fossil fuel. Technologies are not yet commercially practical because they are still in a state of emergence and more time is needed to achieve technical maturation. The power density given by supercapacitors based on CNT was seen as an alternative way of replacing conventional batteries. Supercapacitors with extended use time might, over time, reduce the amount of waste disposed of in the environment. Crucial developments are still under way to meet the long-term challenge of environmental protection. The involvement of CNTs in green nanocomposite design in green materials supports the LCA principle which promotes the reduction, reuse and recyclability of raw materials. Continuous development of CNTs from low-cost sources for large-scale application has taken a closer step to overcoming the high cost of synthesis problem. Given the safety consequences arising from widespread use of CNTs, the safest way to prevent exposure to CNTs in humans and the environment is source reduction.

KEYWORDS

- **carbon nanotubes**
- **environment**
- **wastewater disposal**
- **air pollution control**
- **biotechnology**
- **renewable energy**
- **supercapacitors**

REFERENCES

1. Abraham, J. K.; Philip, B.; Witchurch, A.; Varadan, V. K.; Reddy, C. C. A compact Wireless Gas Sensor Using a Carbon Nanotube/PMMA Thin Film Chemiresistor. *Smart Mate. Struct.* **2004,** *13,* 1045–1049.
2. Agboola, A. E.; Pike, R. W.; Hertwig, T. A.; Lou, H. H. Conceptual Design of Carbon Nanotube Processes. *Clean Technol. Environ. Policy* **2007,** *9,* 289–311.

3. Asuri, P.; Karajanagi, S. S.; Yang, H.; Yim, T.-J.; Kane, R. S.; Dordick, J. S. Increasing Protein Stability through Control of the Nanoscale Environment. *Langmuir* **2006,** *22,* 5833–5836.
4. Balasubramanian, K.; Burghard, M. Chemically Functionalized Carbon Nanotubes. *Small* **2005,** *1,* 180–192.
5. Banerjee, S.; Puri, I. K. Enhancement in Hydrogen Storage in Carbon Nanotubes Under Modified Conditions. *Nanotechnology* **2008,** *19,* 155702.
6. Bethune, D. S.; Kiang, C. H.; de Vries, M. S.; Gorman, G.; Savoy, R.; Vazquez, J.; Beyers, R. Cobalt-catalysed Growth of Carbon Nanotubes with Single-atomic-layer Walls. *Nature* **1993,** *363,* 605–607.
7. Boehm, H. P. Surface Oxides on Carbon and Their Analysis: A Critical Assessment. *Carbon.* **2002,** *40,* 145–149.
8. Bondavalli, P.; Legagneux, P.; Pribat, D. Carbon Nanotubes Based Transistors as Gas Sensors: State of the Art and Critical Review. *Sens. Actuat. B Chem.* **2009,** *140,* 304–318.
9. Brady-Estévez, A. S.; Kang, S.; Elimelech, M. A Single-Walled-Carbon-Nanotube Filter for Removal of Viral and Bacterial Pathogens. *Small.* **2008,** *4,* 481–484.
10. Brown, P.; Takechi, K.; Kamat, P. V. Single-Walled Carbon Nanotube Scaffolds for Dye-Sensitized Solar Cells. *J. Phys. Chem. C 112,* 4776–4782.
11. Cantalini, C.; Valentini, L.; Lozzi, L.; Armentano, I.; Kenny, J. M.; Santucci, S. NO2 Gas Sensitivity of Carbon Nanotubes Obtained by Plasma Enhanced Chemical Vapor Deposition. *Sens. Actuat. B Chem.* **2003,** *93,* 333–337.
12. Capek, I. Dispersions, Novel Nanomaterial Sensors and Nanoconjugates Based on Carbon Nanotubes. *Adv. Coll. Interf. Sci.* **2009,** *150,* 63–89.
13. Cortes, P.; Deng, S.; Smith, G. B. The Adsorption Properties of Bacillus atrophaeusSpores on Single-Wall Carbon Nanotubes. *J. Sens.* **2009,** *2009,* 1–6.
14. Dai, H.; Rinzler, A. G.; Nikolaev, P.; Thess, A.; Colbert, D. T.; Smalley, R. E. Single-wall Nanotubes Produced by Metal-Catalyzed Disproportionation of Carbon Monoxide. *Chem. Phys. Lett. 260,* 471–475.
15. Di Francia, G.; Alfano, B.; La Ferrara, V. Conductometric Gas Nanosensors. *J. Sens.* **2009,** *2009,* 1–18.
16. Dodziuk, H.; Dolgonos, G. Molecular Modeling Study of Hydrogen Storage in Carbon Nanotubes. *Chem. Phys. Lett.* **2002,** *356,* 79–83.
17. Eatemadi, A.; Daraee, H.; Karimkhanloo, H.; Kouhi, M.; Zarghami, N.; Akbarzadeh, A.; Abasi, M.; Hanifehpour, Y.; Joo, S. Carbon Nanotubes: Properties, Synthesis, Purification, and Medical Applications. *Nanoscale Res. Lett.* **2014,** *9,* 393.
18. Fagan, S. B.; Santos, E. J. G.; Souza Filho, A. G.; Mendes Filho, J.; Fazzio, A. Ab Initio Study of *2,3,7,*8-tetrachlorinated Dibenzo-p-dioxin Adsorption on Single Wall Carbon Nanotubes. *Chem. Phys. Lett.* **2007,** *437,* 79–82.
19. Fischback, M. B.; Youn, J. K.; Zhao, X.; Wang, P.; Park, H. G.; Chang, H. N.; Kim, J.; Ha, S. Miniature Biofuel Cells with Improved Stability Under Continuous Operation. *Electroanalysis* **2006,** *18,* 2016–2022.
20. Fornasiero, F.; Park, H. G.; Holt, J. K.; Stadermann, M.; Grigoropoulos, C. P.; Noy, A.; Bakajin, O. Ion Exclusion by Sub-2-nm Carbon Nanotube Pores. *Proc. Natl. Acad. Sci.* **2008,** *105,* 17250–17255.

21. Gao, Z.; Bandosz, T. J.; Zhao, Z.; Han, M.; Qiu, J. Investigation of Factors Affecting Adsorption of Transition Metals on Oxidized Carbon Nanotubes. *J. Hazard. Mater.* **2009,** *167,* 357–365.
22. Gong, J.; Sun, J.; Chen, Q. Micromachined Sol–gel Carbon Nanotube/SnO2 Nanocomposite Hydrogen Sensor. Sens. *Actuat. B Chem.* **2008,** *130,* 829–835.
23. Hadjipaschalis, I.; Poullikkas, A.; Efthimiou, V. Overview of Current and Future Energy Storage Technologies for Electric Power Applications. *Renew. Sustain. Energ. Rev.* **2009,** *13,* 1513–1522.
24. Hsieh, S.-H.; Horng, J.-J. Adsorption Behavior of Heavy Metal Ions by Carbon Nanotubes Grown on Microsized Al2O3 Particles. *J. Univ. Sci. Technol. Beijing Miner. Metall. Mater.* **2007,** *14,* 77–84.
25. Hummer, G.; Rasaiah, J. C.; Noworyta, J. P. Water Conduction through the Hydrophobic Channel of a Carbon Nanotube. *Nature.* **2001,** *414,* 188–190.
26. Iijima, S. Helical Microtubules of Graphitic Carbon. Nature. *354,* 56–58.
27. Iyakutti, K.; Kawazoe, Y.; Rajarajeswari, M.; Surya, V. Aluminum Hydride Coated Single-Walled Carbon Nanotube as a Hydrogen Storage Medium. *Int. J. Hydrogen Energ.* **2009,** *34,* 370–375.
28. Janssen, R. A. J.; Hummelen, J. C.; Sariciftci, N. S. Polymer–Fullerene Bulk Heterojunction Solar Cells. *MRS Bull.* **2011,** *30,* 33–36.
29. Jia, Y.; Wei, J.; Wang, K.; Cao, A.; Shu, Q.; Gui, X.; Zhu, Y.; Zhuang, D.; Zhang, G.; Ma, B.; Wang, L.; Liu, W.; Wang, Z.; Luo, J.; Wu, D. Nanotube-Silicon Heterojunction Solar Cells. *Adv. Mater.* **2008,** *20,* 4594–4598.
30. Journet, C.; Maser, W. K.; Bernier, P.; Loiseau, A.; de la Chapelle, M. L.; Lefrant, S.; Deniard, P.; Lee, R.; Fischer, J. E. Large-scale Production of Single-Walled Carbon Nanotubes by the Electric-Arc Technique. *Nature* **1997,** *388,* 756–758.
31. Kang, S.; Herzberg, M.; Rodrigues, D. F.; Elimelech, M. Antibacterial Effects of Carbon Nanotubes: Size Does Matter! *Langmuir* **2008,** *24,* 6409–6413.
32. Li, Q.; Mahendra, S.; Lyon, D. Y.; Brunet, L.; Liga, M. V.; Li, D.; Alvarez, P. J. J. Antimicrobial Nanomaterials for Water Disinfection and Microbial Control: Potential Applications and Implications. *Water Res.* **2008,** *42,* 4591–4602.
33. Lin, H. F.; Ravikrishna, R.; Valsaraj, K. T. Reusable Adsorbents for Dilute Solution Separation. 6. Batch and Continuous Reactors for the Adsorption and Degradation of *1,*2-Dichlorobenzene from Dilute Wastewater Streams Using Titania as a Photocatalyst. *Separation Purification Technol.* **2002,** *28,* 87–102.
34. Liu, Y.; Li, Y.; Yan, X.-P. Preparation, Characterization, and Application of L-Cysteine Functionalized Multiwalled Carbon Nanotubes as a Selective Sorbent for Separation and Preconcentration of Heavy Metals. *Adv. Funct. Mater.* **2008,** *18,* 1536–1543.
35. Liu, Y.; Lu, S.; Panchapakesan, B. Alignment Enhanced Photoconductivity in Single Wall Carbon Nanotube Films. *Nanotechnology* **2009,** 20, 035203.
36. Logan, B. E.; Hamelers, B.; Rozendal, R.; Schröder, U.; Keller, J.; Freguia, S.; Aelterman, P.; Verstraete, W.; Rabaey, K. Microbial Fuel Cells: Methodology and Technology†. Environ. *Sci. Technol.* **2006,** *40,* 5181–5192.
37. Long, R. Q.; Yang, R. T. Carbon Nanotubes as Superior Sorbent for Dioxin Removal. J. Am. *Chem. Soc.* **2001,** *123,* 2058–2059.
38. Lu, C.; Chiu, H. Adsorption of Zinc(II) from Water with Purified Carbon Nanotubes. Chem. *Eng. Sci.* **2006,** *61,* 1138–1145.

39. Lu, C.; Chiu, H. Chemical Modification of Multiwalled Carbon Nanotubes for Sorption of Zn2+ from Aqueous Solution. *Chem. Eng. J.* **2008,** *139,* 462–468.
40. Lu, J.; Kumar, B.; Castro, M.; Feller, J.-F. Vapour Sensing with Conductive Polymer Nanocomposites (CPC): Polycarbonate-carbon Nanotubes Transducers with Hierarchical Structure Processed by Spray Layer by Layer. Sens. *Actuat. B Chem.* **2009,** *140,* 451–460.
41. Lyu, J.; Kudiiarov, V.; Lider, A. An Overview of the Recent Progress in Modifications of Carbon Nanotubes for Hydrogen Adsorption. *Nanomaterials* **2020,** *10,* 255.
42. Mauter, M. S.; Elimelech, M. Environmental Applications of Carbon-Based Nanomaterials. *Environ. Sci. Technol.* **2008,** *42,* 5843–5859.
43. Mills, A.; Le Hunte, S. An Overview of Semiconductor Photocatalysis. J. Photochem. Photobiol. *A Chem.* **1997,** 108, 1–35.
44. Minteer, S. D.; Liaw, B. Y.; Cooney, M. J. Enzyme-Based Biofuel Cells. *Curr. Opin. Biotechnol.* **2007,** *18,* 228–234.
45. Mitra, B. Environment Friendly Composite Materials: Biocomposites and Green Composites. *Defence Sci. J.* **2014,** *64,* 244–261.
46. Mojumdar, S. C.; Raki, L. Preparation and Properties of Calcium Silicate Hydrate-Poly(Vinyl Alcohol) Nanocomposite Materials. *J. Therm. Analy. Calorimetry* **2005,** *82,* 89–95.
47. Moon, S.-I. L.; Paek, K.-K.; Lee, Y.-H.; Park, H.-K.; Kim, J.-K.; Kim, S.-W.; Ju, B.-K. Bias-Heating Recovery of MWCNT Gas Sensor. *Mater. Lett.* **2008,** *62,* 2422–2425.
48. Morozan, A.; Stamatin, L.; Nastase, F.; Dumitru, A.; Vulpe, S.; Nastase, C.; Stamatin, I.; Scott, K. The Biocompatibility Microorganisms-Carbon Nanostructures for Applications in Microbial Fuel Cells. *Phys. Status Solidi (A)* **2007,** *204,* 1797–1803.
49. Mostafavi, S. T.; Mehrnia, M. R.; Rashidi, A. M. Preparation of Nanofilter from Carbon Nanotubes for Application in Virus Removal from Water. *Desalination* **2009,** *238,* 271–280.
50. Muller, J.; Huaux, F.; Moreau, N.; Misson, P.; Heilier, J.-F.; Delos, M.; Arras, M.; Fonseca, A.; Nagy, J. B.; Lison, D. Respiratory Toxicity of Multi-Wall Carbon Nanotubes. *Toxicol. Appl. Pharmacol.* **2005,** *207,* 221–231.
51. Nepal, D.; Balasubramanian, S.; Simonian, A. L.; Davis, V. A. Strong Antimicrobial Coatings: Single-Walled Carbon Nanotubes Armored with Biopolymers. *Nano Lett.* **2008,** *8,* 1896–1901.
52. Nguyen, L. H.; Phi, T. V.; Phan, P. Q.; Vu, H. N.; Nguyen-Duc, C.; Fossard, F. Synthesis of Multi-Walled Carbon Nanotubes for NH3 Gas Detection. *Phys. E Low-Dimensional Syst. Nanostruct* **2007,** *37,* 54–57.
53. O'Regan, B.; Grätzel, M. A Low-Cost, High-Efficiency Solar Cell Based on Dye-Sensitized Colloidal TiO2 Films. *Nature* **1991,** *353,* 737–740.
54. Ong, Y. T.; Ahmad, A. L.; Zein, S. H. S.; Tan, S. H. A Review on Carbon Nanotubes in an Environmental Protection and Green Engineering Perspective. *Braz. J. Chem. Eng.* **2010,** *27,* 227–242.
55. Palmore, G. T. R.; Whitesides, G. M. Microbial and Enzymatic Biofuel Cells. **1994,** *Am. Chem. Soc. 566,* 271–290.
56. Pasquier, A. D.; Unalan, H. E.; Kanwal, A.; Miller, S.; Chhowalla, M. Conducting and Transparent Single-Wall Carbon Nanotube Electrodes for Polymer-Fullerene Solar Cells. *Appl. Phys. Lett.* **2005,** *87,* 203511.

57. Penza, M.; Rossi, R.; Alvisi, M.; Signore, M. A.; Cassano, G.; Dimaio, D.; Pentassuglia, R.; Piscopiello, E.; Serra, E.; Falconieri, M. Characterization of Metal-Modified and Vertically-Aligned Carbon Nanotube Films for Functionally Enhanced Gas Sensor Applications. *Thin Solid Films* **2009,** *517,* 6211–6216.
58. Qi, P.; Vermesh, O.; Grecu, M.; Javey, A.; Wang, Q.; Dai, H.; Peng, S.; Cho, K. J. Toward Large Arrays of Multiplex Functionalized Carbon Nanotube Sensors for Highly Sensitive and Selective Molecular Detection. *Nano Lett.* **2003,** *3,* 347–351.
59. Raffaelle, R. P.; Landi, B. J.; Evans, C. M.; Castro, S. L.; Bailey, S. G. Quantum Dot-Single Wall Carbon Nanotube Complexes For Polymeric Solar Cells. *Thirty-first IEEE Photovoltaic Specialists Conference,* 2005, 74–77.
60. Rahman, M. Z.; Khan, S. I. Advances in Surface Passivation of C-Si Solar Cells. Mater. Renew. *Sustain. Energ.* **2012,** 1.
61. Reijnders, L. Cleaner Nanotechnology and Hazard Reduction of Manufactured Nanoparticles. *J. Clean. Prod.* **2006,** *14,* 124–133.
62. Ritschel, M.; Uhlemann, M.; Gutfleisch, O.; Leonhardt, A.; Graff, A.; Täschner, C.; Fink, J. Hydrogen Storage in Different Carbon Nanostructures. *Appl. Phys. Lett.* **2002,** *80,* 2985–2987.
63. Savage, N.; Diallo, M. S. Nanomaterials and Water Purification: Opportunities and Challenges. *J. Nanopart. Res.* **2005,** *7,* 331–342.
64. Sayes, C. M.; Liang, F.; Hudson, J. L.; Mendez, J.; Guo, W.; Beach, J. M.; Moore, V. C.; Doyle, C. D.; West, J. L.; Billups, W. E.; Ausman, K. D.; Colvin, V. L. Functionalization Density Dependence of Single-Walled Carbon Nanotubes Cytotoxicity In Vitro. *Toxicol. Lett.* **2006,** *161,* 135–142.
65. Scarselli, M.; Scilletta, C.; Tombolini, F.; Castrucci, P.; De Crescenzi, M.; Diociaiuti, M.; Casciardi, S.; Gatto, E.; Venanzi, M. Photon Harvesting with Multi Wall Carbon Nanotubes. *Superlattices Microstruct.* **2009,** *46,* 340–346.
66. Schaller, R.; Mari, D.; dos Santos, S. M.; Tkalcec, I.; Carreño-Morelli, E. Investigation of Hydrogen Storage in Carbon Nanotube–Magnesium Matrix Composites. *Mater. Sci. Eng. A.* **2009,** *521–522,* 147–150.
67. Sharma, T.; Mohanareddy, A.; Chandra, T.; Ramaprabhu, S. Development of Carbon Nanotubes and Nanofluids Based Microbial Fuel Cell. *Int. J. Hydrogen Energ.* **2008,** *33,* 6749–6754.
68. Sheldon, R. A. Enzyme Immobilization: The Quest for Optimum Performance. *Adv. Synth. Cataly.* **2007,** *349,* 1289–1307.
69. Simon, P.; Gogotsi, Y. Materials for Electrochemical Capacitors. *Nat. Mater.* **2008,** *7,* 845–854.
70. Sitharaman, B.; Shi, X.; Walboomers, X. F.; Liao, H.; Cuijpers, V.; Wilson, L. J.; Mikos, A. G.; Jansen, J. A. In Vivo Biocompatibility of Ultra-Short Single-Walled Carbon Nanotube/Biodegradable Polymer Nanocomposites for Bone Tissue Engineering. Bone **2008,** *43,* 362–370.
71. Srivastava, S. Sorption Of Divalent Metal Ions from Aqueous Solution By Oxidized Carbon Nanotubes and Nanocages: A Review. *Adv. Mater. Lett.* **2013,** *4,* 2–8.
72. Stafiej, A.; Pyrzynska, K. Solid Phase Extraction of Metal Ions Using Carbon Nanotubes. *Microchem. J.* **2008,** *89,* 29–33.
73. Suehiro, J.; Zhou, G.; Hara, M. Detection of Partial Discharge in SF6 Gas Using a Carbon Nanotube-Based Gas Sensor. *Sens. Actuat. B Chem.* **2005,** *105,* 164–169.

74. Tahaikt, M.; El Habbani, R.; Ait Haddou, A.; Achary, I.; Amor, Z.; Taky, M.; Alami, A.; Boughriba, A.; Hafsi, M.; Elmidaoui, A. Fluoride Removal from Groundwater by Nanofiltration. *Desalination* **2007,** *212,* 46–53.
75. Tsai, H.-Y.; Wu, C.-C.; Lee, C.-Y.; Shih, E. P. Microbial Fuel Cell Performance of Multiwall Carbon Nanotubes on Carbon Cloth as Electrodes. *J. Power Sources* **2009,** *194,* 199–205.
76. Türker, A. R. New Sorbents for Solid-Phasc Extraction for Metal Enrichment. CLEAN–Soil, *Air Water* **2007,** *35,* 548–557.
77. Ueda, T.; Bhuiyan, M. M. H.; Norimatsu, H.; Katsuki, S.; Ikegami, T.; Mitsugi, F. Development of Carbon Nanotube-Based Gas Sensors for NOx Gas Detection Working at Low Temperature. Phys. E Low-Dimension. Syst. *Nanostruct.* **2008,** *40,* 2272–2277.
78. Ulbricht, R.; Lee, S. B.; Jiang, X.; Inoue, K.; Zhang, M.; Fang, S.; Baughman, R. H.; Zakhidov, A. A. Transparent Carbon Nanotube Sheets as 3D Charge Collectors in Organic Solar Cells. *Sol. Energ. Mater. Sol. Cells* **2007,** *91,* 416–419.
79. Vaudreuil, S.; Labzour, A.; Sinha-Ray, S.; Mabrouk, K. E.; Bousmina, M. Dispersion Characteristics and Properties of Poly(Methyl Methacrylate)/Multi-Walled Carbon Nanotubes Nanocomposites. *J. Nanosci. Nanotechnol.* **2007,** *7,* 2349–2355.
80. Wang, X.; Lu, J.; Xing, B. Sorption of Organic Contaminants by Carbon Nanotubes: Influence of Adsorbed Organic Matter. *Environ. Sci. Technol.* **2008,** *42,* 3207–3212.
81. Wang, Y.; Deng, W.; Liu, X.; Wang, X. Electrochemical Hydrogen Storage Properties of Ball-Milled Multi-Wall Carbon Nanotubes. *Int. J. Hydrogen Energ.* **2009,** *34,* 1437–1443.
82. Watanabe, K. Recent Developments in Microbial Fuel Cell Technologies for Sustainable Bioenergy. *J. Biosci. Bioeng.* **2008,** *106,* 528–536.
83. Wei, C.; Dai, L.; Roy, A.; Tolle, T. B. Multifunctional Chemical Vapor Sensors of Aligned Carbon Nanotube and Polymer Composites. *J. Am. Chem. Soc.* **2006,** *128,* 1412–1413.
84. Wick, P.; Manser, P.; Limbach, L.; Dettlaffweglikowska, U.; Krumeich, F.; Roth, S.; Stark, W.; Bruinink, A. The Degree and Kind of Agglomeration Affect Carbon Nanotube Cytotoxicity. *Toxicol. Lett.* **2007,** *168,* 121–131.
85. Xu, D.; Tan, X.; Chen, C.; Wang, X. Removal of Pb(II) from Aqueous Solution by Oxidized Multiwalled Carbon Nanotubes. *J. Hazard. Mater.* **2008,** *154,* 407–416.
86. Yang, K.; Zhu, L.; Xing, B. Adsorption of Polycyclic Aromatic Hydrocarbons by Carbon Nanomaterials. Environ. *Sci. Technol.* **2006,** *40,* 1855–1861.
87. Yu, F.; Wu, Y.; Ma, J.; Zhang, C. Adsorption of Lead on Multi-Walled Carbon Nanotubes with Different Outer Diameters and Oxygen Contents: Kinetics, Isotherms and Thermodynamics. *J. Environ. Sci.* **2003,** *25,* 195–203.
88. Zacharia, R.; Rather, S.-u.; Hwang, S. W.; Nahm, K. S. Spillover of Physisorbed Hydrogen from Sputter-Deposited Arrays of Platinum Nanoparticles to Multi-Walled Carbon Nanotubes. *Chem. Phys. Lett.* **2007,** *434,* 286–291.
89. Zhang, Q.; Huang, J.; Wei, F.; Xu, G.; Wang, Y.; Qian, W.; Wang, D. Large Scale Production of Carbon Nanotube Arrays on the Sphere Surface from Liquefied Petroleum Gas at Low Cost. *Chin. Sci. Bull.* **2007,** *52,* 2896–2902.
90. Zhang, W.-D.; Zhang, W.-H. Carbon Nanotubes as Active Components for Gas Sensors. *J. Sens.* **2009,** *2009,* 1–16.

91. Zhou, L. J.; Tan, J. G.; Hu, B. H.; Feng, H. L. [Ultrastructural Study of Sclerotic Dentin in Non-Carious Cervical Lesions Disposed by a Total-Etching Dentin Adhesive]. *Beijing Da Xue Xue Bao Yi Xue Ban* **2004,** *36,* 319–321.
92. Zhou, M.; Deng, L.; Wen, D.; Shang, L.; Jin, L.; Dong, S. Highly Ordered Mesoporous Carbons-Based Glucose/O_2 Biofuel Cell. *Biosens. Bioelectron.* **2009,** *24,* 2904–2908.
93. Zhu, H.; Zeng, H.; Subramanian, V.; Masarapu, C.; Hung, K.-H.; Wei, B. Anthocyanin-Sensitized Solar Cells Using Carbon Nanotube Films as Counter Electrodes. *Nanotechnology* **2008,** *19,* 465204

CHAPTER 4

CNTs as New Emerging Lubricant Additives for Enhancing Energy Efficiency

AVINASH V. BORGAONKAR[1*] and SHITAL B. POTDAR[2]

[1]*Department of Mechanical Engineering, National Institute of Technology Warangal, Warangal, Telangana, India*

[2]*Department of Chemical Engineering, National Institute of Technology Warangal, Warangal, Telangana, India*

[*]*Corresponding author. E-mail: avi.borgaonkar@gmail.com*

ABSTRACT

Almost 30% of the total energy is lost in machinery and engine parts due to wear and frictional loss. The semi-solid and liquid lubricants used in machinery and engine parts to minimize friction and wear are unable to sustain at high load and high operating temperatures. To improve the tribological performance and load wear index, antiwear (AW) and extreme pressure (EP) additives are dispersed into lubricating oil. These additives provide cushioning effect in between the contacting surfaces by forming thin films, in turn preventing severe wear and seizure. In the past two decades, researchers tried different additives, such as graphite, fullerenes, nanoonions, CNT, and graphene due to their extraordinary properties. The sp2 C-C bond of CNTs made them a strongest candidate among solid materials. Because of stress-free surface modification and larger surface area compared with other existing inorganic nanomaterials, the use of CNTs aggrandize the attention of researchers for tribological applications.

4.1 INTRODUCTION

In the recent published literature, it has been reported that friction and wear are the major reasons toward failure of machineries and their components as well as systems used in military, aerospace, and industrial applications. In automobile engines, almost 5% of the total energy produced is lost to overcome frictional resistance.[1] In concern with this issue, there is necessity to formulate novel lubricants with doping of various nanoadditives which helps to improve tribological performance. Undoubtedly, this will have immediate impact on energy efficiency, sustainability, and increased mechanical performance.

The research in the field of energy efficient solutions is attributed toward the development of novel nanomaterials, instead of use of the conventional nanoadditives. Presently, different types of nanomaterials are under examination as base oil additives. The nanomaterials have been evolved as a relatively new class of lubricant additives in the current development of lubricant additives. Nanolubricants mainly contain base oil or fully formulated lubricant with suspended colloidal solid particles within them. The tribological tests can be performed employing different test rigs like pin-on-disc, four-ball tester, ball-on-disc, ball-on-plate, block-on-ring, twin-disc test rig, etc. This section represents an overview on the tribological properties of nanomaterials type, compatibility of base oil/nanomaterial additive, optimum wt.% of nanoparticles, and environmental tests conditions.

The different test rigs used to carry out the tribological analysis are represented in Figure 4.1 (Tables 4.1–4.6).

4.2 ELEMENTS OF NANOLUBRICANTS

4.2.1 LUBRICANT CLASSIFICATION

4.2.1.1 VEGETABLE AND ANIMAL OIL

These are extracted from the fats of vegetables and animals. These oils have a tendency to stick on the contacting surfaces effectively at higher loads and high temperatures due to good oiliness. These oils are used for the purpose of lubrication. But with humid air or water, they easily undergo oxidation, and petroleum-based lubricants are blended with these agents to get enhanced oiliness.

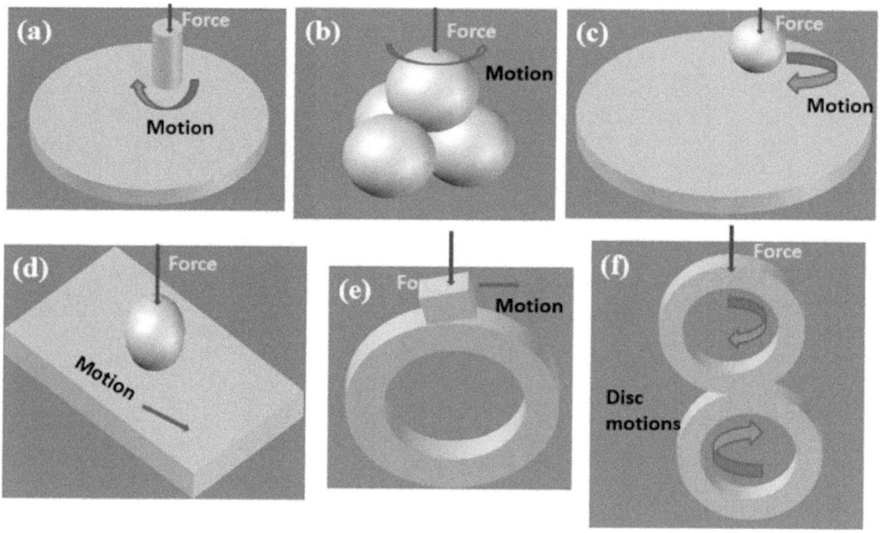

FIGURE 4.1 Different test rigs used for tribological analysis (a) pin-on-disc test rig, (b) four-ball test rig, (c) ball-on-disc test rig, (d) ball-on-flat tester, (e) block-on-ring test rig, (f) twin-disc test rig.

4.2.1.2 PETROLEUM OILS OR MINERAL

These oils are having about 12–50 atoms of carbon. They are available in abundance and are economical. Due to their stability under working conditions, they are extensively used. It has less oiliness compared with animal and vegetable oils.

4.2.1.3 BLENDED OILS

The oils do not have all the required properties that are required for a lubricant, hence they are added with sufficient additives to enhance their properties and for better performance. The additives that are added to lubricating to enhance the properties of oils are named as blended oils.

TABLE 4.1 Tribological Test Carried Out Using Pin-on-disc Test Rig.

Particles	Base oil	Findings	References
Surface-modified MWCNTs	Pure liquid paraffin	Effective membrane of MWCNTs-dispersed lubricant developed on the surface leads to reduction in friction and wear	[2]
MWCNTs	Lithium grease	At lower speed (150 RPM), 0.5 wt.% addition of MWCNTs reduces COF by 42.6%.	[3]
Cu nanoparticles	Commercial lubricant (500SN)	Heat generated during friction brings about the deposition of the Cu nanoparticles, which results in mending effect	[4]
Ni-containing SWCT and carbon nanoonions	PAO oil	The presence of a Ni metallic core inside CNTs have shown significant improvement in tribological performance under boundary lubrication conditions	[5]
Polyphenylene sulfide (PPS) composites filled with CuO particles	Water	The agglomeration of CuO nanoparticles takes place on the contacting substrate surface provides protection from wear	[6]
Nanodiamonds	Ethylene glycol	The decrease in friction and wear was observed with both evenly distributed and agglomerated nanodiamonds	[7]
Ag-MoS_2 nanoparticles	25 cst 75W140 Gear oil and PAO with 8 cst, 25 cst, and 40 cst viscosity	Ag–MoS_2 nanolubricant has significantly reduces COF and wear Besides, Ag nanoparticles prolonged the life of pure MoS_2 nanoparticles at higher temperatures	[8]
Cu nanoparticles	Mineral and synthetic ester base oils	The Cu-dispersed mineral oil exhibits reduced COF and enhanced antiwear characteristics at 0.3% wt	[9]
Titanium oxide (TiO_2) nanoparticles	Mineral-based multigrade engine oil	The TiO_2-dispersed engine oil shows significant reduction in COF and wear rate	[10]

TABLE 4.2 Tribological Testing Using Four-Ball Test Rig.

Particles	Base oil	Findings	References
TiO_2, Al_2O_3, CuO, and MWCNTs	Metkool, Metkut, Montgomery, Ecodraw	CuO-dispersed Metkut lubricant shows reduction in wear scar diameter and friction coefficient by 86% and 7% respectively	[11]
Long and short MWCNTs	EP 140 gear oil	The stability of the MWCNTs-dispersed lubricant improved significantly. This leads to enhanced tribological and extreme pressure properties	[12, 13, 15]
CNTs	Calcium grease	With 3 wt.% CNT addition friction is reduced by about 50%, the wear scar diameter (WSD) decreased to 32% and the EP properties enhance about 38%	[14]
Surface coated by LaF_3 nanoparticles	Liquid paraffin	Enhanced load-carrying capacity and antiwear property have been observed with LaF_3 dispersed oil. Besides significant reduction in friction was observed as compared with ZDDP	[16]
MoS_2 and WS_2 nanoparticles	EP 140 grade transmission oil	The 0.5 wt.% nanoparticles dispersed lubricant enhances tribological properties and EP properties lubricant	[17]
Tetrafluorobenzoic acid-modified TiO_2 nanoparticles	Liquid paraffin	The nanoparticles-dispersed lubricant enhances tribological properties and EP properties lubricant	[18]
WS_2, chloroparaffins (T302), and MoS_2	Petrol engine oil	The WS_2-dispersed petrol engine oil exhibits excellent tribological properties	[19]
CuO, ZnO, and ZrO_2 nanoparticles	PAO oil	CuO and ZnO nanoparticles exhibited best EP behavior whereas ZrO_2 shown worst behavior	[20]
Surface-modified Cu nanoparticles	50 CC oil	A thin Cu film having low elastic modulus and hardness was observed on the contacting surface resulting in the enhanced frictional and wear properties; especially at higher oil temperature	[21]
Modified graphene platelets (MGP)	350SN base oil	The load-carrying capacity and wear resistance significantly improved with 0.075 wt.%. MGP dispersed oil. The friction has been significantly reduced with MGP dispersed oil compared with pure base oil	[22]

TABLE 4.3 Tribological Test Carried Out Using Ball-on-disc Test Rig.

Particles	Base oil	Findings	References
MWCNTs, MWCNTs with surfactant (Triton X-100)	Mineral oil (0.2228 cPs)	The nano lubricant containing MWCNT dispersed with surfactant exhibits minimum wear and COF	[24]
Rhenium-doped inorganic fullerene MoS_2 nanoparticles	Poly alpha olefin PAO oil	In the case of both boundary and mixed lubrication regimes, great improvement in the tribological properties have been observed with formulated oil under severe operating conditions	[25]
Surface-modified Palladium (Pd) nanoparticles	Liquid paraffin+ Tetrabutylammonium acetate (TBA) as surfactant	The Pd-nanoparticles protected by a surfactant layer exhibits better lubrication of metallic contacts possessing lower values of electrical resistivity. The formed protective layer on the ball surface heals the wear tracks developed on the disc surface. This resulted into enhanced load-carrying capacity and reduction in COF.	[26]
MWNTs	PAO oil	The MWNTs-dispersed oil reduces the wear rate by 5–9 times and friction coefficient by two times.	[27]
TiO_2 nanoadditive	Water	The TiO_2-dispersed water lubricant exhibits enhanced antifriction and antiwear properties	[28]
Silver nanoparticles	polyethylene-glycol (PEG) lubricant	The addition of silver nanoparticles shows significant reduction in wear	[29]
MoS_2 nanoparticles	Castor oil	The MoS_2 nanoparticle-dispersed lubricant reduces the possibility of asperities direct contact, resulting in the reduction of COF and adhesive wear	[30]
Composite Si3N4/SiC nanoparticles	Mineral oil mixed with nanoparticles of diamond	The lubricant tends to decrease the COF by more than 40% and also significantly decreases the wear coefficient	[31]
TiO_2 nanoparticles	Sodium dodecyl benzene sulfonate (SDBS) + glycerol + water	The formulated nanolubricant with 4 wt.% of TiO_2 and 0.4 wt.% of SDBS exhibited enhanced frictional and wear properties by reducing friction coefficient and wear rate up to 70.5% and 84.3%, respectively	[32]
h-BN nanoparticles	ester base oil	The lubricant reduces COF by 25%, scar width by 9%, scar depth by 14% and transversal area by 22% all with respect to pure oil	[33]

TABLE 4.4 Tribological Test Carried Out Using Ball-on-flat Tester.

Particles	Base oil	Findings	References
Inorganic fullerene-like (IF) nanoparticles of WS_2	Mineral oil SN 90, SN 150 and bright stock	The COF has been reduced by 50% in the mixed lubrication regime with the use of formulated oil	[34]
WS_2 (MoS_2) inorganic nanotubes and fullerene nanoparticles	Paraffin oil	WS_2-dispersed lubricant results into reduced COF	[35]
		This lubricant shows significant reduction in the run-in period. The friction and wear properties also have been significantly improved	[36]
CuO, TiO_2, and Diamond nanoparticles	PI-SF engine oil and base oil	The CuO nanoparticles-dispersed API-SF engine oil and the base oil decreases the COF by 18.4% and 5.8%, respectively, and reduces the worn scar depth by 16.7% and 78.8%, respectively in comparison with pure base oil	[37]
MoS_2 and SiO_2 nanoparticles	EOT5 engine oil	The MoS_2-dispersed lubricant exhibits excellent stability compared with SiO_2 nanolubricants. Also, MoS_2-dispersed lubricant exhibited significant improvement in the tribological properties and load-carrying capacity	[38]
Fluorographene (FG)	GTL-8-based lubricant oil	The FG-dispersed oil with 0.3 mg/mL reduces the COF and the wear rate by 35% and 90%, respectively	[39]
Perfluoropolyether (PFPE) nanolubricant		The COF of the counter-disc bonded with PEPE film decreases linearly with increase in the surface coverage	[40]
α-pinene tribo-polymers		The vapor of α-pinene and pinnae produces a polymer-like film during sliding between the contacting surfaces. This developed film play a role of an effective lubricant in the absence of continuous supply of the organic vapor	[41]
α-FeOOH nanorods	PAO lubricating oils	Closely packed nanorods has significantly reduces wear and friction. COF decreases by 22%	[42]
MoS_2, SiO_2, ND, CNTs, α-Al_2O_3, and ZrO_2).	Palm oil	The Al_2O_3 nanofluid exhibits significant antifriction property. The COF reduced by 20%	[43]

TABLE 4.5 Tribological Test Carried Out Using Block-on-ring Test Rig.

Particles	Base oil	Findings	References
Nanodiamond, onion-like carbon (OLC), (SWCNTs/MWCNTs) or nanographene platelets	Polyalphaolefin (PAO) oil and PAO + molybdenum dialkyldithiophosphate (MoDDP)	The polishing capability of nanodiamonds along with the protective action of MoDDP greatly reduces the COF and wear rate in comparison with pure PAO	[44]
Modified nanodiamond (ND)	SAE10W engine oil	ND dispersed lubricant helps to improve the frictional and wear properties and reduces the oil temperature. The antiwear mechanism attributed due to developed hard and porous layer between the contacting surfaces	[45]
Ni nanoparticles	PAO6 oil	Ni nanoparticles dispersed oil helps to improve the tribological properties and also improved the load-carrying capacity	[46]
Carbon-coated Cu nanoparticles	PAO6 oil	The carbon-coated Cu nanoparticles dispersed oil leads to enhance the load-carrying capacity and reduces the wear	[47]
Multilayer graphene impregnated with copper (MLG-Cu), and polyaniline (MLG-PANI)	Commercial motor oil	The MLG-Cu with 0.5 wt.% and 2 wt.% exhibits reductions in the COF and wear up to 43% and 63%, respectively	[48]
Nanodiamond-PTFE-MoDDP	Polyalphaolefins and engine oils	Significant reductions in the COF and wear have been observed. With the DND dispersed oil, 35% reduction in surface roughness has been observed compared with pure oil	[49]
CuO nanoparticles	Polyalphaolefin (PAO6)	Due to CuO-dispersed oil reduction in the COF and wear have been observed. This is attributed because of the third body and tribosinterization mechanism.	[50]
Nanodiamond	Commercial oil (CPC circulation oil R68, Taiwan)	3 wt.% addition of the ND exhibits significant reduction in the friction compared with the base oil	[51]

TABLE 4.6 Tribological Test Carried Out Using Twin-Disc Test Rig.

Particles	Base oil	Findings	References
Graphite Carbon nanofibres, CNTs	Mineral oil (Super gear EP220, SK, Korea)	Nanooils with the fibrous particle additives (e.g., CNFs/CNTs) showed higher COF than the nanooils with spherical nanoparticle additives, since the fibrous nanoparticles have higher aspect ratios	[52]
Functionalized SWCNTs/MWCNTs modified with carboxylic acid	Oil and water	Reduced in both COF and wear caused due to the formation of an amorphous carbon film transferred from the CNT's	[53]
Graphite nanoparticles	Industrial gear oil (viscosity 220 cSt)	The graphite-dispersed lubricant exhibits enhanced tribological properties	[54]
CuO nanoparticles	Mineral oil + sodium oleate (SOA) surfactant	The CuO/SOA nanoparticle additives reduce the friction. The effect of nanoparticles is more influential at higher loads and concentrations. The steady-state temperature of the lubricant is lower in the presence of nanoparticles	[55]
Carbon nanopearls (CNPs) and MWCNTs	Mixture of a synthetic ester resin and a rosin ester derived from pine tar, and linseed oil	Biolubricant additivation with raw unpurified MWCNTs and CNPs for enhanced tribological performance	[56]
Fullerene nanoparticles	Mineral oil-based lubricant	In the friction test, the friction coefficient of the disc specimen immersed in the nanooil was significantly lower than that of the disc specimen immersed in the mineral oil	[57]
Sulfur and phosphorus	Environmentally acceptable lubricant (EAL)-based oil and Commercial heavy truck gearbox lubricant	The EAL additives reacted with the surface to a greater extent than the commercial lubricant additives. An increase in the boundary layer, using the EAL, which itself needs to be sheared off, can explain the higher friction	[58]
Antioxidants, dispersants, detergent, EP, AW and FM compounds + 5% of MoS_2	Two commercial greases (LiCaM220 and LiM460)	The lithium-thickened grease (LiM50) showed lower friction values than the calcium-thickened grease (CaM50). LiM50 provides significant wear reduction while CaM50 generates high wear	[59]

Functions of lubricant

- To absorb heat generated between two mating surfaces.
- To reduce friction and wear of the contacting surfaces.
- To restrict the expansion of the contacting surfaces due to heat generated.
- To smoothen the relative motion.
- To reduce maintenance cost.
- To reduce power losses in turns minimizes energy losses.

Desirable features of lubricating oil

- It should possess high boiling point and low-freezing point.
- It should possess optimum viscosity for proper functioning.
- It should possess high oxidation and high heat resistance.
- It should possess anticorrosive properties.
- It should possess higher stability.

4.2.1.4 GREASE OR SEMI-SOLID LUBRICANTS

Greases are the semi-solid lubricants that are obtained by mixing lubricating oil with thickening agents. The principal component in the greases is lubricating oil and it can be from synthetic hydrocarbon or petroleum oil which is of high to low viscosity. The thickeners may be of soaps and non-soap. At lower speed grease, it can support a heavier load. Grease has high internal resistance when compared with that of lubricating oils, lubricating oils can efficiently disperse heat from the bearings when compared with that of grease. So greases work at a relatively lower temperature.

4.2.1.5 NANOMATERIALS

At present, research on nanomaterials is of great scientific interest because of huge potential in mechanical, biomedical, optical, and electronic field applications. Material with a size between 1 and 100 nm is classified as nanomaterial, while the behavior of material relies on its size. Hence, as the size is reduced to nanoscale and as the quantity of the atoms on the surface of the material becomes significant, changes in the properties of materials are witnessed.

There are a number of nanoparticles that can be utilized as grease-added substances. These are listed as follows.

Molybdenum disulfide, carbon nanotubes, Tungsten disulfide, polymers, zinc oxide, gold or silver nanoparticles, dendrimers, liposomes, graphene nanoplatelets, quantum dabs, iron oxide, perflurocarbon, micelles, ferritin, lanthanide particles.

Nanoparticles have surprising optical properties so that they are used in electrons to produce quantum effects. Due to differences in densities the nanoparticles can be suspended in solvents. The surface interaction has a high impact on the suspension, which results in either sinking in or floating on a liquid. At elevated temperatures, the volume of the nanoparticles plays a major role in diffusion. There is a vast development in the last 10 years and has many applications in the industrial, medical, and electronics field.

4.2.2 SELECTION OF BASE LUBRICANT

A lot of research is being done by many researchers to understand the effect of MWCNTs as additives in various lubricating oils. Lubricating oils provide the best wear reduction and friction coefficient and so is the best choice for an engine application. However, normal lubricants are not sufficient as in automotive systems due to their poor wear, friction, and load-bearing capacity, heat transfer coefficient and corrosion properties. So, in addition to the lubricating oil, a small number of other ingredients including corrosion inhibitors antifriction antiwear, and other additives are added to the lubricating oil.

Traditional lubricant typically contains 73–80 wt.% of aliphatic hydrocarbons, 11%–15% of monoaromatic hydrocarbons, 4%–8% polyaromatic hydrocarbons, 2%–5% diaromatic hydrocarbons. Approximately, 20% of the lubricants contain additives like zinc diaryl, zinc dithiophosphate, molybdenum disulfide and other organic metallic compounds. These lubricants gained international popularity for having an unsurpassed property for extended time intervals. The formulation process used for preparing a nanolubricant is depicted in Figure 4.2.

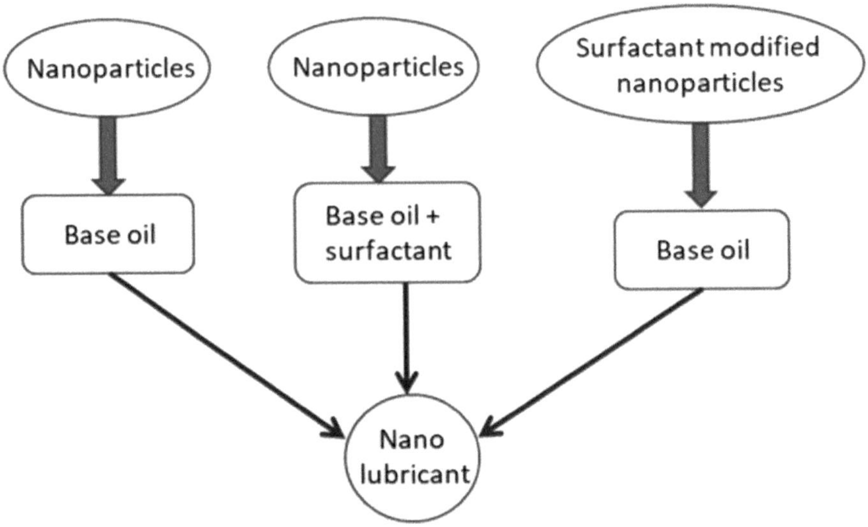

FIGURE 4.2 Formulation of nanolubricant.

4.3 LUBRICATION MECHANISMS

The lubrication mechanism analysis is an important step to understand the tribological behavior of nanoparticles. Various types of mechanisms have been suggested by researchers employing the surface analysis techniques to elaborate the lubrication enhancement by the nanoparticle-suspended lubricating oil. These mechanisms reported in the literature listed as ball-bearing effect,[37,60,61] the development of protective film,[62–64] mending effect[4] and polishing effect.[65] These mechanisms have been briefly elaborated as follows.

4.3.1 BALL-BEARING EFFECT

The quasi-spherical and spherical nanoparticles are assumed to function like minute ball bearings which roll in between the contacting surfaces. Such shapes of nanoparticles lead to change the sliding friction toward a combination of sliding and rolling friction. Such type of mechanism occurs in the case of tribological pair having steady low-load conditions between the shearing surfaces to manage the shape and rigidity of the nanoparticles. The ball-bearing mechanism is shown in Figure 4.3.

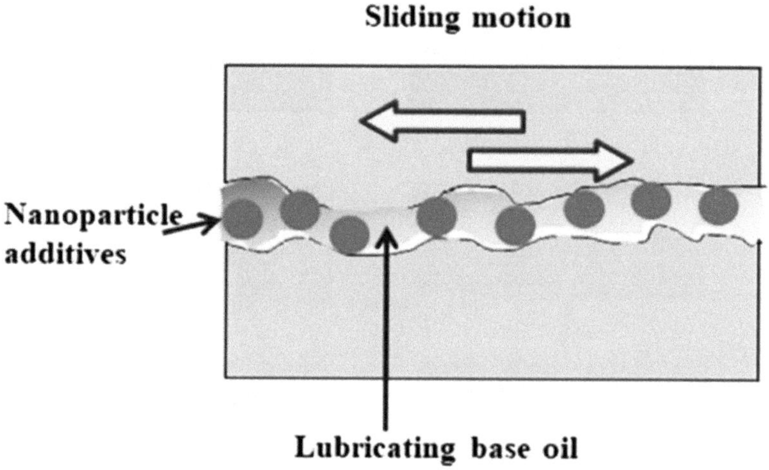

FIGURE 4.3 Schematic representation of ball-bearing mechanism with nanolubricant.

4.3.2 DEVELOPMENT OF PROTECTIVE FILM

The developed tribofilms on the contacting surfaces govern the tribological behaviors of the contacting surfaces. This development of the film is initialized due to chemical reaction between the substrate material, base oil, and the additives under the set operating conditions.[46] The formation of a durable tribofilm with the addition of Mo, S, and P elements reduces the friction, wear, and seizure for MoS_2 nanoparticles dispersed IN the lubricant.[35,58] Figure 4.4 shows schematically the tribofilm development, which protects the surfaces from crack propagation by reducing the friction between the contacting asperities.

4.3.3 MENDING EFFECT

The self-repairing effect or mending effect is analyzed by the deposition of the nanoparticles on the contacting surfaces and wear compensation. In this phenomenon, the nanoparticles get deposited on the comparatively soft surface to reduce abrasion.[66] Nanoparticles dispersed in the base oil have the tendency to fill grooves and asperities of the frictional surface. Figure 4.5 represents schematically the filled up asperities and grooves with nanoparticles on the frictional surface.

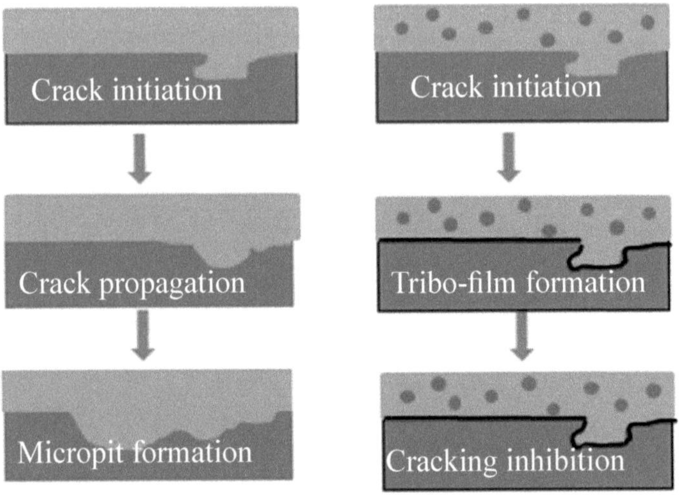

FIGURE 4.4 Schematic representation of the development of tribofilm with nanolubricant.

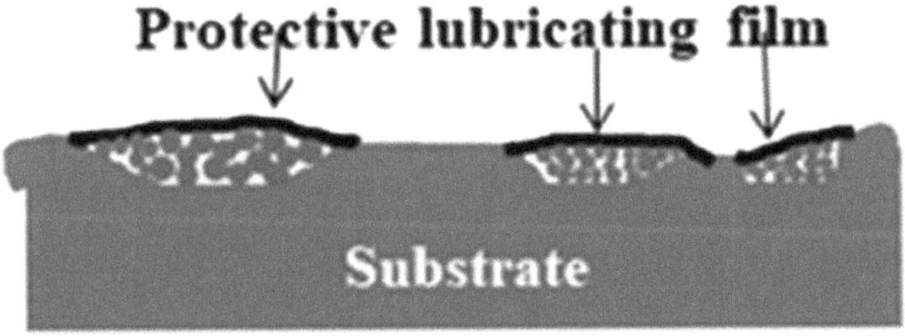

FIGURE 4.5 Schematics of nanoparticles filled up asperities and grooves on frictional surface.

4.3.4 POLISHING EFFECT

This is also defined as smoothing effect. In this phenomenon, it is hypothesized that the roughness of the mating surfaces is reduced by nanoparticle-assisted abrasion. The nanoparticles fill the rough asperities, which may act as reservoirs for nanolubricants within the contact, as depicted schematically in Figure 4.6. The process of filling up the rough asperities termed as the smoothing process. This polishing process or artificial smoothing resulted into improved tribological performance mainly because of the decreased surface roughness.[67]

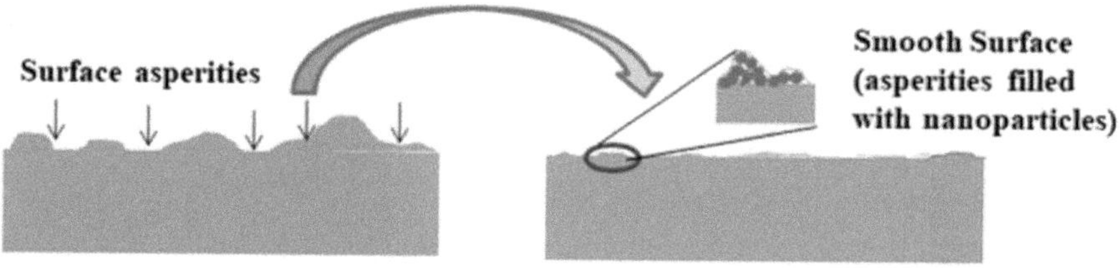

FIGURE 4.6 Schematics of polishing/smoothing effect due to filling up of rough valleys of contacting surfaces with nanoparticles.

4.4 SUMMARY

The referred literature reported that CNTs have been popularly used as a nanomaterial in order to improve the tribological performance of the contacting surfaces. Figure 4.7 shows that the carbon nanotubes (CNTs) are the potential candidate for various tribological applications.

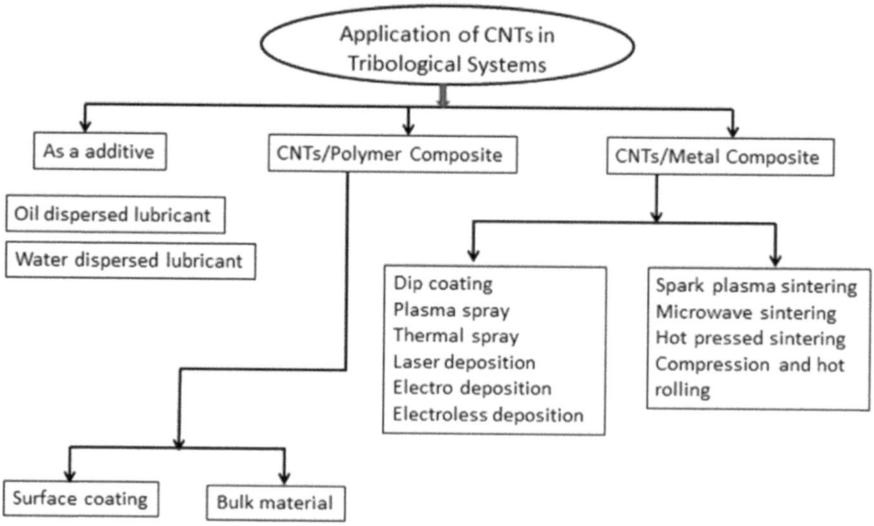

FIGURE 4.7 Application of CNTs in tribological systems.

The usage of different nanoparticles-dispersed lubricants helps to reduce friction and wear of the contacting surfaces having different relative motions. The reduction in friction in wear resulted into energy saving and enhanced efficiency of the tribological applications. The literature review represented in Table 4.7 shows that CNTs possessing the interesting antifriction and antiwear behavior make them suitable candidates as additives to improve the tribological performance. The Percentage (%) contribution is determined on the basis of nanoparticles repetitively used in cited articles in the current section.

TABLE 4.7 Different Nano Materials Additives Used for the Tribological Studies.

Sr. No.	Nanomaterial additives	Percentage (%) Contribution	References (from year 2000–2020)
1	LaF_3	1.2	[16]
2	Ni	2.4	[23,46]
3	TiO_2	7.1	[10,11,18,28,32,37]
4	Perfluoropolyether PFPE	1.2	[40]
5	Cu	5.9	[4,9,21,47,48]
6	WS_2	3.6	[17,19,34]
7	MWNTs	19.05	[2,3,11–15,24,27,43,44,52,53, 6]
8	WS_2/MoS_2	2.4	[35,36]
9	Chloroparaffins	1.2	[19]
10	MoS_2	7.1	[17,19,30,38,43,59]
11	PPS–CuO	1.2	[6]
12	CuO	5.9	[11,20,37,50,55]
13	ZnO	1.2	[20]
14	ZrO_2	2.4	[20,43]
15	Palladium (Pd)	1.2	[26]
16	Nanodiamond	7.1	[7,37,43,44,49,51]
17	Ni-SWCNT	1.2	[5]
18	Carbon nanoonions	3.6	[5,44,56]
19	Graphite	4.8	[13,39,52,54]
20	Fullerene	2.4	[25,57]
21	Modified graphene platelets (MGP)	2.4	[22,44]
22	Ag–MoS_2	2.4	[8,44]
23	Al_2O_2	2.4	[11,43]
24	Nanodiamond-PTFE	1.2	[49]
25	Composite Si_3N_4/SiC	1.2	[31]
26	SiO_2	2.4	[38,43]
27	α-pinene tribo-polymers	1.2	[41]
28	Polyaniline MLG-PANI	1.2	[48]
29	α-FeOOH nanorods	1.2	[42]
30	h-BN	1.2	[33]
31	Sulfur and phosphorus	1.2	[58]

Due to sulfur and phosphorus content, traditional lubricants lead to health and environmental hazards. These sensitive issues can be tackled by replacing conventional base oils by biobased base stocks dispersed with different nanomaterials that will suit for the specific applications. In this regard, the dispersion stability and the tribological compatibility of various nanomaterials as additives to biobased base stocks are required to be investigated for different applications.

KEYWORDS

- **energy**
- **friction**
- **MWCNTs**
- **SWCNTs**
- **tribological properties**
- **wear**

REFERENCES

1. Holmberg, K.; Erdemir, A. Influence of Tribology on Global Energy Consumption, Costs and Emissions. *Friction* **2017,** *5* (3), 263–284.
2. Chen, C. S.; Chen, X. H.; Xu, L. S.; Yang, Z.; Li, W. H. Modification of Multi-Walled Carbon Nanotubes with Fatty Acid and Their Tribological Properties as Lubricant Additive. *Carbon* **2005,** *43* (8), 1660–1666.
3. El-Adly, R. A.; Hussein, M. F.; Mohamed, A. Preparation and Evaluation of Lubricating Greases Based on Carbon Nanotube. *Int. J. Adv. Pharm. Biol. Chem.* **2015,** *4* (3), 675–684.
4. Liu, G.; Li, X.; Qin, B.; Xing, D.; Guo, Y.; Fan, R. Investigation of the Mending Effect and Mechanism of Copper Nano-Particles on a Tribologically Stressed Surface. *Tribol. Lett.* **2004,** *17* (4), 961–966.
5. Joly-Pottuz, L.; Vacher, B.; Le Mogne, T.; Martin, J. M.; Mieno, T.; He, C. N.; Zhao, N. Q. The Role of Nickel in Ni-Containing Nanotubes and Onions as Lubricant Additives. *Tribol. Lett.* **2008,** *29* (3), 213–219.
6. Cho, M. H.; Bahadur, S. Friction and Wear of Polyphenylene Sulfide Composites Filled with Micro and Nano CuO Particles in Water-Lubricated Sliding. *Tribol. Lett.* **2007,** *27* (1), 45–52.

7. Elomaa, O.; Oksanen, J.; Hakala, T. J.; Shenderova, O.; Koskinen, J. A Comparison of Tribological Properties of Evenly Distributed and Agglomerated Diamond Nanoparticles in Lubricated High-load Steel–steel Contact. *Tribol. Int.* **2014,** *71*, 62–68.
8. Zhang, W.; Demydov, D.; Jahan, M. P.; Mistry, K.; Erdemir, A.; Malshe, A. P. Fundamental Understanding of the Tribological and Thermal Behavior of Ag–MoS_2 Nanoparticle-based Multi-component Lubricating System. *Wear* **2014,** *288*, 9–16.
9. Borda, F. L. G.; de Oliveira, S. J. R.; Lazaro, L. M. S. M.; Leiroz, A. J. K. Experimental Investigation of the Tribological Behavior of Lubricants with Additive Containing Copper Nanoparticles. *Tribol. Int.* **2018,** *117*, 52–58.
10. Laad, M.; Jatti, V. K. S. Titanium Oxide Nanoparticles as Additives in Engine Oil. *J. King Saud Univ.-Eng. Sci.* **2018,** *30* (2), 116–122.
11. Pena-Paras, L.; Taha-Tijerina, J.; Garcia, A.; Maldonado, D.; Gonzalez, J. A.; Molina, D.; Palacios, E.; Cantu, P. Antiwear and Extreme Pressure Properties of Nanofluids for Industrial Applications. *Tribol. Trans.* **2014,** *57* (6), 1072–1076.
12. Chebattina, K. R. R.; Srinivas, V.; Rao, N. M. Effect of Size of Multiwalled Carbon Nanotubes Dispersed in Gear Oils for Improvement of Tribological Properties. *Advances in Tribology* **2018,** *V2018*, 1–13.
13. Bhaumik, S.; Prabhu, S.; Singh, K. J. Analysis of Tribological Behavior of Carbon Nanotube Based Industrial Mineral Gear Oil 250 cSt Viscosity. *Adv. Tribol.* **2014,** 1–8.
14. Kamel, B. M., Mohamed, A., El Sherbiny, M.; Abed, K. A. Tribological Behaviour of Calcium Grease Containing Carbon Nanotubes Additives. *Indus. Lubric. Tribol.* **2016,** *68* (6), 723–728.
15. Khalil, W.; Mohamed, A.; Bayoumi, M.; Osman, T. A. Tribological Properties of Dispersed Carbon Nanotubes in Lubricant. *Fullerenes Nanotubes Carbon Nanostruct.* **2016,** *24* (7), 479–485.
16. Zhou, J.; Wu, Z.; Zhang, Z.; Liu, W.; Dang, H. Study on an Antiwear and Extreme Pressure Additive of Surface Coated LaF3 Nanoparticles in Liquid Paraffin. *Wear* **2001,** *249* (6), 333–337.
17. Srinivas, V.; Thakur, R. N.; Jain, A. K.; Saratchandra Babu, M. Tribological Studies of Transmission Oil Dispersed with Molybdenum Disulfide and Tungsten Disulfide Nanoparticles. *J. Tribol.* **2017,** *139* (4), 1–6.
18. Ye, W.; Cheng, T.; Ye, Q.; Guo, X.; Zhang, Z.; Dang, H. Preparation and Tribological Properties of Tetrafluorobenzoic Acid-Modified TiO_2 Nanoparticles as Lubricant Additives. *Mater. Sci. Eng. A* **2003,** *359* (2), 82–85.
19. Chen, S. H. I.; Da-Heng, M. A. O.; Hao, F. E. N. G. Study on Tribological Characteristics of Ultrafine WS_2 Powders in Petrol Engine Oil [J]. *China Tungsten Indust.* **2006,** 1–3.
20. Battez, A. H.; González, R.; Felgueroso, D.; Fernández, J. E.; del Rocío Fernandez, M.; García, M. A.; Penuelas, I. Wear Prevention Behaviour of Nanoparticle Suspension Under Extreme Pressure Conditions. *Wear* **2007,** *263* (12), 1568–1574.
21. Yu, H.; Xu, Y.; Shi, P. Tribological Properties and Lubricating Mechanisms of Cu Nanoparticles in Lubricant. *Trans. Nonferrous Met. Soc. China* **2008,** *18* (1) 636–641.
22. Lin, J.; Wang, L.; Chen, G. Modification of Graphene Platelets and Their Tribological Properties as a Lubricant Additive. *Tribol. Lett.* **2011,** *41* (1), 209–215.
23. Qiu, S.; Zhou, Z.; Dong, J.; Chen, G. Preparation of Ni Nanoparticles and Evaluation of Their Tribological Performance as Potential Additives in Oils. *J. Trib.* **2001,** *123* (3), 441–443.

24. Lijesh, K. P.; Muzakkir, S. M.; Hirani, H. Experimental Tribological Performance Evaluation of Nano Lubricant Using Multi-Walled Carbon Nano-Tubes (MWCNT). *Int. J. Appl. Eng. Res.* **2015,** *10* (6), 14543–14550.
25. Yadgarov, L.; Petrone, V.; Rosentsveig, R.; Feldman, Y.; Tenne, R.; Senatore, A. Tribological Studies of Rhenium Doped Fullerene-Like MoS_2 Nanoparticles in Boundary, Mixed and Elasto-Hydrodynamic Lubrication Conditions. *Wear* **2013,** *297* (1–2), 1103–1110.
26. Kolodziejczyk, L.; Martinez-Martinez, D.; Rojas, T. C.; Fernandez, A.; Sanchez-Lopez, J. C. Surface-Modified Pd Nanoparticles as a Superior Additive for Lubrication. *J. Nanopart. Res.* **2007,** *9* (4), 639–645.
27. Kalin, M.; Kogovsek, J.; Remskar, M. Mechanisms and Improvements in the Friction and Wear Behavior Using MoS_2 Nanotubes as Potential Oil Additives. *Wear* **2012,** *280*, 36–45.
28. Wu, H.; Zhao, J.; Cheng, X.; Xia, W.; He, A.; Yun, J. H.; Huang, S.; Wang, L.; Huang, H.; Jiao, S.; Jiang, Z. Friction and Wear Characteristics of TiO_2 Nano-additive Water-based Lubricant on Ferritic Stainless Steel. *Tribol. Int.* **2018,** *117*, 24–38.
29. Neuffer, H.; Ghaednia, H.; Jackson, R. Wear Volume Analysis Using a Nano-lubricant for Ball-on-disk Testing. *Tribol. Lubric. Technol.* **2014,** *2*, 1–3.
30. Yu, R.; Liu, J.; Zhou, Y. Experimental Study on Tribological Property of MoS_2 Nanoparticle in Castor Oil. *J. Tribol.* **2019,** *141* (10), 1–5.
31. Charoo, M. S.; Wani, M. F. Friction and Wear Properties of Nano-Si_3N_4/Nano-SiC Composite Under Nanolubricated Conditions. *J. Adv. Ceram.* **2016,** *5* (2), 145–152.
32. Wu, H.; Jia, F.; Li, Z.; Lin, F.; Huo, M.; Huang, S.; Sayyar, S.; Jiao, S.; Huang, H.; Jiang, Z. Novel Water-based Nanolubricant with Superior Tribological Performance in Hot Steel Rolling. *Int. J. Extreme Manuf.* **2020,** *2* (2), 1–11.
33. Lineira del Rio, J. M.; Guimarey, M. J.; Comunas, M. J.; Lopez, E. R.; Prado, J. I.; Lugo, L.; Fernandez, J. Tribological and Thermophysical Properties of Environmentally-Friendly Lubricants Based on Trimethylolpropane Trioleate with Hexagonal Boron Nitride Nanoparticles as an Additive. *Coatings* **2019,** *9* (8), 1–17.
34. Greenberg, R.; Halperin, G.; Etsion, I.; Tenne, R. The Effect of WS_2 Nanoparticles on Friction Reduction in Various Lubrication Regimes. *Tribol. Lett.* **2004,** *17* (2), 179–186.
35. Rapoport, L.; Fleischer, N.; Tenne, R. Applications of WS_2 (MoS_2) Inorganic Nanotubes and Fullerene-like Nanoparticles for Solid Lubrication and for Structural Nanocomposites. *J. Mater. Chem.* **2005,** *15* (18), 1782–1788.
36. Perfiliev, V.; Moshkovith, A.; Verdyan, A.; Tenne, R.; Rapoport, L. A New Way to Feed Nanoparticles to Friction Interfaces. *Tribol. Lett.* **2006,** *21* (2), 89–93.
37. Wu, Y. Y.; Tsui, W. C.; Liu, T. C. Experimental Analysis of Tribological Properties of Lubricating Oils with Nanoparticle Additives. *Wear* **2007,** *262* (8), 819–825.
38. Xie, H.; Jiang, B.; He, J.; Xia, X.; Pan, F. Lubrication Performance of MoS_2 and SiO_2 Nanoparticles as Lubricant Additives in Magnesium Alloy-steel Contacts. *Tribol. Int.* **2016,** *93*, 63–70.
39. Ci, X.; Zhao, W.; Luo, J.; Wu, Y.; Ge, T.; Shen, L.; Gao, X.; Fang, Z. Revealing the Lubrication Mechanism of Fluorographene Nanosheets Enhanced GTL-8 Based Nanolubricant Oil. *Tribol. Int.* **2019,** *138*, 174–183.

40. Choi, J.; Kawaguchi, M.; Kato, T. The Surface Coverage Effect on the Frictional Properties of Patterned PFPE Nanolubricant Films in HDI. *IEEE Trans. Magnet.* **2003,** *39* (5), 2492–2494.
41. He, X.; Barthel, A. J.; Kim, S. H. Tribochemical Synthesis of Nano-Lubricant Films from Adsorbed Molecules at Sliding Solid Interface: Tribo-Polymers from α-pinene, Pinane, and n-decane. *Surf. Sci.* **2016,** *648*, 352–359.
42. Zin, V.; Agresti, F.; Barison, S.; Litti, L.; Fedele, L.; Meneghetti, M.; Fabrizio, M. Effect of External Magnetic Field on Tribological Properties of Goethite (a-FeOOH) Based Nanofluids. *Tribol. Int.* **2018,** *127*, 341–350.
43. Wang, Y.; Li, C.; Zhang, Y.; Li, B.; Yang, M.; Zhang, X.; Guo, S.; Liu, G.; Zhai, M. Comparative evaluation of the lubricating properties of vegetable-oil-based nanofluids between frictional test and grinding experiment. *J. Manuf. Proc.* **2017,** *26*, 94–104.
44. Nunn, N.; Mahbooba, Z.; Ivanov, M. G.; Ivanov, D. M.; Brenner, D. W.; Shenderova, O. Tribological properties of polyalphaolefin oil modified with nanocarbon additives. *Diamond Relat. Mater.* **2015,** *54*, 97–102.
45. Puzyr, A. P.; Burov, A. E.; Selyutin, G. E.; Voroshilov, V. A.; Bondar, V. S. Modified Nanodiamonds as Antiwear Additives to Commercial Oils. *Tribol. Trans.* **2012,** *55* (1), 149–154.
46. Chou, R.; Battez, A. H.; Cabello, J. J.; Viesca, J. L.; Osorio, A.; Sagastume, A. Tribological Behavior of Polyalphaolefin with the Addition of Nickel Nanoparticles. *Tribol. Int.* **2010,** *43* (12), 2327–2332.
47. Viesca, J. L.; Battez, A. H.; González, R.; Chou, R.; Cabello, J. J. Antiwear Properties of Carbon-Coated Copper Nanoparticles Used as an Additive to a Polyalphaolefin. *Tribol. Int.* **2011,** *44* (8), 829–833.
48. Ramon-Raygoza, E. D.; Rivera-Solorio, C. I.; Gimenez-Torres, E.; Maldonado-Cortes, D.; Cardenas-Aleman, E.; Cue-Sampedro, R. Development of Nanolubricant Based on Impregnated Multilayer Graphene for Automotive Applications: Analysis of Tribological Properties. *Powder Technol.* **2016,** *302*, 363–371.
49. Shenderova, O.; Vargas, A.; Turner, S.; Ivanov, D. M.; Ivanov, M. G. Nanodiamond-based Nanolubricants: Investigation of Friction Surfaces. *Tribol. Trans.* **2014,** *57* (6), 1051–1057.
50. ChuBattez, A. H.; Viesca, J. L.; Gonzalez, R.; Blanco, D.; Asedegbega, E.; Osorio, A. Friction Reduction Properties of a CuO Nanolubricant Used as Lubricant for a NiCrBSi Coating. *Wear* **2010,** *268* (2), 325–328.
51. Chu, H. Y.; Hsu, W. C.; Lin, J. F. The Anti-scuffing Performance of Diamond Nano-particles as an Oil Additive. *Wear* **2010,** *268* (8), 960–967.
52. Hwang, Y.; Lee, C.; Choi, Y.; Cheong, S.; Kim, D.; Lee, K.; Lee, K.; Kim, S. H. Effect of the Size and Morphology of Particles Dispersed in Nano-oil on Friction Performance between Rotating Discs. *J. Mech. Sci. Technol.* **2011,** *25*, 2853–2857.
53. Cornelio, J. A. C.; Cuervo, P. A.; Hoyos-Palacio, L. M.; Lara-Romero, J.; Toro, A. Tribological Properties of Carbon Nanotubes as Lubricant Additive in Oil and Water for a Wheel–Rail System. *J. Mater. Res. Technol.* **2016,** *5* (1), 68–76.
54. Lee, C. G.; Hwang, Y. J.; Choi, Y. M.; Lee, J. K.; Choi, C.; Oh, J. M. A Study on the Tribological Characteristics of Graphite Nano Lubricants. *Int. J. Prec. Eng. Manuf.* **2009,** *10* (1), 85–90.

55. Ghaednia, H.; Jackson, R. L.; Khodadadi, J. M. Experimental Analysis of Stable CuO Nanoparticle Enhanced Lubricants. *J. Exp. Nanosci.* **2015,** *10* (1), 1–18.
56. Perez-de Brito, A. F.; Ponce, S.; Perez-Robles, J. F.; Higuera-Ciapara, I.; Toro, A.; Esparza, R.; Medina, D. I.; Villaseñor-Ortega, F.; Luna-Barcenas, G. Linseed and Complex Rosin Ester Oils Additivated with MWCNTs and Nanopearls for Gears/Wheel-Rail Systems. *Eur. J. Lipid Sci. Technol.* **2020,** *122* (2), 1–15.
57. Lee, K.; Hwang, Y.; Cheong, S.; Choi, Y.; Kwon, L.; Lee, J.; Kim, S. H. Understanding the Role of Nanoparticles in Nano-oil Lubrication. *Tribol. Lett.* **2009,** *35* (2), 127–131.
58. Ellen, B.; Zhu, Y.; Söderberg, A. Study of Surface Roughness on Friction in Rolling/Sliding Contacts: Ball-on-Disc versus Twin-Disc. *Tribol. Lett.* **2020,** *68* (2), 1–15.
59. Biazon, L.; Ferrer, B. P.; Toro, A.; Cousseau, T. Correlations between Rail Grease Formulation and Friction, Wear and RCF of a Wheel/Rail Tribological Pair. *Tribol. Int.* **2020,** 1–10.
60. Rapoport, L.; Leshchinsky, V.; Lvovsky, M.; Nepomnyashchy, O.; Volovik, Y.; Tenne, R. Mechanism of Friction of Fullerenes. *Indust. Lubric. Tribol.* **2002,** *54* (4), 171–176.
61. Chinas-Castillo, F.; Spikes, H. Mechanism of Action of Colloidal Solid Dispersions. *J. Tribol.* **2003,** *125* (3), 552–557.
62. Ginzburg, B.; Shibaev, L.; Kireenko, O.; Shepelevskii, A.; Baidakova, M.; Sitnikova, A. Antiwear Effect of Fullerene C_{60} Additives to Lubricating Oils. *Russian J. Appl. Chem.* **2002,** *75* (8), 1330–1335.
63. Hu, Z. S.; Lai, R.; Lou, F.; Wang, L. G.; Chen, Z. L.; Chen, G. X., et al. Preparation and Tribological Properties of Nanometer Magnesium Borate as Lubricating Oil Additive. *Wear* **2002,** *252* (6), 370–374.
64. Xiaodong, Z.; Xun, F.; Huaqiang, S.; Zhengshui, H. Lubricating Properties of Cyanex 302–modified MoS_2 Microspheres in Base Oil 500SN. *Lubric. Sci.* **2007,** *19* (1), 71–79.
65. Sui, T.; Song, B.; Zhang, F.; Yang, Q. (2015). Effect of Particle Size and Ligand on the Tribological Properties of Amino Functionalized Hairy Silica Nanoparticles as an Additive to Polyalphaolefin. *J. Nanomater.* **2015,** *2015*, 1–9.
66. Song, X.; Zheng, S.; Zhang, J.; Li, W.; Chen, Q.; Cao, B. Synthesis of Monodispersed $ZnAl_2O_4$ Nanoparticles and Their Tribology Properties as Lubricant Additives. *Mater. Res. Bull.* **2012,** *47* (12), 4305–4310.
67. Kheireddin, B. A. Tribological Properties of Nanoparticle-Based Lubrication Systems. PhD thesis, Texas A&M University, 2013.

CHAPTER 5

Carbon Nanotubes in the Agricultural Sector

ELVIN THOMAS*

School of Environmental Sciences, Mahatma Gandhi University Priyadarsini Hills P.O. Kottayam, Kerala, India

**Corresponding author. E-mail: elvinthomas1992@gmail.com*

ABSTRACT

Carbon nanotubes (CNTs), owing to their unique properties, have attracted widespread interest, leading to its large-scale production worldwide. This chapter presents the major events that led up to the discovery of CNTs in 1991 by Sumio Iijima. It then briefly discusses the three popular production techniques of CNTs namely, the arc-discharge method, the laser-ablation method, and the chemical vapor deposition method, with each technique having its merits as well as demerits. The extraordinary mechanical, electrical, vibrational, thermal, and chemical properties of the materials are explored to gain a better understanding of its potential applications. In the agricultural sector, scientists are fascinated by the ability of CNTs to improve crop growth and yield. Some of the proposed applications of CNTs in the sector include nanofertilizers, antimicrobial agents, nanopesticides, nanosensors, and vectors for gene delivery. Despite the numerous positive impact CNTs have on the agricultural sector, the existing literature also reveals some of the negative effects of CNT exposure on plants as well as soil microbial communities. As such, with the current knowledge base developed using laboratory studies, it is difficult to predict the outcome of CNT exposure in the natural environment. This calls forth the need for reliable and accurate investigation on the impact of different CNTs on a wider range of crops, pathogens, soils, and other nontarget organisms, thereby helping in minimizing the knowledge gap between experimental and environmental conditions, before the widespread commercialization of CNTs.

5.1 INTRODUCTION

Nanotechnology, a field of research and innovation focused on the development of materials and devices on the scale of atoms and molecules, continues to emerge as the scientific beacon of the future and is one of the areas of intense research and development in modern science. Due to their extraordinary optical, mechanical, electrical, and thermal properties, carbon nanomaterials (CNMs) a class of engineered nanomaterials, which has evolved significantly over the past two decades have found diverse applications worldwide. While an overwhelming majority of CNMs are engineered or artificially synthesized, in nature, CNMs exist only in negligible quantities. Theoretically, its production can be performed in unlimited quantities as long as the raw materials for its synthesis are available. The different members of the CNM family include fullerenes, nanocones, nanoonions, nanohorns, nanobeads, nanofibers, nanodiamonds, graphene, carbon dots, and carbon nanotubes (CNTs).[1] Among the different CNMs available, CNTs owing to their tunable physical properties with respect to the graphene layers, their functionalization, and chirality have attracted attention in multiple disciplines including physics, chemistry, and material sciences. An ideal CNT structure can be illustrated by considering a seamless cylinder composed of one or more layers of graphene with open or closed ends (composed of a highly curved fullerene-like hemisphere).[2] While perfect CNTs have all their carbons bonded in a hexagonal lattice except at their ends, the mass production of CNTs often introduces pentagons, hexagons, and other defects in its sidewalls resulting in less-desired properties. CNTs based on the number of concentric layers of rolled graphene sheets are further categorized into single-wall carbon nanotubes (SWCNTs) and multiwall carbon nanotubes (MWCNTs). For SWCNTs, the outer diameter is typically in the range of 0.8–2 nm, whereas, for the MWCNTs, it is in the range 5–20 nm, although diameters can sometimes exceed 100 nm. Also depending on the desired application, the length of these CNTs varies from 100 nm to several centimeters.[3]

5.2 HISTORY

5.2.1 PRE-1991 TRACE EVIDENCE OF CARBON NANOTUBES

As early as in 1880, Thomas A. Edison filed a patent for the manufacture of coiled carbon resistors, developed by pyrolyzing spirally coiled Japanese Kyoto bamboo filament, which when heated ohmically provided a filament to be used in an early model of an incandescent light bulb.[4] In 1886, a patent was filed by T.V. Hughes and C.R. Chambers of England for the manufacture of hair-like carbon filaments, by the destructive distillation of a gaseous carbon compound, for electric lighting.[5] However, further research on carbon filaments proceeded slowly, as sturdy tungsten filaments soon replaced carbon filaments in electric light bulbs. Several other reports on carbon filaments were also presented at the French Academy of Sciences.[6,7] These studies were mainly based on vapor-grown carbon filaments, where carbon filaments were developed through the thermal decomposition of hydrocarbons. Also, none of these reports could establish the presence of CNTs due to the limitations in the resolution of the then-available microscopic tools. But with the invention of the transmission electron microscope (TEM) in the late 1930s, significant progress was achieved in understanding the morphology and inner texture of the carbon filaments. The first evidence (TEM images) for the tubular nature of carbon filaments is believed to have appeared in 1952 in the Journal of Physical Chemistry of Russia, in which Radushkevich and Lukyanovich reported the hollow nature of the carbon filaments and their nanometer-sized diameter.[8,9] Figures in their publication, clearly depicted carbon filaments with a continuous inner cavity, thereby exhibiting tubes with a diameter in the range of 50 nm.[9] Also, the tubes imaging 15–20 graphenes, obviously exhibited a multiwalled tubular structure.[8] However, due to the cold war and the Russian language barrier, Western scientist's limited access to Russian scientific journals resulted in this discovery receiving no widespread attention. At that time with the TEM resolution only in the nanometer range, the arrangement of graphene in the tube walls could not be distinguished further. However, in the following years, the textures of various multiwall tubular carbon structures (concentric, herringbone, bamboo, platelet, etc.) were deduced using the diffraction studies. It must also be noted that most of the research on carbon filaments and nanotubes were used to understand their growth mechanisms to prevent their formation during the processing in the coal and steel industry, as well as in the coolant channels of nuclear reactors.[8]

The mechanical and elastic properties of carbon fibers were explored in the late 1950s when space and aircraft industry was searching for strong and stiff lightweight fibers that could be used for building composite materials that had superior mechanical properties. This search led to the development of carbon fibers based on organic polymer precursors mainly rayon, polyacrylonitrile (PAN), and later mesophase pitch.[10] It also stimulated the development of graphite whiskers, which exhibited high degrees of flexibility and tensile strengths up to 2000 kg/mm2.[11] Though the dc carbon arc technique used to produce the graphite whiskers is quite similar to the arc evaporation method for CNT synthesis, the graphite whiskers showed major differences when compared with the CNTs. The whiskers had a recoverable length of up to 3 cm and a diameter ranging from a fraction of a micron to over 5 μ.[11] Also, unlike the CNTs, the graphite whiskers had a scroll-like structure. In an attempt to synthesize more crystalline filamentous carbon materials, the 1960s and early 1970s witnessed the synthesis of carbon fibers via a catalytic chemical vapor deposition (CVD) method. Concurrently worldwide, research continued to focus on the control of the processes for the synthesis of vapor-grown carbon fibers as well as polymer-based carbon fibers with the desire to achieve increased strength and modulus as well as high thermal conductivity.[12,13] Carbon fibers formed via the thermal decomposition of a mixture of benzene and hydrogen at 1100°C were found to have diameters less than several 100 μ and also have a length of up to 25 cm. Scanning electron microscope (SEM) images of these fibers depicted concentric sheets of carbon set around the fiber axis similar to the annual tree rings. Located in the center along the fiber axis, high-resolution electron microscope observations revealed the innermost region of the fibers to be composed of a hollow tube, about 10 nm in diameter.[14]

During the late 1970s and early 1980s, Iijima of NEC Research laboratory in Japan had been studying carbon films obtained from an evacuated arc-evaporated apparatus when he came across some unusual structures in the resulting films. These structures upon further investigation consisted of discrete graphite particles, made up of concentric closed shells tightly curved around a central cavity. Though using the tilting experiment Ijima confirmed the tubular nature of one of these structures, he did not explore them further suggesting that the curved layers might be due to sp3 bonding.[15]

Howard G. Tennent, in 1987, was granted a US Patent for the production of "cylindrical discrete carbon fibrils" characterized by a constant diameter between 3.5 and 70 nm. The method involved the catalytic decomposition of a variety of carbon-containing gasses (e.g., hydrocarbons and acetone) on a metal surface.[16] In 1985, Harry Kroto of the University of Sussex and Richard Smalley of Rice University Houston, came together at Rice and along with a group of students and colleagues while doing a series of experiments on the vaporization of graphite in an atmosphere of helium accidentally discovered fullerenes (C60), which is a soccer ball (isohedral)-shaped molecule composed of 60 carbon atoms bonded together in pentagons and hexagons. This discovery marked the beginning of a new era in carbon science and extended way beyond the boundaries of academic chemical physics. However, due to the minuscule amount of C60 produced in the Kroto-Smalley experiments, further progress in the research in this area was slow, until in 1990, a far simpler technique was developed which produced C60 in bulk quantities. This was achieved by Wolfgang Krätschmer of the Max Planck Institute at Heidelberg, Donald Huffman of the University of Arizona, and their co-workers by using a simple carbon arc to vaporize graphite, again in an atmosphere of helium. The soot settled on the walls of the vessel were then dispersed in benzene and dried to produce plate-like crystals of fullerite, composed of 90% C60 and 10% C70.[17] This discovery that macroscopic amounts of C60 could be produced by a simple arc evaporation apparatus accessible to any laboratory gave a new impetus to research in carbon science. Later in December 1990, at a carbon–carbon composites workshop, Smalley envisioned the existence of a tubular fullerene or bucky tube made by elongating a C60 molecule.[18] Following up, at a fullerene workshop in Philadelphia in August of 1991, Mildred Dresselhaus, an American nanotechnologist at the MIT, during her oral presentation on the symmetry of the CNTs proposed how the CNTs would be capped at either end by fullerene hemisphere, and also suggested on how the zone folding could explain the electron–photon dispersion relations of such structures.[19] The pre-1991 carbon literature has several other examples mentioning the occurrence of nanotube like structures. In 1991, while studying C60 crystals using high-resolution electron microscopy, Su Wang and Peter Buseck of the Arizona State University reported the presence of large fullerenes with elongated ellipsoid shape and containing over 100 carbon atoms.[20] Several other scientists too anticipated the discovery of CNTs. For instance, in 1990, theoretical studies describing small cylindrical fullerene molecules were carried out by Patrick Fowler of Exeter University, England.[21] Fine tubules of carbon were known almost two centuries ago, but they all had a much less perfect structure and greater diameter than the now known CNTs. Hence, these findings should in no way diminish the importance of Iijima's work in 1991, which was the first to appreciate fully the nature and importance of CNTs (Figure 5.1).

5.2.2 THE DISCOVERY OF C60-RELATED CARBON NANOTUBES

Fascinated by the Krätschmer–Huffman Nature paper and believing it might contain some interesting structures, Iijima made a detailed TEM study of the soot produced by their technique. Disappointing him, the initial high-resolution TEM studies from such soot contained mostly short-range completely amorphous structures. Thereafter, he turned his attention to the hard, cylindrical deposits on the graphite cathode after arc evaporation and were amazed by a whole range of novel graphitic structures. The scientific breakthrough in CNTs research came in 1991 when Iijima imaged using TEM the cathodic soot and came upon the multiwall tubular carbon structures, consisting of up to several tens of graphitic shells with adjacent shell separation of 0.34 nm and diameters of 1 nm. In an argon-filled vessel (100 Torr), the graphitic carbon needles were grown on the negative end of the carbon electrode used for the dc arc-discharge evaporation of carbon.[22] This discovery then prompted fullerene scientists all over the world to examine the used graphite cathodes, previously discarded as junk. Working in the same laboratory as Iijima, another significant advance was made by Thomas Ebbesen and Pulickel Ajayan when they serendipitously discovered a method for preparing CNTs in gram quantities. They found that increasing the pressure of helium in the arc evaporation chamber helped improve the yields of CNTs in the cathodic soot. The CNTs thereby formed had diameters in the range of 2–20 nm and length up to several micrometers.[23] The increased availability of CNTs using this method accelerated the pace of research worldwide. The symmetrical and electronic properties of the fullerene tubules were studied by American theoreticians at the Massachusetts Institute of Technology[24] and the Naval Research Laboratory, Washington DC,[25] respectively. The CVD technique earlier used for the preparation of carbon filaments and fibers was used for the first time, in 1993, to grow MWCNTs, by the Mexican physicist, José Yacamán and his colleagues of the National Autonomous University of Mexico.[26] The CVD technique for MWCNTs was subsequently improved and optimized over the years, and in 1996, large-scale synthesis of CNTs was achieved by this technique using iron as catalyst.[27]

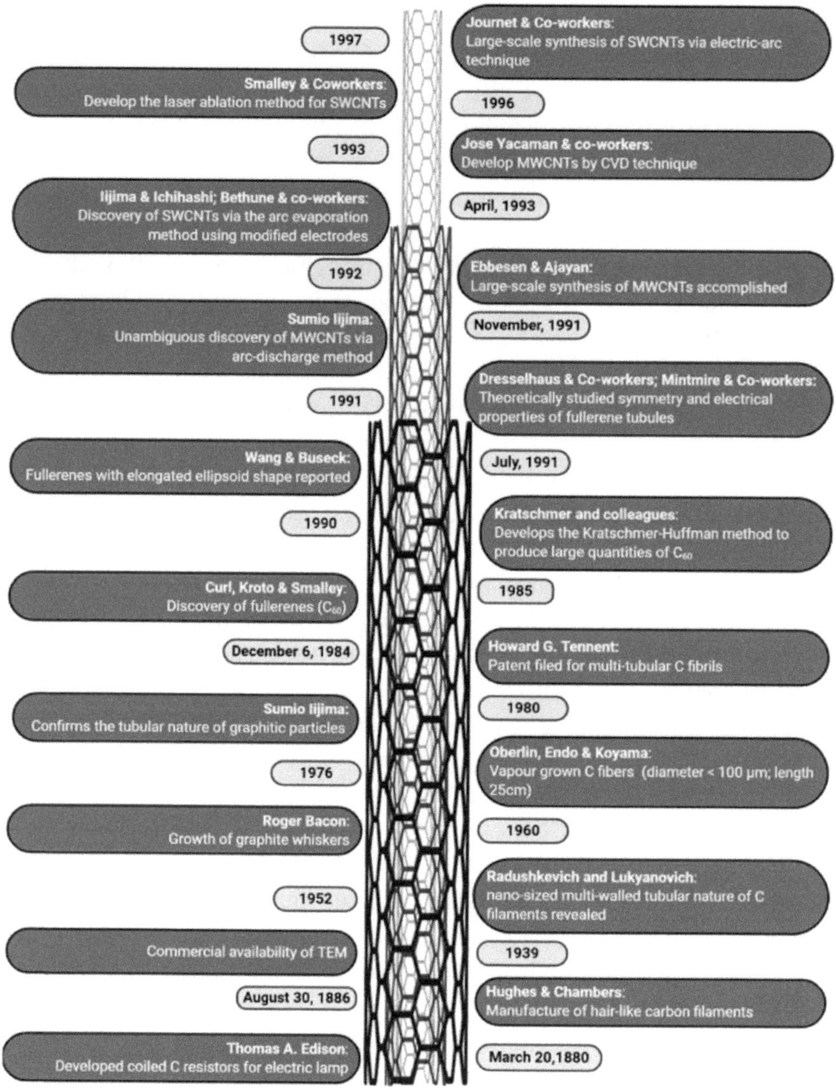

FIGURE 5.1 Timeline depicting the early history and discovery of carbon nanotubules. (Abbreviations used: C – carbon, C60 – fullerenes, TEM – Transmission electron microscope, MWCNTs – Multiwalled carbon nanotubules, SWCNTs – Single-walled carbon nanotubules, CVD – Chemical vapor deposition)[1]

Source: Figure created with BioRender.com.

In 1993, Iijima and Toshinari Ichihashi of NEC Research laboratory in Japan[28] as well as D.S. Bethune and his colleagues[29] of the IBM Almaden Research Center in California independently developed in their laboratories single-walled tubular carbon structures with diameters of 1–1.2 nm. Iijima reported the arc-discharge synthesis of CNTs, in a chamber filled with a mixture of 10 Torr methane and 40 Torr argon gasses. The cathode of the apparatus had a shallow dip to hold a small piece of iron during the evaporation. The use of argon, methane, and iron in this method was crucial for the synthesis of single-walled tubular carbon structures.[28] D.S. Bethune and colleagues, on the other hand, used thin electrodes with bored holes filled with a mixture of pure powdered metals (iron, nickel, or cobalt) and graphite as their anode. TEM images of the specimens thereby obtained revealed that only cobalt-catalyzed nanotubes had single-atomic layer walls.[29] The discovery of SWCNTs was an important milestone in the CNT's research as they appeared to have structures similar to that of ideal nanotubes. Smalley's group in 1996, developed an alternate method for the preparation of SWCNTs, which involved the laser vaporization of graphite with small amounts of nickel and cobalt at 1200°C. The laser-ablation method resulted in a high yield of SWCNTs with unusually uniform diameters (5–20 nm). Unlike the SWCNTs formed by the arc evaporation method, these highly uniform tubes tended to form aligned bundles, which led Smalley to christen the nanotube bundles, "ropes."[30] Subsequent works, however, suggested that the ropes may be less homogenous than originally thought. The electric-arc technique usually used to generate fullerenes and MWCNTs, though is a cheaper and easier to implement method, previously had led to only low yields of SWCNTs. However, in 1997, this technique was used to generate large quantities of SWCNTs, which had properties similar to that of the laser-ablation method.[31] Since the discovery of CNTs by Iijima in 1991, the study has progressed rapidly and various methods have been developed over the years for the fabrication of CNTs from different carbon precursors, with the popular methods recognized as the arc-discharge method, the laser-ablation method, and the CVD method. The comparative study of the different methods and its merits and demerits are summarized in Table 5.1.

5.3 PROPERTIES OF CARBON NANOTUBES

CNTs, the most widely studied allotrope of carbon, with their sp^2 hybridization can be single-, double-, or multiwalled. Despite their simple chemical

TABLE 5.1 A Comparative Study of the Major Methods Used in the Synthesis of Carbon Nanotubes.

Method	Arc Discharge	Laser Vaporization	Chemical Vapor Deposition
Conditions	Current (50–100 A) Voltage (25–60 V) Plasma temperature (>3000°C) Inert gas-filled vessel (~100 Torr) Gap between graphite electrodes (1–4 mm) Water-cooled trap	Laser beam (Nd: YAG or CO2 laser) External furnace temperature (1200°C) Inert gas-filled vessel pressure (~500 Torr) Conical water-cooled copper collector	Temperature (550°C–1000°C) Metal nanoparticle crystal Reactive gasses such as H_2 or inert gasses such as Ar/N_2 Substrate
Feedstock	Graphite	Graphite	Hydrocarbons (methane, carbon monoxide, acetylene, etc.)
SWCNTs	Co-vaporization of graphite and metal (Ni/Y and Co/Ni catalysts)	Transition metals such as Ni, Fe, or Co added in small amounts to the carbon target	Higher temperature (900°C–1200°C) yield SWCNTs
Yield	~30%–90%	~70%–90%	~20%–100%
Advantages	• Highly crystalline with very few defects • Short reaction (micro to milliseconds) • Cost-efficient with respect to laser-ablation method • Perfectly straight CNTs obtained	• High quality CNTs • Minimal defects and contaminants with respect to arc-discharge CNTs • Short reaction (micro to milliseconds)	• Simple and economic process • High yield • Utilizes a variety of hydrocarbons at any state • Low temperature, ambient pressure • Better control over growth parameters • CNTs grow in a variety of forms • Possible to scale-up the process
Disadvantages	• Little control over the alignment (chirality) of CNT • Purification of obtained CNTs is a must	• Expensive • High maintenance with low production • Branched CNTs produced	• Structurally more defective • Long-time reaction (minutes to hours)
References	[32,33]	[32]	[34]

composition and atomic bonding configuration, CNTs exhibit an extreme variety of richness in their structure and properties, most of which are derived from the graphene layers they are composed of. CNTs with their high surface area and large aspect ratio, show remarkable anisotropic properties in their mechanical strength, electrical and thermal conductivity, all of which find applications in nanoelectronics, nanopackages, nanotransistors, nanopassives, nanoantennas as well as fabricants for nanointerconnects.[35] To comprehend the potential applications of CNTs in various fields, it is necessary to understand their different properties which will be discussed in brief in the following sections.

5.3.1 ELECTRICAL PROPERTIES

In addition to being either single- or multiwalled, depending on the angle of rolling of the graphene sheets and the radius of the closing cylinder, CNTs can assume any of the three forms: achiral armchair, achiral zig-zag, or chiral type. The chirality or the rolled-up direction of the graphene sheets depends uniquely on the chiral indices (*n and m*). While in the achiral armchair type ($n = m$), the two sides of each hexagon are directed perpendicular to the axis of the CNT, in the achiral zig-zag type (*n or m* = 0), the two sides of each hexagon remain parallel to the axis of the CNTs. On the other hand, in the chiral type, any pair of the sides of each hexagon lies at an angle different from 0 to 90 degree to the axis of CNT.[36] Depending upon their different structures, CNTs can exhibit metallic or semiconducting properties. Upon satisfying the condition $n - m = 3i$, (where i is an integer), the CNT is described as metallic or highly conducting in nature, and if not, it is considered semi-metallic. At all times, it is found that the arm-chair CNTs have a metallic nature, whereas the zig-zag ones have a semi-metallic or semiconducting nature. It must also be noted that a natural mix of CNTs, statistically speaking, will have 1/3 of its chirality metallic and 2/3 semiconducting.[37] With the carbon atom in CNTs being sp^2 hybridized, the fourth valence electron of each carbon atom remains free and delocalized all over the atom contributing to the electrical nature of the CNTs. CNTs have got extremely low resistance, which is found to occur only when an electron collides with the few defects in the crystalline structure through which it passes. Such collisions due to a defect in the crystal structure or an impurity deflects the electrons from its path. However, the very small diameter and high aspect ratio of CNTs prevents the electrons from getting scattered easily.[38]

For instance, SWCNTs produced via the condensation of a laser vaporized carbon–nickel–cobalt mixture was found to have a single-rope resistivity of $< 10^{-4}$ Ω-cm at 300 K.[30] With the nature of electron transport in SWCNTs being ballistic, they are often described as quantum wires. The high field transport properties of SWCNTs are critical to the high power operations of nanoelectronics. It has been found that by reducing the length of the nanotube to ~ 10 nm, metallic SWCNTs were able to carry unprecedented 100 μA current per tube., which corresponds to a current density of ~4 × 109A/cm^2. This current density is three times higher than that tolerable by a typical metal before breakdown.[39] On the other hand, the MWCNTs have a fairly diffusive or quasi-ballistic nature of electron transport. The strong carbon–carbon covalent bonding and the lack of surface dangling bonds are the major reasons for the advanced properties of CNTs which make it useful in future electronics. The fact that CNTs depict their metallic behavior even in tubes with 1 nm in diameter is interesting. Unlike most one-dimensional materials which exhibit a Peierls instability, that opens an energy gap and causes the material to undergo a metal to insulator transition, CNTs due to their cylindrical structure and strong carbon–carbon bond do not show Peierls instability even at the lowest measured temperature.[40] CNTs due to their electronic properties such as high carrier mobility, high current density, good switching characteristics, and scalability can be used in transistors and other switching applications. Recently, CNTs also find application as electron field emitters, as emissions can be obtained even at lower threshold voltage. Other applications of CNTs include their use as interconnect material, photovoltaics, sensors, X-ray generation, vacuum microelectronics, computer circuits, and micron-scale on-chip triodes at a high frequency (>200 MHz).[41] Capable of overcoming most of the problems associated with the conventional copper and aluminum electrical wires, CNTs due to their lightweight and high electrical as well as mechanical performance, also find potential applications in electrical wiring.[38]

5.3.2 MECHANICAL PROPERTIES

Considering the mechanical properties, CNTs are among the stiffest and strongest materials so far measured. Their stiffness is often attributed to the graphene sheets which are stiff in the in-plane direction, with an in-plane modulus of 1.06 Tpa.[42] With an exceptionally high Young's modulus in the terapascal range (~1TPA),[43] CNTs have high strength and extraordinary flexibility as well as resilience. Young's modulus values reported are found to vary throughout the literature as some authors use the entire cross-sectional area that the CNT occupies, while others use only a ring of the thickness of the graphene interlayer spacing (3.4 Å). Their excellent mechanical properties are a result of the sp^2 hybridization of the individual carbon atoms. The graphitic sp^2 bond in CNTs is found to be 33% stronger than the sp3 bond in diamond, which is the hardest substance in nature. Recent studies report that the strength of the CNTs depends on their chiral structure, with near-arm chair CNTs with small diameter exhibiting the highest tensile strength.[44] The cylindrical shape of graphene sheets in CNTs provides the structural stability to CNTs even under compression. CNTs are found to remain stable up to very high temperatures, approx. close to the melting points of graphite which is 4000 K.[45] While they are very strong materials in the axial direction, in the radial direction, CNTs are found to be soft.[46] The cylindrical symmetry of the tubes gets destroyed due to the van der Waals forces between adjacent nanotubes, which can deform them substantially.[47] This elasticity of CNTs in the radial direction is important, especially during the formation of nanocomposites. The flexibility and compliance of CNTs for various applications vary depending on the geometric parameters of the nanotubes, which can be tailored. CNTs, due to their flexibility, can be repeatedly bent up to 90 degree without causing any break or damage in their structure.[40] Although upon compression and bending, they develop kinks or ripples and also flatten into deflated ribbon while under torsion, intriguingly, they still can reversibly restore their original shape. While most of the known hard materials fail at a strain of 1% or less, it is predicted that the failure strain of CNTs can be as high as ~15%.[48] This sets them apart from the conventional carbon fibers and other currently known fibers which are prone to fracture when bent or twist. The intrinsic tensile strength of CNTs along with its lightweight structures has encouraged the construction of space elevators, which requires ~63 GPa. Also, because of their extraordinary mechanical properties as well as high aspect ratios, CNTs find potential application as filler in polymer-based composites to produce ultrahigh performance structural materials, and also as alternatives to the traditional electrical actuators for different microscopic and macroscopic applications.[49] Since the demonstration by Ajayan and Iijima that CNTs could be filled with molten lead, and thereby be used as molds for the fabrication of nanowires (less than 2 nm in diameter), the scientific world has been attracted to the idea of using

CNTs as "molecular containers."[50] After this discovery, several controlled methods have been developed for the opening and filling of the tubes with a wide range of materials including biological ones. Such filled CNTs with their fascinating properties may find important applications in catalysis, magnetic recording, nuclear medicines, or as a biological sensors. Other mechanical applications of CNTs include the development of oscillators achieving higher speed than other technologies, development of superhydrophobic cotton fabric,[45] nanoropes with better load transfer mechanisms, filling of nanotubes thereby providing an efficient nanoscale storage system, and nanoelectromechanical systems.[51]

5.3.3 THERMAL PROPERTIES

CNTs, due to their low-temperature specific heat and thermal conductivity, serve as an ideal material for the study of 1-D quantization of the phonon band structure as well as thermal management, both on macroscale and microscale.[41] CNTs are known to have a thermal conductivity at least twice that of diamonds, which was previously considered to be the best thermal conductors. The thermal conductivity and specific heat of CNTs are determined primarily by phonons. At room temperature, the thermal conductivity of isolated SWCNTs was speculated to be as high as 6600 W/mK, while that for a single MWNT was reported to be 3000 W/mK.[52] However, different forms of CNTs such as the SWNT films aligned in the magnetic field as well as the annealed SWNT fiber spun from sulfuric acid suspension were found to have very low thermal conductivity. Although some CNTs are metallic and thereby could conduct thermal energy via electrons, the transport primarily rests on the vibration of atoms, that is, the transport of thermal energy through a nanotube is often mediated by the phonon conduction mechanism. The two main mechanisms which hinder the transport of the otherwise efficient phonon transport in CNTs include topological defects and the removal of extraneous contaminants. While topological defects such as simple vacancies are found to decrease the mean free path of phonons via scattering, the removal of contaminants often affects the structure of CNTs, thereby introducing local thermal resistance areas.[53] Also, bundled CNTs have thermal conductivity an order of magnitude lower than that of the isolated CNTs.[52] The length of the CNTs also plays a crucial role in thermal conductivity, with thermal conductivity increasing with length up to the mean free path of phonons, which for MWCNTs is found to be 500 nm and longer than that for SWCNTs. The temperature at which thermal conductivity is measured is also important, as it is found that increasing the temperature increases the thermal conductivity up to a certain limit. Both the temperatures inside as well as outside the nanotubes are found to affect the thermal conductivity. Considering their chirality, the zig-zag CNTs have the highest thermal conductivity, followed by the arm-chair CNTs and finally, the chiral nanotubes have the lowest thermal conductivity. However, the thermal conductivity of all three types of CNTs shows similar temperature dependence.[40] Thus, the thermal properties observed in these tubes are directly associated with their small size and unique structure.[52] Since a wide range of parameters is to be optimized to obtain the expected thermal conductivity, in theory, it is possible to tailor materials as per need to be either thermally insulating or conducting. Taking into consideration the thermal properties of CNTs, nanotubes find applications in mainly two areas: thermal interface devices, which require high thermal conductivity, and thermoelectric devices, for which thermal conductivity is undesirable.[53]

5.3.4 VIBRATIONAL PROPERTIES

The one-dimensional confinement of electronic and phonon states in CNTs has a prominent effect not only on the electronic band structure but also on their vibrational properties. Among the common spectroscopic methods used for the investigation of vibrations in solids such as the inelastic light scattering, far-infrared spectroscopy, and neutron scattering. Raman scattering is the most prominent method used. The Raman spectra of CNTs provide detailed information about the vibrational properties of the nanotubes including their phonon and electronic structure, as well as information regarding their defects. The spectra are sensitive to the chiral indices (n,m) and provide information on the radial breathing mode (RBM) which corresponds to the radial expansion–contraction of the CNT, the G band corresponding to planar vibrations of neighboring atoms moving in opposite directions, the D band originating from the structural defects and its second-order-related harmonic G' band. Of these four features, the frequency of RBM appears to be the most sensitive to the diameter of the CNTs.[54] The presence of CNTs can be confirmed by the combined spectrum of the RBM and the double-peak structure just below 1600 cm1. However,

RBM alone does not provide convincing evidence for the presence of CNTs as there are many materials which have Raman peaks between 100 and 300 cm−1. The RBM is unique to SWCNTs and is found to be consistent with tube contraction and expansion, in which all the carbon atoms in a CNT move in the radial direction synchronously in an effect similar to that of breathing. However, MWCNTs lack the RBM mode, as their outer tubes restrict the breathing mode. The characteristic Raman line shape of the G band helps distinguish between metallic and semiconducting CNTs, that is, for metallic CNTs – one Lorentzian and one Breit–Wigner–Fano line dominates the line shape, whereas for semiconducting CNTs, two Lorentzian features dominate the line shape.[38] Also, based on its sensitivity to the electronic energy state, the G band can be utilized for monitoring changes in the energy state due to the environment. A broad peak around 1540 cm−1 (less pronounced in isolated tubes) is representative of metallic CNTs.[55] The D band, also known as the defect band or the disorder band, ranges from 1250 to 1450 cm−1 and is indicative of disorders in the structure of the CNTs. The ratio of the intensity of the D band to the G band is used to characterize the chemical functionalization of the CNT as well as to assess its quality. It is found that the functionalization of the sidewall damages the nanotube by increasing the intensity of the D band. In MWCNTs, the multilayer configuration and greater disorder in structure result in a more prominent D band. The load transfer between the SWNT and the matrix is monitored using the G' band, which is usually in the range of 2500–2900 cm-1. Thus, except for the RBM mode and the D band, the MWNT and SWCNTs have very similar Raman spectra (Figure 5.2). Change in bond length or angles due to externally applied stress or temperature results in a shift in the phonon frequency at fixed excitation energy. Important modes of vibrations observed in SWCNTs are the acoustic and optical modes. An acoustic phonon corresponds to a vibration in the crystal lattice in which all moving components are of the same mass. While longitudinal acoustic mode depicts atomic displacement parallel to the axis of the nanotube, transverse acoustic mode is doubly degenerated and shows an atomic displacement which is perpendicular to the axis of the nanotube.[38] Thus, the phonon properties of CNTs are important as they show their unique one-dimensional behaviors and special characteristics, such as the twist acoustic behaviors, the phonon subbands, and the radial breathing mode. Hence the analysis of phonon properties is a very important field of continuous investigation crucial for revealing the distinct mechanics of the molecular-sized nanomechanical oscillator. However, mechanics at the molecular scale possesses rich dynamics, unexplored to a large extent due to lack of an efficient way to detect motion in real-time. Currently, this field of miniature mechanical oscillators is rapidly evolving with emerging applications, including the fundamental tests of quantum mechanics, signal processing, and biological detection.[56]

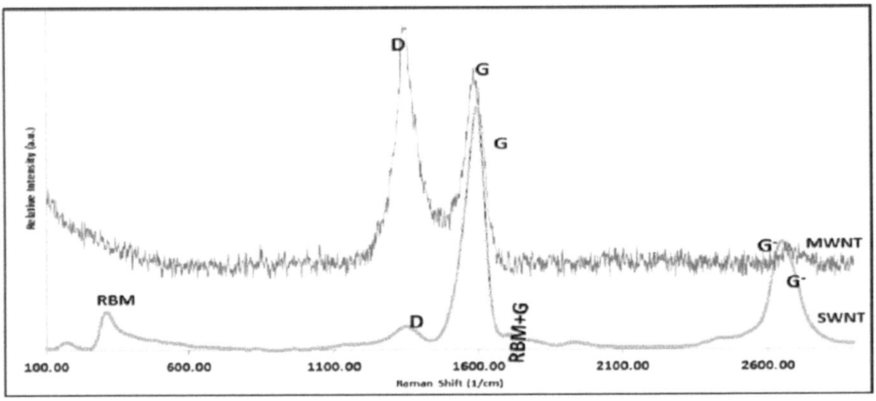

FIGURE 5.2 Comparative Raman spectra of SWCNTs and MWCNTs (100–2600 cm−1).
Source: Reprinted with permission from Ref. 57 (Endnote 2). https://creativecommons.org/licenses/by/3.0/

5.3.5 CHEMICAL PROPERTIES

CNTs have a formidable sorptive character, which find their application as potential water decontaminating, metal sorption, and aromatic compounds removal materials. Solubility is an important property concerning the processibility of a substance. It enhances the degree of reactivity and homogenous dispersion of a substance in a solvent and also has an impact on its purification process. The inadequate solubility of CNTs in water or any other aqueous or nonaqueous medium raises a great limitation to their applicability. The agglomeration of the tubes and their hydrophobic nature leads to difficulty in their dispersion in water, different solvents, as well as polymeric mediums. CNTs can be dispersed in some solvents via sonication, however, the interruption of the process results in immediate precipitation. Chemical functionalization of CNTs helps in creating new variants for the state-of-the-art nanotechnologies and bionanotechnologies. Functionalization also helps tailor their properties for other applications, such as fuel sources, battery components, catalysis, and also environmental remediation. Modification and functionalization of CNTs also give rise to completely new nanomaterials with unique spintronic properties as a starting base for the development of new components in bionanosensors, nanoelectronic, photonics, switches, and resistors. The three fundamental approaches for modifying CNTs include solubilization, dispersion, and polyelectrolyzation. With their surfaces having a highly unique electrostatic character arranged in a nonperiodic manner, pristine CNTs are mostly insoluble in any solvent but are found to form suspensions in toluene, diethyl ether, *N-N*-dimethylformamide (DMF), ethyl acetate, acetone, acetonitrile, tetrahydrofuran (THF), etc. For this to happen, their surface of interaction must be made more accessible to various reactants by devising the right solvents, which would preserve the chemical surface of the modified CNTs. Given their population of π-electrons not counter-balanced by aromatic hydrogens, CNTs are neither regarded as ordinary polar, aromatic, or lipophilic compounds. The curvature of the CNTs plays an important role in the sorption and dispersion potential of CNTs in the media. For instance, when compared with the low curvature CNTs with a large diameter, high curvature CNTs with small diameter have a lower sorption efficiency toward large planar aromatic compounds. In turn, The high curvature CNTs with small diameters were found to have an increasing sorption efficiency for smaller polycyclic aromatic hydrocarbons. A multitude of surfactants (such as sodium dodecyl sulfate (SDS), sodium dodecylbenzene sulfonate, benzalkonium chloride, etc.) has also been used for the dispersion and purification of CNTs in the aqueous medium. Though ultrasounds (US) are also commonly used for the dispersion of CNT bundles into solutions or suspensions of individual CNTs, they are found to induce local defects on their structure resulting in the breaking or shortening of CNTs. The dispersion of CNTs in room temperature ionic liquids is a promising technique for applications that require pristine CNTs. Recently, indefinitely stable dispersions of CNTs were developed by conversion of SWCNTs to polyelectrolytes, composed of charged CNTs (SWCNTn−) dissolved in a strong polar aprotic solvent.[58] The functionalization of CNTs introduces a functional group onto its sidewalls or ends, thereby changing their solubility, physicochemical properties, and chemical reactivity. Based on the location of the functional group, functionalization can be exohedral or endohedral. While exohedral functionalization introduces a functional group on the outer surface of the CNTs, endohedral functionalization fills the nanotubes with atoms or molecules at their defective sites. Functionalization can also be classified based on linkage-type into chemical (covalent) or physical (noncovalent) functionalization. Chemical functionalization refers to the covalent linkage of a functional group onto the CNT via a change in hybridization. One of the major advantages of chemical functionalization is the efficiency that can be achieved based on the high degree of chemical modification. It attaches the functional groups to the tips or the side-walls of CNT in a nonreversible manner, resulting in new properties such as better dispersion in polar solvents (alcohol, water, or acetone). On the other hand, physical functionalization is a noninvasive mode of functionalization based on hydrophobic, van der Waals or π–π interactions. In this mode of functionalization, as the electronic structure of the nanotubes remains intact, the original optical and electronic properties of the CNTs are almost completely preserved, which is essential for several applications. However, the physical functionalization of CNTs is difficult not only to control but also to characterize due to its weakly bound molecules which causes the desorption resulting in the aggregation of the partially functionalized CNTs.[38] Thus, the characterization of the surface chemistry and structure of CNTs is necessary to understand not only their modifications but also their potential applications.

5.4 CARBON NANOTUBES IN THE AGRICULTURAL SECTOR

Sustainable food production is under serious threat and is of global concern due to the growing population, global warming, urbanization, and limited availability of natural resources, such as fertile land, water, energy, nutrients, and quality seeds. The intensification of agriculture and the injudicious use of fertilizers has led to land degradation causing increased pressure on the remaining land. Agriculture is a crucial part of economic growth and its products influence most aspects of human life such as food, feed, feedstocks for bio-based products, textiles, fuel, furniture, etc. Thus, agricultural development is considered to be one of the most powerful tools to end extreme poverty, boost shared prosperity as well as feed the growing population expected to reach 9.7 billion by 2050. Advances in nanotechnology have the potential to revolutionalize and meet the future global needs of the agricultural sector. The different applications of nanotechnology in the agricultural sector include nanofertilizers, nanopesticides, nanoscale biosensors, and nanoparticle-based plant gene transfer. It is found to improve the utilization of nutrients by plants and also help in the fast and accurate detection as well as the management of diseases. The following sections will discuss the potential applications of CNTs in the agricultural sector.

5.4.1 NANOFERTILIZERS

Some common practices of fertilizer application in conventional agriculture include the broadcasting, side-dressing, and banding of fertilizers as well as the foliar application by spraying. However, these techniques often result in considerable losses of nutrients via evaporation and leaching. Such leaching of fertilizers leads to a lower fraction of the applied fertilizer being available to the target crop, thereby reducing its productivity. For instance, it is found that the nitrogen use efficiency by crop plants is low when compared with the amount of nitrogen applied to the soil. Almost 50%–70% of the nitrogen applied as conventional fertilizers, in the form of plant nutrient formulations with dimensions greater than 100 nm, is lost via the long-term incorporation of mineral nitrogen into soil organic matter by soil microorganisms, the leaching of water-soluble nitrates, and also via the emissions of gaseous ammonia and nitrogen oxides.[59] One of the possible solutions to this problem is suggested to be nanofertilizers. Nanofertilizers exploit the nanoscale porous domains on plant surfaces and deliver nutrients to crops in any one of the following three ways. The nutrients can be delivered as an emulsion or particles of nanoscale dimensions, or can be coated as a thin polymer film protecting a nanomaterial or can be encapsulated within it. Nanofertilizers, based on plant nutrient requirement, are classified into macro-nanofertilizers, micro-nanofertilizers, and nanoparticulate fertilizers.[60] Some of the proposed benefits of using nanofertilizers include reduced fertilizer use and increased agricultural yields through optimized nutrient management. The highly controlled release of fertilizers and site-specific characteristics of nanofertilizers avoid leaching and losses of nutrients to unintended targets,[61] thereby reducing the amount and cost of fertilizers spent by farmers. Further, the reduction in the amount of fertilizers used would also help in improving the overall soil quality.

The ability of water-soluble CNTs to traverse both plant cell wall and cell membrane holds great potential to function as nanotransporters to deliver different cargoes into different plant cell organelles[62] and thus may be used as carriers to enhance the nutrient uptake by plant cells.[63] In a study conducted by Yatim and his colleagues,[61] various amounts of sonicated functionalized and nonfunctionalized MWCNTs were stirred with urea fertilizers for 6 h at 150 rpm and at room temperature to create both functionalized and nonfunctionalized MWNT-urea fertilizers. A comparative study on the role of these nanofertilizers on the growth of a local paddy variety revealed the functionalized nanofertilizers to function better in terms of growth and yield when compared with nonfunctionalized MWNT-urea fertilizers. The functionalization of CNTs highly disperses them in water and thereby increases their solubility and softness allowing them to pass through regular channels in the plant cell without inflicting any physical damage.[64] CNTs loaded with nitrogen, phosphorus, and potassium by functioning as a slow-release fertilizer reduce the utilization of high concentrations of fertilizers usually used in conventional agriculture. Such easily and inexpensively developed macro- and micro-nano fertilizers could revolutionize the agricultural sector.[65] Carbon-based nanofertilizers were also found to promote the growth, improve the quality, as well as increase the yield of different vegetable crops.[66] In all of the cases mentioned, the fertilizer is composed of an amendment of mineral and organic fertilizer along with the CNTs, which act as a fertilizer synergist to reduce nutrient losses and consequently improve nutrient availability and plant growth.

In addition to functioning as macro- and micro-nanofertilizers, CNTs also function as nanoparticulate fertilizers like titanium dioxide and silicon dioxide, by showing plant growth-promoting activities.[60] Exposure of valuable crop seeds (like tomatoes) to MWCNTs is found to dramatically increase its germination and growth rates via the penetration of the thick seed coat and promotion of water uptake.[67] The insertion of SWCNTs into extracted plant chloroplasts was found to enhance maximum electron transport rates and to promote their photosynthetic activity three times higher than that of the control chloroplasts without SWCNTs in their lipid envelope.[68] CNTs were also found to be promising materials for stimulating the growth and regeneration of date palm in vitro cultures. The results of the study showed that although the number of embryos decreased when compared with the control, CNTs did increase the number of germinated embryos as well as roots. During the elongation stage, a significant increase in the shoot length and leaf number was also observed. CNTs also facilitated the adsorption or transportation of nutrients, such as nitrogen, phosphorus, potassium, and calcium, while inhibiting that of sodium.[69] CNTs were also found to enhance biomass, flowering, fruit yield, as well as improve the medicinal properties of the plants.[70,71] Even though several studies are highlighting the positive effects of CNTs on plants,[72–74] it must also be noted that many researchers have reported CNTs to adversely affect the plant growth in some instances.[75–77] For instance, in a 15-day hydroponic study, zucchini plants exposed to MWCNTs showed a decrease in the plant biomass by 60% when compared with control plants.[78] However, in another study, MWCNTs stimulated seed germination, increased the root length, shoot length, and biomass of young seedlings of sorghum and switchgrass.[79] Thus, the effect of CNT-based nanofertilizers varies from plant to plant, their growth stages, the chemical composition, and the concentration of the nanotube used, and hence, extensive research must be done before application of the same in the field.

5.4.2 ANTIMICROBIAL AGENTS

Recently, antimicrobial agents have found huge applications in the agricultural sector in their use in crops, aquamarine, poultry production, and livestock rearing. As humans and animals are both prone to pathogens in agricultural products, it has turned out to be essential to include antimicrobial agents in them. Researchers have found CNTs to have great antibacterial and antifungal properties, which can be exploited in the agricultural sector for plant disease control. The antimicrobial property of CNTs is found to depend on various factors, such as their diameter, length, density, surface functional molecules, aspect ratio, and purity of the nanotubes. The CNTs by forming close associations with the membrane surface of microorganisms change the integrity of the cell membrane by inflicting severe damage and cause the leakage of all intracellular components from the cell matrix. In some cases, CNTs via entering into the microorganism nuclei, induces the formation of reactive oxygen species which results in DNA damage and eventual cell death. Metal toxicity caused by the impurity associated with CNTs is also a possible mechanism for microbial inactivation.[80,81] SWCNTs were found to be more efficient than MWCNTs as bactericidal agents due to their smaller diameter. Although in the presence of both the CNTs, the bacteria expressed high levels of stress-related genes, the magnitude, and the quantity of the same was higher in the presence of SWCNTs.[82] Studies have demonstrated the ability of dual-functional CNT-modified filters to remove undesirable pathogens from water sources and inhibit the activities of captured Gram-positive bacteria in one single filtration step.[83] The antifungal activity of pristine and functionalized MWCNTs have also been studied and have found functionalized MWCNTs to exhibit higher antifungal activity than pristine MWCNTs. The antifungal activity of functionalized MWCNTs was found to be higher than that of various other nanoparticles and native antimicrobial agents, such as silver nanoparticles, carbohydrate nanoparticles, dermaseptin, nigrocins, copper, and zinc-doped hydroxyapatite nanopowders, etc.[84] The coating of CNTs with metallic nanoparticles[85] and covalent functionalization[86] are among the different methods proposed to increase the antimicrobial activity of CNTs. Furthermore, adsorption of nanotubes to microbial membranes was found to be greater when the CNTs were functionalized with positively charged agents.[84,87]

CNTs were also found to be effective against both plant bacterial[88] and fungal pathogens.[89,90] For example, the fungal pathogen *Fusarium graminearum* is known worldwide to affect wheat, corn, and other crops causing the plant disease *Fusarium* head blight (FHB). Currently, very few effective methods exist for the control of the same and so the disease is of great concern in the agricultural sector. However, it was demonstrated that both pristine and functionalized MWCNTs have an effective and durable inhibitory effect against the pathogen. The MWCNTs functionalized with –COOH, $-NH_{2,}$ and –OH were found to show enhanced inhibition of spore germination during the growth cycle preventing the formation of mature fungal mycelium, thus interrupting the pathogen reproduction and terminating infection cycle.[89] Bacterial wilt is another plant disease caused by the bacterial pathogen *Ralstonia solanacearum* and is found to affect a large number of crops during the growing season and also during the post-harvest storage period. Though numerous chemical measures against the pathogen exist, with use, most of the pathogens have developed defense mechanisms against such antimicrobial agents and have built resistant strains. The current integrated management strategies, such as pathogen-free transplants, and crop rotation using nonhost cover crops have failed to control *R. solanacearum*. SWCNTs dispersion was found to destroy the cellular membrane of *R. solanacearum* and release the cytoplasm content, thereby showing strong antibacterial activity. Since they exhibit multipoint antibacterial mechanisms, it is less susceptible for the bacteria to develop resistance to CNT-based antimicrobial agents.[88] Despite the exceptional antimicrobial properties of CNTs, very few studies have documented the antibacterial and antifungal properties of CNTs against different plant pathogens, which have turned resistant to conventional management practices. Thus, future studies may focus on other agricultural pathogenic microbes to extend the spectrum of CNTs' antimicrobial activities thereby helping eliminate severe agricultural problems caused by them.

5.4.3 NANOCIDES

Currently worldwide, approximately 2 million tons of pesticides are used of which 47.5% comprises herbicides, 29.5% are insecticides, 17.5% are fungicides and 5.5% are other pesticides.[91] Though pesticides contribute significantly to enhanced crop production, the impact of their usage on the environment cannot be unseen. Over time, they are found to accumulate in different compartments of the environment (water, air, soil, and biota), thereby posing a severe threat to human beings. In addition to environmental pollution, conventional pesticide applications also cause economic loss, as most of the applied fertilizers are lost by leaching, evaporation, and degradation. However, the encapsulation of pesticides via nanotechnology helps in reducing its utilization through the highly controlled and target-specific measured release of sufficient amounts of pesticides over a period of time, thereby obtaining the fullest biological efficacy. Such encapsulated agrochemicals are protected from being degraded and hence exhibit improved stability and use efficiency.[92] The high surface area to volume ratio of these nanoscale delivery systems also enhance the bioavailability of these pesticides and thus lowers the application rate. While conventional pesticides face the obstacle of water insolubility and hence are dissolved in organic solvents or organic-based emulsions, downsizing of these agrochemical suspensions helps lower the toxicity due to the elimination of organic solvents.[93] Encapsulation of water-soluble dithiocarbamate fungicides, mancozeb, and zineb, into functionalized MWCNTs, exhibited higher toxicity against *Alternaria alternata* fungi compared with the bulk pesticide.[94] Apart from being carriers of chemical pesticides, CNTs, due to their antifungal and antibacterial properties, alone are promising materials for the development of nanocides.[95]

5.4.4 NANOSENSORS

Nanosensors are instruments incorporated with receptor transducers which upon recognizing a signal from biological and chemical analytes of similar size, responds using a physical sign. They are analytical devices, having at least one sensing dimension less than 100 nm, used for monitoring physicochemical properties of areas otherwise difficult to reach.[96] The transducers are of four major types: electrochemical (electrodes), thermal (heat-sensitive sensors), optical (optrodes), and mass (surface acoustic wave devices, or piezoelectric crystals). Among these, the electrochemical nanosensors are the most common sensors according to the numerous studies which have led to a better understanding of their biointeraction and detection processes. In comparison to other commercially available conventional sensors, nanosensors due to their high surface to volume ratio, and excellent luminescence properties show many advantages, such as high stability, fast response time, low detection limit (up to nanograms or lesser) and high sensitivity.[33] The development of nanosensors has already proven its potential in the agricultural sector facilitating the real-time monitoring of crop and field conditions, as well as for the detection of pollutants, pathogens, and other environmental stressors. In addition to the physical monitoring of soil temperature, humidity, quality as well as fertility, they also help monitor the soil microorganisms. Other major implications of nanosensors in agriculture include detection of viability and shelf life of seeds, response sensors for irrigation and safety in precision agriculture, and detection of residual pesticides, toxins, and fertilizers.[97] Such real-time monitoring helps prevent the excessive use of fertilizers and pesticides, which thereby reduces its production cost and also minimizes environmental contaminations. Through energy-efficient and environment-friendly approaches, the application of nanosensors helps in the conversion of conventional agriculture practices into sustainable smart agriculture practices.

CNTs due to their nanoscale dimensions, high strength, high electrical, and thermal conductivity as well as high specific surface area exhibit enormous potential as nanosensors for numerous applications in the agriculture sector. Due to their high sensitivity, a few molecules are quite sufficient to change the electrical properties (chemical to electrical transduction) of the nanotubes. The great surface area of the CNTs also provides sufficiently large spaces for the interaction with the sensed molecules. CNT-based nanosensors for measurement and monitoring of crop and soil health have also been described. Recently, scientists have developed H_2O_2 selective SWNT-based nanosensors for decoding wound signaling in plants by the spatiotemporal monitoring of the wound-induced H_2O_2 signaling pathways in different plants. The CNTs embedded in the plant leaves via the lipid exchange envelope penetration method would then be able to detect chemical signals that are produced when a plant is damaged. Termed plant nanobionics, this approach enables the interfacing of plant defense pathways to portable electronic devices as simple as smartphones at a remote distance. Such sensors could assist agricultural scientists in deciphering the response of plants to different types of stress and pathogens, thereby helping them to develop new strategies to improve crop yield.[98,99] MWCNTs have also been used as humidity sensors, due to the linear response of their conductivity to temperature and humidity.[100,101] Most CNT-based sensors are field-effect transistors (FET) and are widely used to detect gasses. Studies have demonstrated the integration of CNT-graphite electronic devices on to surfaces of live plants and insects, enabling real-time, wireless sensing of toxic gases in the environment.[102] Monitoring of carbon dioxide levels within greenhouses or other controlled growth environments using CNT-based carbon dioxide sensors can be useful in achieving an optimal environment for plant growth.

Several studies have exhibited the ability of CNT-based nanosensors in the detection of pathogenic microorganisms. For example, the rapid and accurate detection of Staphylococcal enterotoxin B (SEB), a toxin produced by the Gram-positive bacterium *Staphylococcal aureus*, in the linear range 0.5–83.5 ng/mL was made possible by a conductometric immune-biosensor method based on the immobilization of horseradish peroxidase (HRP)-labeled SEB antibody (HRP-anti-SEB) onto nano-gold/chitosan-MWCNT functionalized biorecognition interface.[103] In another study, liposome coated CNTs were used for the electrochemical detection of ultratrace (10^{-16} g) levels of cholera toxin, secreted by the bacterium *Vibrio cholera*.[104] CNTs have also been used for the detection of agricultural pathogens, such as the Chili leaf curl virus that causes chili leaf curl disease in chili pepper. For this, the electrocatalytic potential of MWCNTs-based zinc nanocomposites was exploited for the recognition and discrimination of DNA biomolecules. The high selectivity and specificity of this sensor with its complementary and noncomplementary DNA can be modified for the in-field detection of other viruses in crops.[105]

In another study, CNTs (50–100 nm diameter) anchoring highly electrolytic agglomerates of copper nanoparticles were used for the detection of viruses in the sap extracted from cotton leaves.[106] Gold and chitosan nanoparticle-reinforced CNTs were used for the detection of quinoline, a secondary metabolite produced in oil palm trees within 24 h upon attack by the fungus *Ganoderma boninense*, thus enabling the early detection of fungal attack in oil palm plantations.[107]

The intensification of agriculture has led to an increase in pesticide use because of its positive impact on crop yield. Although they have significantly increased crop yield, and reduced postharvest loss, their accumulation in different compartments of the environment including edible parts of plants poses a severe threat to human health. Several methods, such as gas chromatography (GC), high-performance liquid chromatography (HPLC), gas chromatography with mass spectrometry (GC/MS), and liquid chromatography with mass spectrometry (LC/MS) exist for the ultratrace detection of pesticide residues. However, in addition to being expensive and time-consuming, the requirement of well-trained personnel, as well as the inconvenience for in situ application, largely restricts further applications of these methods. A number of simple, sensitive, and cheap techniques have been developed in recent years, that use different electrode architectures designed with pristine or functionalized CNTs for the detection of various pesticides. Electrochemical sensors for pesticide detection have been classified into seven different groups based on the CNT technique used, namely, CNT sensors, phthalocyanine, molecularly imprinted polymers (MIP), ionic liquid, metallic nanoparticles, β-cyclodextrin, and fullerene (C60)-functionalized carbon nanotubes.[108] For instance, a sensitive electrochemical sensing strategy was developed for the detection of methyl parathion residues in soil and water samples, by combining the electrochemical reduction of Ellman's reagent with acetylcholinesterase (AChE) inhibition. The biosensor consisted of immobilized AChE on MWCNTs–chitosan nanocomposites modified glassy carbon electrode (GCE). The favorable biocompatibility and excellent conductivity of MWCNTs–chitosan nanocomposites, along with the fast response, and low detection limit, provided high sensitivity to the sensor, enabling the detection of trace amounts of pesticide residues.[109] In another study, direct determination of the insecticide Cyromazine in environmental samples was made possible using a technique involving MWCNT paste electrodes.[110] In some studies, a noticeable improvement in the detection of the analytical signals was observed upon a combined use of different CNT-based sensors.[111] CNTs help not only in the detection of pesticides but also are found to be efficient in trapping them. Though pristine CNTs are quite inefficient, defective CNTs (i.e., transition metal atom incorporated nitrogen-doped CNTs) due to the strong bonding between the aromatic group of pesticide and the transition metal atom of the defective CNT, are efficient in trapping hazardous pesticides.[112]

Nanosensors aid farmers in precision farming by the accurate analysis of nutrients and pesticides in the soil as well as help maximize the water use efficiency for smart agriculture. Deployment of networks of wireless CNT-based nanosensors throughout the cultivated fields could assure comprehensive and real-time monitoring of crop growth, as well as provide data that could help agricultural scientists suggest excellent management practices that avoid an overdose of agricultural inputs. Although covering agricultural fields with multiple such sensors could provide a full picture of the soil and crop health, currently, the majority of these nanosensors are still in the development stage and their large-scale development has not been established yet. Some of the major challenges in the upscaling nanosensors revolve around the environment safety and human health risks associated with the use of such sensors, the assessment of which requires an entire life cycle analysis of the product. Up-scaling the production of nanosensors, would also require the long-term testing for the calibration and validation of these devices.[113]

5.4.5 VECTORS FOR GENE THERAPY

Intending to address the increasing demand for food and energy, plant biotechnology has revolutionized conventional agricultural practices. Recent advancements in the discipline have succeeded in improving crop yields as well as tolerance against environmental stresses. Exogenous biomolecule delivery into plant cells is difficult due to the complex cellular uptake mechanism through its rigid cell wall, which poses as a dominant transport barrier. Conventional methods of DNA delivery suffer several limitations, such as poor cellular penetration, low transformation efficiencies, tissue damage, unavoidable or uncontrolled DNA integration with the genome of host species, and limitations in host species selection. However, with the ability to act as "molecular containers," CNTs have attracted much attention from biotechnologists for the targeted delivery of molecules into in vivo cells overcoming most of these limitations. Both SWCNTs[62] and MWCNTs[63] are found capable to penetrate the intact plant cell wall and cell membrane to deliver molecular cargoes into different cell organelles. Among these, SWCNTs exhibit several characteristics that make them optimal candidates for plasmid DNA delivery into intact plant cell: (i) efficient plant cell wall transversal due to their high aspect ratio and extraordinary tensile strength, (ii) optimal activity of delivered cargoes via the protection of cargo biomolecule from cellular degradation and metabolism, and (iii) biocompatibility. Polyethylenimine-functionalized SWCNTs enabled efficient plasmid DNA delivery into a variety of nonmodal plant species and thereby resulted in high protein expression levels without transgene integration. This method can also be expanded to efficiently transfer linear DNA amplicons into crop plants, in a nontoxic manner, without external mechanical aids. The protocol also enables rapid high throughput screening of mature plants so as to identify genotypes that resulted in desired phenotypes. Functionalized SWCNT-based DNA delivery also aids the maximization of the synthesis of plant-mediated natural products as well as enables the mapping and optimization of the biosynthetic pathways in plants.[114] Recently, chitosan-complexed positively charged SWCNTs were used as carriers to protect and deliver the plasmid DNA to the chloroplasts by exploiting the lipid exchange envelope penetration mechanism. This mode of gene delivery was found to be easy to perform, cost-effective, and applicable to mature plants of different species, thus enabling its widespread application in plant biology and bioengineering studies.[115] Posttranscriptional gene silencing can also be achieved by the direct delivery of small interfering RNA (siRNA) using SWCNTs. The CNT-based delivery of siRNA resulted in high slicing efficiency in intact plant cells.[116] During the line development for commercial crops, with multiple rounds of selection and breeding done before their seeds are brought to the market, there should not be any trace of CNTs left in the lines. Thus, with edible plants constituting generations not directly exposed to CNTs, this could be suggested as a commercially viable approach for crop transformation.[115]

5.5 OUTLOOK

Despite the numerous applications of CNTs in the agricultural sector, its safety and biocompatibility remain controversial. In addition to the positive effects of CNTs, a review of literature also reveals the adverse effects of CNT exposure on the plant as well as soil microbes. For instance, a hydroponic study revealed the dose-dependent toxicity of MWCNTs exposure in mustard. While at low concentrations (2.3×10^{-3} mg/mL) MWCNTs showed no toxicity, at high concentrations (46×10^{-3} mg/mL) they hindered the germination and growth of mustard seeds. Furthermore, oxidized-MWCNTs exhibited more negative effects than pristine ones.[117] In another study, culturing rice cell suspensions (*Oryza sativa* L.) with MWCNTs resulted in a hypersensitive response cascade that increased the concentration of reactive oxygen species and decreased the cell viability.[118] In the case of Arabidopsis T87 suspension cells, toxic effects of exposure to MWCNTs included a decrease in cell chlorophyll content, viability, dry weight, and superoxide dismutase activity. Moreover, it was found that as the diameter of the agglomerates of MWCNTs decreased, the toxicity increased sharply.[119] Soil microbial communities by playing a major role in the decomposition of organic matter nutrient cycling, and its symbiotic relationships with terrestrial plant species, have a direct impact on the soil quality. So, any factor that affects the soil microbial community also affects the soil quality and vice versa, thus making the protection of soil microbial diversity and biomass of utmost importance and a major challenge in the agricultural sector. Unlike plants, only limited information is available on the interactions of soil microbes with CNTs. Although few studies have reported neutral or positive biological effects [120] of CNTs on the soil microbial community, in general, many studies consider them to be toxic.[2,121] CNTs are found to interact strongly with bacterial cell membranes via electrostatic interactions and thereby disrupt cell membranes integrity by physical damage or oxidative stress.[122] In addition to inactivating *Escherichia coli* and *Staphylococcus epidermis*, SWCNTs were also found to inactivate beneficial soil microorganisms like *Pseudomonas aeruginosa*, and *Bacillus subtilis* as well as diverse microbial communities in river and wastewater effluents.[123] In vivo studies have reported the introduction of MWCNTs into the abdominal cavity of mice to show asbestos-like pathogenic behavior, raising concerns about the potential impact of widespread industrialization of such nanoparticles.[124] Thus, the cytotoxic behavior of nanotubes appears to be situational, depending very much on the degree and type of functionalization, the occurrence of metal-catalyst particles remaining from synthesis, and the aggregation state of the CNTs. With the behavior of CNTs in the complex environmental matrices remaining largely unknown, it becomes quite difficult to predict to what extent the described in vitro properties will be manifested upon releasing CNT-related products into the environment. As such, with the current knowledge base, the widespread field application of CNTs in the agriculture sector is not yet possible. This calls forth a thoroughly reliable and accurate investigation of the impact of different CNTs on a wider range of crops, pathogens, and soils. Also, for environmental risk assessment, the impact of CNT-related products on nontarget organisms must be evaluated before the commercialization of products. Although this will increase the complexity of the studies, it will provide information about the impact of CNTs in a realistic scenario, thereby decreasing the current knowledge gap between experimental and environmental conditions.

KEYWORDS

- **nanofertilizers**
- **nanocides**
- **carbon nanotubes**
- **nanosensors**
- **toxicity**
- **agriculture**

ENDNOTES

1. Image created with BioRender.com
2. http://creativecommons.org/licenses/by/3.0

REFERENCES

1. Sharon, M.; Sharon, M. *Carbon Nano Forms and Applications*; McGraw-Hill Education: New York, 2010.
2. Mukherjee, A.; Majumdar, S.; Servin, A. D.; Pagano, L.; Dhankher, O. P.; White, J. C. Carbon Nanomaterials in Agriculture: A Critical Review. *Front. Plant Sci.* **2016**, *7* (FEB2016), 1–16. https://doi.org/10.3389/fpls.2016.00172.

3. De Volder, M. F. L.; Tawfick, S. H.; Baughman, R. H.; Hart, A. J. Carbon Nanotubes: Present and Future Commercial Applications. *Science (80-.).* **2013**, *339* (6119), 535–539. https://doi.org/10.1126/science.1222453.
4. Edison, T. A. Manufacture of Filaments for Incandescent Electric Lamps. 470, 925, March 15, 1892.
5. Hughes, T. V.; Chambers, C. R. Manufacture of Carbon Filaments, 1889.
6. Pélabon, C.; Pélabon, H. Sur Une Variété de Carbone Filamenteux. *C. R. Acad. Sci. Paris* **1903**, *137*, 706–708.
7. Schützenberger, P.; Schützenberger, L. Sur Quelques Faits Relatifs Ál'histoire Du Carbone. *C. R. Acad. Sci. Paris* **1890**, *111*, 774–778.
8. Monthioux, M.; Kuznetsov, V. L. Who Should Be Given the Credit for the Discovery of Carbon Nanotubes? *Carbon N. Y.* **2006**, *44* (9), 1621–1623. https://doi.org/10.1016/j.carbon.2006.03.019.
9. Radushkevich, L. V.; Lukyanovich, V. M. O. Strukture Ugleroda, Obrazu- Jucegosja Pri Termiceskom Razlozenii Okisi Ugleroda Na Zeleznom Kontakte. *Zurn Fis. Chim* **1952**, *26*, 88–95.
10. Dresselhaus, M. S.; Dresselhaus, G.; Avouris, P. *Carbon Nanotubes: Synthesis, Structure, Properties, and Applications*; Springer Verlag: Berlin Heidelberg, 2001.
11. Bacon, R. Growth, Structure, and Properties of Graphite Whiskers. *J. Appl. Phys.* **1960**, *31* (2), 283–290. https://doi.org/10.1063/1.1735559.
12. Endo, M.; Koyama, T.; Hishiyama, Y. Structural Improvement of Carbon Fibers Prepared from Benzene. *Jpn. J. Appl. Phys.* **1976**, *15* (11), 2073–2076. https://doi.org/10.1143/JJAP.15.2073.
13. Tibbetts, G. G. Vapor-Grown Carbon Fibers: Status and Prospects. *Carbon N. Y.* **1989**, *27* (5), 745–747. https://doi.org/10.1016/0008-6223(89)90208-X.
14. Oberlin, A.; Endo, M.; Koyama, T. High Resolution Electron Microscope Observations of Graphitized Carbon Fibers. *Carbon N. Y.* **1976**, *14* (2), 133–135. https://doi.org/10.1016/0008-6223(76)90124-X.
15. Iijima, S. High Resolution Electron Microscopy of Some Carbonaceous Materials. *J. Microsc.* **1980**, *119* (1), 99–111. https://doi.org/10.1111/j.1365-2818.1980.tb04081.x.
16. Tennent, H. G. Carbon Fibrils, Method for Producing Same and Compositions Containing Same. 4,663,230, 1987.
17. Krätschmer, W.; Lamb, L. D.; Fostiropoulos, K.; Huffman, D. R. Solid C60: A New Form of Carbon. *Nature* **1990**, *347*, 354–358. https://doi.org/10.1038/347354a0.
18. Smalley, R. E. Formation and Properties of C60 and the Fullerenes. *Natl. Institu. Standards Technol.* 1990.
19. Dresselhaus, M. S. Oral Presentation at Fullerene Workshop, 1991.
20. Wang, S.; Buseck, P. R. Packing of C_{60} Molecules and Related Fullerenes in Crystals: A Direct View. *Chem. Phys. Lett.* **1991**, *182* (1), 1–4. https://doi.org/10.1016/0009-2614 (91)80093-D.
21. Fowler, P. W. Carbon Cylinders: A Class of Closed-Shell Clusters. *J. Chem. Soc. Faraday Trans.* **1990**, *86* (12), 2073–2077. https://doi.org/10.1039/FT9908602073.
22. Iijima, S. Helical Microtubules of Graphitic Carbon. *Nature* **1991**, *354*, 56–58. https://doi.org/10.1038/354056a0.
23. Ebbesen, T. W.; Ajayan, P. M. Large-Scale Synthesis of Carbon Nanotubes. *Nature* **1992**, *358*, 220–222. https://doi.org/10.1038/358220a0.

24. Dresselhaus, M. S.; Dresselhaus, G.; Saito, R. Carbon Fibers Based on C60 and Their Symmetry. *Phys. Rev. B* **1992**, *45* (11), 6234–6242. https://doi.org/10.1103/PhysRevB.45.6234.
25. Mintmire, J. W.; Dunlap, B. I.; White, C. T. Are Fullerene Tubules Metallic? *Phys. Rev. Lett.* **1992**, *68* (5), 631–634. https://doi.org/10.1103/PhysRevLett.68.631.
26. José-Yacamán, M.; Miki-Yoshida, M.; Rendón, L.; Santiesteban, J. G. Catalytic Growth of Carbon Microtubules with Fullerene Structure. *Appl. Phys. Lett.* **1993**, *62* (2), 202–204. https://doi.org/10.1063/1.109315.
27. Li, W. Z.; Xie, S. S.; Qian, L. X.; Chang, B. H.; Zou, B. S.; Zhou, W. Y.; Zhao, R. A.; Wang, G. Large-Scale Synthesis of Aligned Carbon Nanotubes. *Science (80-.)* **1996**, *274* (5293), 1701–1703. https://doi.org/10.1126/science.274.5293.1701.
28. Iijima, S.; Ichihashi, T. Single-Shell Carbon Nanotubes of 1-Nm Diameter. *Nature* **1993**, *363*, 603–605. https://doi.org/10.1038/363603a0.
29. Bethune, D. S.; Kiang, C. H.; de Vries, M. S.; Gorman, G.; Savoy, R.; Vazquez, J.; Beyers, R. Cobalt-Catalysed Growth of Carbon Nanotubes with Single-Atomic-Layer Walls. *Nature* **1993**, *363*, 605–607. https://doi.org/10.1038/363605a0.
30. Thess, A.; Lee, R.; Nikolaev, P.; Dai, H.; Petit, P.; Robert, J.; Xu, C.; Lee, Y. H.; Kim, S. G.; Rinzler, A. G.; et al. Crystalline Ropes of Metallic Carbon Nanotubes. *Science (80-.)* **1996**, *273* (5274), 483–487. https://doi.org/10.1126/science.273.5274.483.
31. Journet, C.; Maser, W. K.; Bernier, P.; Loiseau, A.; Lamy de la Chapelle, M.; Lefrant, S.; Deniard, P.; Lee, R.; Fischer, J. E. Large-Scale Production of Single-Walled Carbon Nanotubes by the Electric-Arc Technique. *Nature* **1997**, *388*, 756–758. https://doi.org/10.1038/41972.
32. Das, R.; Shahnavaz, Z.; Ali, M. E.; Islam, M. M.; Abd Hamid, S. B. Can We Optimize Arc Discharge and Laser Ablation for Well-Controlled Carbon Nanotube Synthesis? *Nanoscale Res. Lett.* **2016**, *11* (1), 1–23. https://doi.org/10.1186/s11671-016-1730-0.
33. Patel, D. K.; Kim, H.; Dutta, S. D.; Ganguly, K.; Lim, K.-T. Carbon Nanotubes-Based Nanomaterials and Their Agricultural and Biotechnological Applications. *Materials (Basel)* **2020**, *13* (7), 1–28. https://doi.org/10.3390/ma13071679.
34. Manawi, Y. M.; Ihsanullah; Samara, A.; Al-Ansari, T.; Atieh, M. A. A Review of Carbon Nanomaterials' Synthesis via the Chemical Vapor Deposition (CVD) Method. *Materials (Basel)* **2018**, *11* (5), 1–36. https://doi.org/10.3390/ma11050822.
35. Morris, J. E.; Iniewski, K. *Graphene, Carbon Nanotubes, and Nanostructures*; CRC Press, Taylor & Francis Group: Boca Raton, London, 2013.
36. Maniecki, T.; Shtyka, O.; Mierczynski, P.; Ciesielski, R.; Czylkowska, A.; Leyko, J.; Mitukiewicz, G.; Dubkov, S.; Gromov, D. Carbon Nanotubes: Properties, Synthesis, and Application. *Fibre Chem.* **2018**, *50* (4), 297–300. https://doi.org/10.1007/s10692-019-09979-2.
37. Kaushik, B. K.; Majumder, M. K. Carbon Nanotube: Properties and Applications. In *Carbon Nanotube Based VLSI Interconnects. SpringerBriefs in Applied Sciences and Technology*; Springer Verlag: New Delhi, 2015; pp 17–37. https://doi.org/10.1007/978-81-322-2047-3_2.
38. Gupta, N.; Gupta, S. M.; Sharma, S. K. Carbon Nanotubes: Synthesis, Properties and Engineering Applications. *Carbon Lett.* **2019**, *29* (5), 419–447. https://doi.org/10.1007/s42823-019-00068-2.

39. Javey, A.; Qi, P.; Wang, Q.; Dai, H. Ten- to 50-Nm-Long Quasi-Ballistic Carbon Nanotube Devices Obtained without Complex Lithography. *Proc. Natl. Acad. Sci. USA* **2004**, *101* (37), 13408–13410. https://doi.org/10.1073/pnas.0404450101.
40. O'Connell, M. J. *Carbon Nanotubes Properties and Applications*; Taylor & Francis: London, 2006.
41. Ibrahim, K. S. Carbon Nanotubes-Properties and Applications: A Review. *Carbon Lett.* **2013**, *14* (3), 131–144. https://doi.org/10.5714/cl.2013.14.3.131.
42. Kelly, B. T. *Physics of Graphite*; Applied Science: London, 1981.
43. Treacy, M. M. J.; Ebbesen, T. W.; Gibson, J. M. Exceptionally High Young's Modulus Observed for Individual Carbon Nanotubes. *Nature* **1996**, *381* (6584), 678–680. https://doi.org/10.1038/381678a0.
44. Takakura, A.; Beppu, K.; Nishihara, T.; Fukui, A.; Kozeki, T.; Namazu, T.; Miyauchi, Y.; Itami, K. Strength of Carbon Nanotubes Depends on Their Chemical Structures. *Nat. Commun.* **2019**, *10*, 1–7. https://doi.org/10.1038/s41467-019-10959-7.
45. Jorio, A.; Dresselhaus, G.; Dresselhaus, M. S. *Carbon Nanotubes Advanced Topics in the Synthesis, Structure, Properties and Applications*; Springer: Berlin Heidelberg, 2008. https://doi.org/10.1071/978-3-540-72865-8.
46. Yu, M. F.; Files, B. S.; Arepalli, S.; Ruoff, R. S. Tensile Loading of Ropes of Single Wall Carbon Nanotubes and Their Mechanical Properties. *Phys. Rev. Lett.* **2000**, *84* (24), 5552–5555. https://doi.org/10.1103/PhysRevLett.84.5552.
47. Ruoff, R. S.; Tersoff, J.; Lorents, D. C.; Subramoney, S.; Chan, B. Radial Deformation of Carbon Nanotubes by Van Der Waals Forces. *Nature* **1993**, *364* (6437), 514–516. https://doi.org/10.1038/364514a0.
48. Meyyappan, M. *Carbon Nanotubes Science and Applications*; 2005.
49. Baughman, R. H.; Cui, C.; Zakhidov, A. A.; Iqbal, Z.; Barisci, J. N.; Spinks, G. M.; Wallace, G. G.; Mazzoldi, A.; De Rossi, D.; Rinzler, A. G.; et al. Carbon Nanotube Actuators. *Science (80-.)* **1999**, *284* (5418), 1340–1344. https://doi.org/10.1126/science.284.5418.1340.
50. Ajayan, P. M.; Iijima, S. Capillarity-Induced Filling of Carbon Nanotubes. *Nature* **1993**, *361* (6410), 333–334. https://doi.org/10.1038/361333a0.
51. Qian, D.; Wagner, G. J.; Liu, W. K.; Yu, M. F.; Ruoff, R. S. Mechanics of Carbon Nanotubes. *Appl. Mech. Rev.* **2002**, *55* (6), 533. https://doi.org/10.1115/1.1490129.
52. Hone, J.; Llaguno, M. C.; Biercuk, M. J.; Johnson, A. T.; Batlogg, B.; Benes, Z.; Fischer, J. E. Thermal Properties of Carbon Nanotubes and Nanotube-Based Materials. *Appl. Phys. A Mater. Sci. Process.* **2002**, *74*, 339–343. https://doi.org/10.1007/s003390201277.
53. Kumanek, B.; Janas, D. Thermal Conductivity of Carbon Nanotube Networks: A Review. *J. Mater. Sci.* **2019**, *54*, 7397–7427. https://doi.org/10.1007/s10853-019-03368-0.
54. Costa, S.; Borowiak-Palen, E.; Kruszyñska, M.; Bachmatiuk, A.; Kaleńczuk, R. J. Characterization of Carbon Nanotubes by Raman Spectroscopy. *Mater. Sci. Pol.* **2008**, *26* (2), 433–441.
55. Reich, S.; Thomsen, C.; Maultzsch, J. *Carbon Nanotubes: Basic Concepts and Physical Properties*; Wiley-VCH Verlag: Darmstadt, 2007. https://doi.org/10.1002/9783527618040.
56. Barnard, A. W.; Zhang, M.; Wiederhecker, G. S.; Lipson, M.; McEuen, P. L. Real-Time Vibrations of a Carbon Nanotube. *Nature* **2019**, *566*, 89–93. https://doi.org/10.1038/s41586-018-0861-0.

57. Abdulrazzak, F. H.; Hussein, F. H. Photocatalytic Hydrogen Production on Nanocomposite of Carbon Nanotubes and TiO_2. *J. Phys. Conf. Ser.* **2018**, *1032*, 1–12. https://doi.org/10.1088/1742-6596/1032/1/012056.
58. Manzetti, S.; Gabriel, J.-C. P. Methods for Dispersing Carbon Nanotubes for Nanotechnology Applications: Liquid Nanocrystals, Suspensions, Polyelectrolytes, Colloids and Organization Control. *Int. Nano Lett.* **2019**, *9* (1), 31–49. https://doi.org/10.1007/s40089-018-0260-4.
59. Monreal, C.; McGill, W. B.; Nyborg, M. Spatial Heterogeneity of Substrates: Effects on Hydrolysis, Immobilization and Nitrification of Urea-N. *Can. J. Soil Sci.* **1986**, *66* (3), 499–511. https://doi.org/10.4141/cjss86-050.
60. Chhipa, H. Nanofertilizers and Nanopesticides for Agriculture. *Environ. Chem. Lett.* **2017**, *15*, 15–22. https://doi.org/10.1007/s10311-016-0600-4.
61. Yatim, N. M.; Shaaban, A.; Dimin, M. F.; Yusof, F.; Razak, J. A. Effect of Functionalised and Non-Functionalised Carbon Nanotubes-Urea Fertilizer on the Growth of Paddy. *Trop. Life Sci. Res.* **2018**, *29* (1), 17–35. https://doi.org/10.21315/tlsr2018.29.1.2.
62. Liu, Q.; Chen, B.; Wang, Q.; Shi, X.; Xiao, Z.; Lin, J.; Fang, X. Carbon Nanotubes as Molecular Transporters for Walled Plant Cells. *Nano Lett.* **2009**, *9* (3), 1007–1010. https://doi.org/10.1021/nl803083u.
63. Serag, M. F.; Kaji, N.; Gaillard, C.; Okamoto, Y.; Terasaka, K.; Jabasini, M.; Tokeshi, M.; Mizukami, H.; Bianco, A.; Baba, Y. Trafficking and Subcellular Localization of Multiwalled Carbon Nanotubes in Plant Cells. *ACS Nano* **2011**, *5* (1), 493–499. https://doi.org/10.1021/nn102344t.
64. Singh, A.; Bhati, A.; Gunture; Tripathi, K. M.; Sonkar, S. K. Nanocarbons in Agricultural Plants: Can Be a Potential Nanofertilizer? In *Nanotechnology in Environmental Science*; Hussain, C. M., Mishra, A. K., Eds.; Wiley-VCH Verlag, 2018; pp 153–190. https://doi.org/10.1002/9783527808854.ch6.
65. Nagib Abdel-Ghany Hasaneen, M.; Mahmoud Mohammad Abdel-Aziz, H.; Moheb Omer, A. Characterization of Carbon Nanotubes Loaded with Nitrogen, Phosphorus and Potassium Fertilizers. *Am. J. Nano Res. Appl.* **2017**, *5* (2), 12–18. https://doi.org/10.11648/j.nano.20170502.11.
66. Jian, L.; Yang-de, Z.; Zhi-ming, Z. The Application Research of Nano-Biotechnology to Promote Increasing of Vegetable Production– 《Hubei Agricultural Sciences》 2009年01期. *Hubei Agric. Sci.* **2009**, *1*, 123–127.
67. Khodakovskaya, M.; Dervishi, E.; Mahmood, M.; Xu, Y.; Li, Z.; Watanabe, F.; Biris, A. S. Carbon Nanotubes Are Able to Penetrate Plant Seed Coat and Dramatically Affect Seed Germination and Plant Growth. *ACS Nano* **2009**, *3* (10), 3221–3227. https://doi.org/10.1021/nn900887m.
68. Giraldo, J. P.; Landry, M. P.; Faltermeier, S. M.; McNicholas, T. P.; Iverson, N. M.; Boghossian, A. A.; Reuel, N. F.; Hilmer, A. J.; Sen, F.; Brew, J. A.; et al. Plant Nanobionics Approach to Augment Photosynthesis and Biochemical Sensing. *Nat. Mater.* **2014**, *13*, 400–408. https://doi.org/10.1038/nmat3890.
69. Taha, R. A.; Hassan, M. M.; Ibrahim, E. A.; Abou Baker, N. H.; Shaaban, E. A. Carbon Nanotubes Impact on Date Palm in Vitro Cultures. *Plant Cell. Tissue Organ Cult.* **2016**, *127* (2), 525–534. https://doi.org/10.1007/s11240-016-1058-6.

70. Husen, A.; Siddiqi, K. S. Carbon and Fullerene Nanomaterials in Plant System. *J. Nanobiotechnology* **2014**, *12*, 1–10. https://doi.org/10.1186/1477-3155-12-16.
71. Khodakovskaya, M. V.; Kim, B. S.; Kim, J. N.; Alimohammadi, M.; Dervishi, E.; Mustafa, T.; Cernigla, C. E. Carbon Nanotubes as Plant Growth Regulators: Effects on Tomato Growth, Reproductive System, and Soil Microbial Community. *Small* **2013**, *9* (1), 115–123. https://doi.org/10.1002/smll.201201225.
72. Cañas, J. E.; Long, M.; Nations, S.; Vadan, R.; Dai, L.; Luo, M.; Ambikapathi, R.; Lee, E. H.; Olszyk, D. Effects of Functionalized and Nonfunctionalized Single-Walled Carbon Nanotubes on Root Elongation of Select Crop Species. *Environ. Toxicol. Chem.* **2008**, *27* (9), 1922–1931. https://doi.org/10.1897/08-117.1.
73. Khodakovskaya, M. V.; De Silva, K.; Biris, A. S.; Dervishi, E.; Villagarcia, H. Carbon Nanotubes Induce Growth Enhancement of Tobacco Cells. *ACS Nano* **2012**, *6* (3), 2128–2135. https://doi.org/10.1021/nn204643g.
74. Wang, X.; Han, H.; Liu, X.; Gu, X.; Chen, K.; Lu, D. Multi-Walled Carbon Nanotubes Can Enhance Root Elongation of Wheat (Triticum Aestivum) Plants. *J. Nanoparticle Res.* **2012**, *14*, 1–10. https://doi.org/10.1007/s11051-012-0841-5.
75. Begum, P.; Ikhtiari, R.; Fugetsu, B.; Matsuoka, M.; Akasaka, T.; Watari, F. Phytotoxicity of Multi-Walled Carbon Nanotubes Assessed by Selected Plant Species in the Seedling Stage. *Appl. Surf. Sci.* **2012**, *262*, 120–124. https://doi.org/10.1016/j.apsusc.2012.03.028.
76. Fathi, Z.; Khavari Nejad, R. A.; Mahmoodzadeh, H.; Satari, T. N. Investigating of a Wide Range of Concentrations of Multi-Walled Carbon Nanotubes on Germination and Growth of Castor Seeds (Ricinus Communis L.). *J. Plant Prot. Res.* **2017**, *57* (3), 228–236. https://doi.org/10.1515/jppr-2017-0032.
77. Tiwari, D. K.; Dasgupta-Schubert, N.; Villaseñor Cendejas, L. M.; Villegas, J.; Carreto Montoya, L.; Borjas García, S. E. Interfacing Carbon Nanotubes (CNT) with Plants: Enhancement of Growth, Water and Ionic Nutrient Uptake in Maize (Zea Mays) and Implications for Nanoagriculture. *Appl. Nanosci.* **2014**, *4*, 577–591. https://doi.org/10.1007/s13204-013-0236-7.
78. Stampoulis, D.; Sinha, S. K.; White, J. C. Assay-Dependent Phytotoxicity of Nanoparticles to Plants. *Environ. Sci. Technol.* **2009**, *43* (24), 9473–9479. https://doi.org/10.1021/es901695c.
79. Pandey, K.; Lahiani, M. H.; Hicks, V. K.; Hudson, M. K.; Green, M. J.; Khodakovskaya, M. Effects of Carbon-Based Nanomaterials on Seed Germination, Biomass Accumulation and Salt Stress Response of Bioenergy Crops. *PLoS One* **2018**, *13* (8), 1–17. https://doi.org/10.1371/journal.pone.0202274.
80. Anzar, N.; Hasan, R.; Tyagi, M.; Yadav, N.; Narang, J. Carbon Nanotube - A Review on Synthesis, Properties and Plethora of Applications in the Field of Biomedical Science. *Sensors Int.* **2020**, *1*, 1–10. https://doi.org/10.1016/j.sintl.2020.100003.
81. Hossain, F.; Perales-Perez, O. J.; Hwang, S.; Román, F. Antimicrobial Nanomaterials as Water Disinfectant: Applications, Limitations and Future Perspectives. *Sci. Total Environ.* **2014**, *466–467*, 1047–1059. https://doi.org/10.1016/j.scitotenv.2013.08.009.
82. Kang, S.; Herzberg, M.; Rodrigues, D. F.; Elimelech, M. Antibacterial Effects of Carbon Nanotubes: Size Does Matter! *Langmuir* **2008**, *24* (13), 6409–6413. https://doi.org/10.1021/la800951v.

83. Dong, X.; Yang, L. Dual Functional Nisin-Multi-Walled Carbon Nanotubes Coated Filters for Bacterial Capture and Inactivation. *J. Biol. Eng.* **2015**, *9*, 1–10. https://doi.org/10.1186/s13036-015-0018-8.
84. Zare-Zardini, H.; Amiri, A.; Shanbedi, M.; Memarpoor-Yazdi, M.; Asoodeh, A. Studying of Antifungal Activity of Functionalized Multiwalled Carbon Nanotubes by Microwave-Assisted Technique. *Surf. Interface Anal.* **2013**, *45* (3), 751–755. https://doi.org/10.1002/sia.5152.
85. Mohan, R.; Shanmugharaj, A. M.; Sung Hun, R. An Efficient Growth of Silver and Copper Nanoparticles on Multiwalled Carbon Nanotube with Enhanced Antimicrobial Activity. *J. Biomed. Mater. Res. Part B Appl. Biomater.* **2011**, *96B* (1), 119–126. https://doi.org/10.1002/jbm.b.31747.
86. Hu, N.; Dang, G.; Zhou, H.; Jing, J.; Chen, C. Efficient Direct Water Dispersion of Multi-Walled Carbon Nanotubes by Functionalization with Lysine. *Mater. Lett.* **2007**, *61* (30), 5285–5287. https://doi.org/10.1016/j.matlet.2007.04.084.
87. Zhou, J.; Qi, X. Multi-Walled Carbon Nanotubes/Epilson-Polylysine Nanocomposite with Enhanced Antibacterial Activity. *Lett. Appl. Microbiol.* **2011**, *52* (1), 76–83. https://doi.org/10.1111/j.1472-765X.2010.02969.x.
88. Wang, X.; Liu, X.; Han, H. Evaluation of Antibacterial Effects of Carbon Nanomaterials against Copper-Resistant Ralstonia Solanacearum. *Coll. Surf. B Biointerf.* **2013**, *103*, 136–142. https://doi.org/10.1016/j.colsurfb.2012.09.044.
89. Wang, X.; Zhou, Z.; Chen, F. Surface Modification of Carbon Nanotubes with an Enhanced Antifungal Activity for the Control of Plant Fungal Pathogen. *Materials (Basel)* **2017**, *10* (12), 1–11. https://doi.org/10.3390/ma10121375.
90. Wang, X.; Liu, X.; Chen, J.; Han, H.; Yuan, Z. Evaluation and Mechanism of Antifungal Effects of Carbon Nanomaterials in Controlling Plant Fungal Pathogen. *Carbon N. Y.* **2014**, *68*, 798–806. https://doi.org/10.1016/j.carbon.2013.11.072.
91. Sharma, A.; Kumar, V.; Shahzad, B.; Tanveer, M.; Sidhu, G. P. S.; Handa, N.; Kohli, S. K.; Yadav, P.; Bali, A. S.; Parihar, R. D.; et al. Worldwide Pesticide Usage and Its Impacts on Ecosystem. *SN Appl. Sci.* **2019**, *1*, 1–16. https://doi.org/10.1007/s42452-019-1485-1.
92. González, P.; González-Melendi, G.; Fernández, R.; Fernández-Pacheco, F.; Coronado, M. J.; Corredor, E.; Testillano, P. S.; Risueño, M. C.; Risueño, R.; Marquina, C.; et al. Nanoparticles as Smart Treatment-Delivery Systems in Plants: Assessment of Different Techniques of Microscopy for Their Visualization in Plant Tissues. *Ann. Bot.* **2008**, *101*, 187–195. https://doi.org/10.1093/aob/mcm283.
93. Sasson, Y.; Levy-Ruso, G.; Toledano, O.; Ishaaya, I. Nanosuspensions: Emerging Novel Agrochemical Formulations. In *Insecticides Design Using Advanced Technologies*; Ishaaya, I., Horowitz, A. R., Nauen, R., Eds., Vol. 9783540469070; Springer: Berlin Heidelberg, 2007; pp 1–39. https://doi.org/10.1007/978-3-540-46907-0_1.
94. Sarlak, N.; Taherifar, A.; Salehi, F. Synthesis of Nanopesticides by Encapsulating Pesticide Nanoparticles Using Functionalized Carbon Nanotubes and Application of New Nanocomposite for Plant Disease Treatment. *J. Agric. Food Chem.* **2014**, *62* (21), 4833–4838. https://doi.org/10.1021/jf404720d.
95. Mohamed, M. A.; Hashim, A. F.; Alghuthaymi, M. A.; Abd-Elsalam, K. A. Nano-Carbon: Plant Growth Promotion and Protection. In *Nanobiotechnology Applications in Plant*

Protection. Nanotechnology in the Life Sciences.; Abd-Elsalam, K., Prasad, R., Eds.; Springer, Cham, 2018; pp 155–188. https://doi.org/10.1007/978-3-319-91161-8_7.
96. Fraceto, L. F.; Grillo, R.; de Medeiros, G. A.; Scognamiglio, V.; Rea, G.; Bartolucci, C. Nanotechnology in Agriculture: Which Innovation Potential Does It Have? *Front. Environ. Sci.* **2016**, *4*, 1–5. https://doi.org/10.3389/fenvs.2016.00020.
97. Srivastava, A. K.; Dev, A.; Karmakar, S. Nanosensors and Nanobiosensors in Food and Agriculture. *Environ. Chem. Lett.* **2018**, *16*, 161–182. https://doi.org/10.1007/s10311-017-0674-7.
98. Lew, T. T. S.; Koman, V. B.; Silmore, K. S.; Seo, J. S.; Gordiichuk, P.; Kwak, S. Y.; Park, M.; Ang, M. C. Y.; Khong, D. T.; Lee, M. A., et al. Real-Time Detection of Wound-Induced H_2O_2Signalling Waves in Plants with Optical Nanosensors. *Nat. Plants* **2020**, *6* (4), 404–415. https://doi.org/10.1038/s41477-020-0632-4.
99. Ren, Q. Q.; Yuan, X. J.; Huang, X. R.; Wen, W.; Zhao, Y. Di; Chen, W. In Vivo Monitoring of Oxidative Burst on Aloe under Salinity Stress Using Hemoglobin and Single-Walled Carbon Nanotubes Modified Carbon Fiber Ultramicroelectrode. *Biosens. Bioelectron.* **2013**, *50*, 318–324. https://doi.org/10.1016/j.bios.2013.07.001.
100. Tsai, J. T. H.; Lu, C. C.; Li, J. G. Fabrication of Humidity Sensors by Multi-Walled Carbon Nanotubes. *J. Exp. Nanosci.* **2010**, *5* (4), 302–309. https://doi.org/10.1080/17458080903513300.
101. Zhao, Z. G.; Liu, X. W.; Chen, W. P.; Li, T. Carbon Nanotubes Humidity Sensor Based on High Testing Frequencies. *Sens. Actuat. A Phys.* **2011**, *168* (1), 10–13. https://doi.org/10.1016/j.sna.2011.03.029.
102. Lee, K.; Park, J.; Lee, M. S.; Kim, J.; Hyun, B. G.; Kang, D. J.; Na, K.; Lee, C. Y.; Bien, F.; Park, J. U. In-Situ Synthesis of Carbon Nanotube-Graphite Electronic Devices and Their Integrations onto Surfaces of Live Plants and Insects. *Nano Lett.* **2014**, *14* (5), 2647–2654. https://doi.org/10.1021/nl500513n.
103. Chen, Z.-G. Conductometric Immunosensors for the Detection of Staphylococcal Enterotoxin B Based Bio-Electrocalytic Reaction on Micro-Comb Electrodes. *Bioprocess Biosyst. Eng.* **2008**, *31* (4), 345–350. https://doi.org/10.1007/s00449-007-0168-2.
104. Viswanathan, S.; Wu, L. C.; Huang, M. R.; Ho, J. A. A. Electrochemical Immunosensor for Cholera Toxin Using Liposomes and Poly(3,4-Ethylenedioxythiophene)-Coated Carbon Nanotubes. *Anal. Chem.* **2006**, *78* (4), 1115–1121. https://doi.org/10.1021/ac051435d.
105. Tahir, M. A.; Hameed, S.; Munawar, A.; Amin, I.; Mansoor, S.; Khan, W. S.; Bajwa, S. Z. Investigating the Potential of Multiwalled Carbon Nanotubes Based Zinc Nanocomposite as a Recognition Interface towards Plant Pathogen Detection. *J. Virol. Methods* **2017**, *249*, 130–136. https://doi.org/10.1016/j.jviromet.2017.09.004.
106. Tahir, M. A.; Bajwa, S. Z.; Mansoor, S.; Briddon, R. W.; Khan, W. S.; Scheffler, B. E.; Amin, I. Evaluation of Carbon Nanotube Based Copper Nanoparticle Composite for the Efficient Detection of Agroviruses. *J. Hazard. Mater.* **2018**, *346*, 27–35. https://doi.org/10.1016/j.jhazmat.2017.12.007.
107. Akanbi, F.; Yusof, N.; Abdullah, J.; Sulaiman, Y.; Hushiarian, R. Detection of Quinoline in G. Boninense-Infected Plants Using Functionalized Multi-Walled Carbon Nanotubes: A Field Study. *Sensors* **2017**, *17* (7), 1–14. https://doi.org/10.3390/s17071538.

108. Wong, A.; Silva, T.; Caetano, F.; Bergamini, M.; Marcolino-Junior, L.; Fatibello-Filho, O.; Janegitz, B. An Overview of Pesticide Monitoring at Environmental Samples Using Carbon Nanotubes-Based Electrochemical Sensors. *C* **2017**, *3* (1), 8. https://doi.org/10.3390/c3010008.
109. Dong, J.; Fan, X.; Qiao, F.; Ai, S.; Xin, H. A Novel Protocol for Ultra-Trace Detection of Pesticides: Combined Electrochemical Reduction of Ellman's Reagent with Acetylcholinesterase Inhibition. *Anal. Chim. Acta* **2013**, *761*, 78–83. https://doi.org/10.1016/j.aca.2012.11.042.
110. Mercan, H.; Inam, R.; Aboul-Enein, H. Y. Square Wave Adsorptive Stripping Voltammetric Determination of Cyromazine Insecticide with Multi-Walled Carbon Nanotube Paste Electrode. *Anal. Lett.* **2011**, *44* (8), 1392–1404. https://doi.org/10.1080/00032719.2010.512684.
111. Herrero, M.; Simõ, C.; García-Cañas, V.; Ibáñez, E.; Cifuentes, A. Foodomics: MS-Based Strategies in Modern Food Science and Nutrition. *Mass Spectrom. Rev.* **2012**, *31* (1), 49–69. https://doi.org/10.1002/mas.20335.
112. Bandyopadhyay, A.; Ghosh, D.; Pati, S. K. Trapping and Sensing of Hazardous Insecticides by Chemically Modified Single Walled Carbon Nanotubes. *Phys. Chem. Chem. Phys.* **2017**, *19*, 24059–24066. https://doi.org/10.1039/c7cp04980c.
113. Zaytseva, O.; Neumann, G. Carbon Nanomaterials: Production, Impact on Plant Development, Agricultural and Environmental Applications. *Chem. Biol. Technol. Agric.* **2016**, *3*, 1–26. https://doi.org/10.1186/s40538-016-0070-8.
114. Demirer, G. S.; Zhang, H.; Goh, N. S.; González-Grandío, E.; Landry, M. P. Carbon Nanotube–Mediated DNA Delivery without Transgene Integration in Intact Plants. *Nat. Protoc.* **2019**, *14*, 2954–2971. https://doi.org/10.1038/s41596-019-0208-9.
115. Kwak, S. Y.; Lew, T. T. S.; Sweeney, C. J.; Koman, V. B.; Wong, M. H.; Bohmert-Tatarev, K.; Snell, K. D.; Seo, J. S.; Chua, N. H.; Strano, M. S. Chloroplast-Selective Gene Delivery and Expression in Planta Using Chitosan-Complexed Single-Walled Carbon Nanotube Carriers. *Nat. Nanotechnol.* **2019**, *14* (5), 447–455. https://doi.org/10.1038/s41565-019-0375-4.
116. Demirer, G. S.; Huan, Z.; Goh, N. S.; Pinals, R. L.; Chang, R.; Landry, M. P. Carbon Nanocarriers Deliver SiRNA to Intact Plant Cells for Efficient Gene Knockdown. *bioRxiv* **2019**, 1–36. https://doi.org/10.1101/564427.
117. Mondal, A.; Basu, R.; Das, S.; Nandy, P. Beneficial Role of Carbon Nanotubes on Mustard Plant Growth: An Agricultural Prospect. *J. Nanopart. Res.* **2011**, *13* (10), 4519–4528. https://doi.org/10.1007/s11051-011-0406-z.
118. Tan, X. ming; Lin, C.; Fugetsu, B. Studies on Toxicity of Multi-Walled Carbon Nanotubes on Suspension Rice Cells. *Carbon N. Y.* **2009**, *47* (15), 3479–3487. https://doi.org/10.1016/j.carbon.2009.08.018.
119. Lin, C.; Fugetsu, B.; Su, Y.; Watari, F. Studies on Toxicity of Multi-Walled Carbon Nanotubes on Arabidopsis T87 Suspension Cells. *J. Hazard. Mater.* **2009**, *170* (2–3), 578–583. https://doi.org/10.1016/J.JHAZMAT.2009.05.025.
120. Shrestha, B.; Acosta-Martinez, V.; Cox, S. B.; Green, M. J.; Li, S.; Cañas-Carrell, J. E. An Evaluation of the Impact of Multiwalled Carbon Nanotubes on Soil Microbial Community Structure and Functioning. *J. Hazard. Mater.* **2013**, *261*, 188–197. https://doi.org/10.1016/j.jhazmat.2013.07.031.

121. Dinesh, R.; Anandaraj, M.; Srinivasan, V.; Hamza, S. Engineered Nanoparticles in the Soil and Their Potential Implications to Microbial Activity. *Geoderma* **2012**, *173–174*, 19–27. https://doi.org/10.1016/j.geoderma.2011.12.018.
122. Jackson, P.; Jacobsen, N. R.; Baun, A.; Birkedal, R.; Kühnel, D.; Jensen, K. A.; Vogel, U.; Wallin, H. Bioaccumulation and Ecotoxicity of Carbon Nanotubes. *Chem. Cent. J.* **2013**, *7* (1), 1–21. https://doi.org/10.1186/1752-153X-7-154.
123. Kang, S.; Mauter, M. S.; Elimelech, M. Microbial Cytotoxicity of Carbon-Based Nanomaterials: Implications for River Water and Wastewater Effluent. *Environ. Sci. Technol.* **2009**, *43* (7), 2648–2653. https://doi.org/10.1021/es8031506.
124. Poland, C. A.; Duffin, R.; Kinloch, I.; Maynard, A.; Wallace, W. A. H.; Seaton, A.; Stone, V.; Brown, S.; MacNee, W.; Donaldson, K. Carbon Nanotubes Introduced into the Abdominal Cavity of Mice Show Asbestos-like Pathogenicity in a Pilot Study. *Nat. Nanotechnol.* **2008**, *3* (7), 423–428. https://doi.org/10.1038/nnano.2008.111.

CHAPTER 6

Carbon Nanotubes as Sensors in Food and Agricultural Applications

BINILA K. KORAH, ANU ROSE CHACKO, SNEHA MATHEW, and BEENA MATHEW*

School of Chemical Sciences, Mahatma Gandhi University, Kottayam, Kerala, India

**Corresponding author. E-mail: beenamathew@mgu.ac.in*

ABSTRACT

Carbon nanotubes (CNTs) are emerging as one of the rising stars in the field of nanosensors due to their fascinating properties, such as good electronic and mechanical properties and high surface-to-volume ratios. As a promising tool for their applications in the field of food and agriculture, CNTs assure a crucial space for the nurture of greener environment. An appreciable improvement in the selectivity, sensitivity, and speed along with their small size, less power consumption, simplicity, and capability to detect complex analytes compared with the traditional methods are the superior traits of CNTs. This chapter is focused around the role of CNTs in the field of agriculture and food safety considering the growing need for development in that direction. Improvement of crop yield, reduction of applied agrochemicals, and improved farming methods using CNTs which are necessary for the realization of a smart agricultural sector within the limited areas and minimum negative environmental impact are discussed in this chapter. Furthermore, the part of CNTs in ensuring food safety is detailed giving special emphasis on quality assurance, prevention of food spoilage, and detection of food-borne pathogens. Concluding with a brief note on the negative impacts of CNTs on environment, we hope this chapter will provide the readers a compilation of the trends and developments of carbon nanotubes as sensors in food and agricultural applications.

6.1 INTRODUCTION

The journey of carbon nanotubes (CNTs) toward attaining a greener environment started long time ago and still it is on its way leaving enduring marks throughout the development. CNTs exhibit unique properties, such as small size and great stability and that make them a favorable material for the scientific community to use them as a tool for accomplishing green applications in environment. Along with the applications involving environmental remediation, CNTs have been proved useful for the upgradation of agriculture which is not only a mean for filling the abdomen of people but is also a part of the economy. Agriculture, which is considered as the backbone of developing countries is now facing a lot of challenges. Increasing population, decrease in land areas suitable for farming, overuse of pesticide-induced environmental problems, variation in climatic conditions, and reduced yield are some of the main challenges faced by the farmers worldwide. A solution for all these issues is actually a giant leap toward greener environment and richer economy. Carbon nanotubes in this scenario have exhibited miracles in accomplishing smart agriculture through improving the crop yield, farming, and reducing agrochemicals. Another closely related field of agriculture is food sector, which needs assistance from the technology for ensuring quality and safety. Food is the basic component and necessity for a healthier community. Carbon nanotubes can monitor the quality of food, provide information regarding the integrity in food packing, prevent food spoilage, and can detect foodborne pathogens.

This chapter starts with a brief introduction on structure, properties, and classification of carbon nanotubes and progresses to the application of carbon nanotubes in agro-food sector (Figure 6.1). Giving special focus on the contributions of CNTs in improving crop yield, farming and reducing agrochemicals, the section deals with the positive impact of CNTs in seed germination and plant growth. It also discusses how CNTs function as nanofertilizers and nanopesticides in molding a greener environment along with a brief overview of CNT-based sensors used in detecting pesticides. The next section is dedicated toward describing the development of CNTs in food-related applications. In this section, the role of CNTs in ensuring quality and safety of food along with their ability of detecting food-borne pathogens is given special attention. It also explains how CNT-based sensors detect taste and smell and how they assure the integrity in packaged food in detail. With a brief note on the negative impacts that the CNTs pose to the environment, the section concludes pointing out the gaps and new research directions. We hope that this chapter will provide the reader a brief understanding of the developments, gaps, and challenges in the area of CNTs in food and agriculture applications.

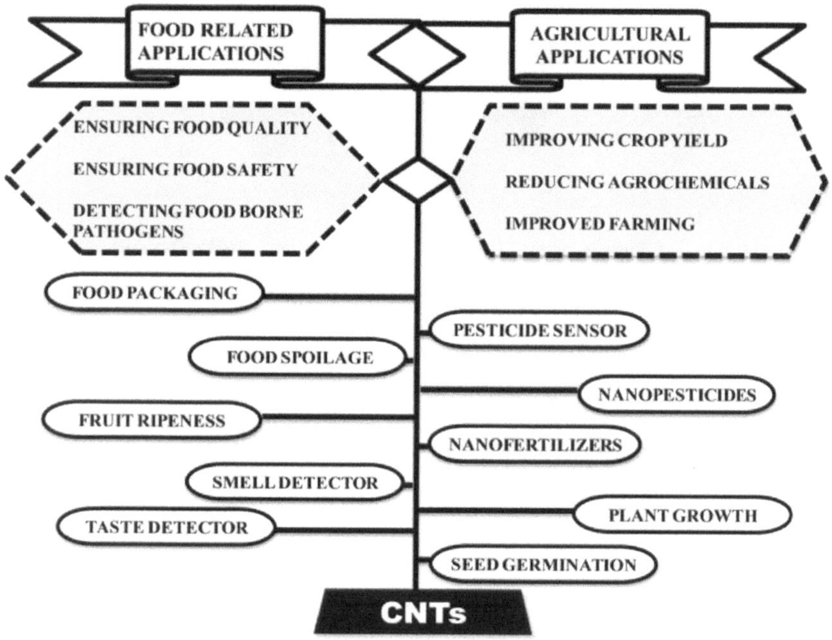

FIGURE 6.1 Schematic representation of the application fields of carbon nanotubes in agricultural and food sectors covered in this chapter.

6.2 STRUCTURE, CLASSIFICATION AND PROPERTIES OF CARBON NANOTUBES

Carbon nanotubes discovered by S. Iijima in 1991 has now emerged as one of the best carbon allotropes in nanotechnology. They are actually rolled graphene sheets with cylindrical geometry and are of several nanometers in diameter. They are classified according to the variations in length, number of layers, diameter, and chirality. The two main groups according to the number of layers are single-walled carbon nanotubes (SWCNTs) and multi-walled carbon nanotubes (MWCNTs) (Figure 6.2).The diameter and length for SWCNTs are around 1–3 nm and a few micrometers, respectively. For MWCNTs, the diameter is around 5–40 nm and length is of 10 μm.

FIGURE 6.2 Inside view of single-walled carbon nanotubes (SWCNTs) and multiwalled carbon nanotubes (MWCNTs).

CNTs have great electronic,[1] thermal,[3] chemical, optical,[2] and mechanical characteristics.[4,5] The exceptional properties of CNTs include rigidity, elasticity, and high strength with higher aspect ratio. Furthermore, high thermal and electrical conductivity than other conductors and high surface area are extraordinary. It is the unusual electronic structure and one-dimensional structure of graphite that is responsible for the high electrical conductivity in CNTs. MWCNTs have remarkable mechanical properties than electrical properties, whereas SWCNTs have significant electrical properties due to their chirality. SWCNTs are again classified based on the way the graphite is wrapped as arm chair, zigzag, and chiral types. Carbon nanotubes are considered as an excellent choice of designing sensors due to the low cost, light weight, and less power consumption.[6] The enhanced surface-to-volume ratio and hollow pipe structure enable the CNTs to exhibit faster response time and high sensitivity.[7] Additionally, they have lower redox potentials, high stability, longer lifetime, and lower surface fouling effects. The excellent sorption capacity is attributed to the high surface active sites to volume ratio and well-arranged distribution in pore size.[8]

6.3 AGRICULTURAL APPLICATIONS OF CNTs

Agriculture is defined as the science or art of cultivating plants which once played a key role toward the development of human civilization. Modern technologies influenced the growth of agriculture through agrochemicals including pesticides and fertilizers, which actually helped in increasing the yield, at the same time, it caused widespread environmental impact. Along with this industrialization, sharp increase in population, nutrient deficiency, reduction in water availability, and changes in climatic conditions affected the development of agriculture sector. It indicates the need for technologies that raise the drowning field of agriculture with one hand and protects the environment with the other. Also, there should be a smart agricultural sector within the limitation of lesser area and lesser environmental impacts in future with increasing population.

There has been a significant interest around the application of nanotechnology in agricultural systems which nowadays has been generalized as "nanoagriculture." The major contributions of modern nanotechnology toward agricultural development include nanomaterial-based plant growth promoters and fertilizers for increasing the productivity of crop.[176–178] Several plant protection products are also available, such as nanomaterial-based pesticides and herbicides.[179–182] Optimization of agricultural practices by the introduction of precision farming[183] and nanoencapsulated fertilizers are the recent advances of nanomaterials.

Carbon nanotubes play different roles to ensure sufficient crop yield with reduction in the usage of agrochemicals. The small size, chemical reactivity, and their capability of penetration into the cells are the superior traits responsible for their widespread use in the field of agriculture. Carbon nanotubes are excellent option for promoting agriculture without causing any damage to the environment. Carbon nanotubes have been widely used for improving the crop yield, farming, and reducing the use of agrochemicals. There are several reports on applying CNTs for seed germination, plant growth, nanopesticides, and nanofertilizers. This section primarily deals with the demonstration of how CNTs were applied in the agricultural field and how it ensured the safety of our green environment.

6.3.1 *IMPROVING CROP YIELD*

The use of carbon nanotubes as a promoter for plant growth and development is an emerging field of immense potential. The unique structures and excellent properties like thermal stability, high electrical conductivity and surface area of carbon nanotubes have been extensively employed in several applications.[9] Improvement in the crop yield is associated with the increase in the plant growth rate and timely germination of seeds. This section deals with the extensive research of CNTs in achieving developments in the area of seed germination and seedling growth. The secret or science behind the increased rate of germination and growth is proposed to be the ability of CNTs to penetrate thick coat of seed thereby activating the water uptake process.[10] SWCNTs are responsible for delivering chemicals to the cell by penetrating the cell wall and are also a key factor in increasing the crop yield.[11,12] At the same time, MWCNTs enhance seed germination by penetrating into the seed coat.[13]

6.3.1.1 SEED GERMINATION

Germination is a physiological process that is initiated by water imbibitions through the membranes. Any method that increases the hydration of seeds is said to improve seed germination process. CNTs can break the dormancy by mechanically bringing out changes in the cell wall. Higher amount of CNTs in soil increases the absorption of water by acting as a catalyst.[14] Different properties of plants such as their growth stages result in changes in attitude toward each carbon nanotubes and their chemical composition. There have been several evidences in the literature pointing the same. Six different crop species, such as radish (*Raphanus sativus*), lettuce, rapeseed (*Brassica napus*), rye, cucumber, and maize showed positive influence of CNTs on seed germination and its growth.[15] There were more studies on carbon nanotubes on plant interaction compared with other nanomaterials (see Table 6.1).

TABLE 6.1 Effect of Carbon Nanotubes on Seed Germination in Different Plants with Different Types and Concentration of CNTs.

Plant species	Carbon nanotubes (CNTs)	Concentration (mg/L)	Medium	Effect on germination	Refs.
Wheat	MWCNTs	20, 50	DI water	Enhances germination	[37]
Maize					
Peanut					
Garlic bulb					
Barley Maize	MWCNTs	50, 100, 200	MS medium	Accelerates germination	[22]
Soybean					
Barley	MWCNTs	25, 50, 100 (spray)	Water	No effect at 25 mg/L, increased germination rate at 100 mg/L	[22]
Maize					
Soybean					
Tomato	MWCNTs	10, 20, 40	MS medium	Accelerates germination	[10]
Rice	CNTs	50, 100, 150	MS medium	Enhances germination speed and rate	[38]
Rice	CNTs	NA	Basal growth medium	Increases seed water content and germination rate	[31]
Tomato Onion	CNTs	10, 20 and 40	Ultra-pure water	Improves germination of tomato and onion	[39]
Turnip Radish					
Radish	MWCNTs	2000	DI water	No effect	[15]
Rape					
Ryegrass Lettuce					
Maize					
Cucumber					

TABLE 6.1 *(Continued)*

Plant species	Carbon nanotubes (CNTs)	Concentration (mg/L)	Medium	Effect on germination	Refs.
Alfalfa Wheat	MWCNTs	40–2560	Agar medium	No effect	[40]
Zucchini	MWCNTs	1000	Hoagland medium	No effect	[41]
Mustard Black lentil	MWCNTs	10, 20, 40	DI water	No effect	[33]
Wheat	oMWCNTs	10, 20, 40, 80, 160	DI water	No effect	[23]

6.3.1.1.1 *Effect of SWCNTs in Germination*

The uptake of SWCNTs through penetration into the plant cell was first explained in plant cell cultures. Intracellular penetration of SWCNTs with a length less than 500 nm was exhibited in cell cultures of tobacco.[16] Fluidic phase endocytosis was proposed to be the mechanism for the penetration of CNTs into hard cell wall and cell membrane which was confirmed by the decrease in penetration in the presence of endocytosis inhibitors. The process was temperature-independent since it does not affect the internalization of SWCNT into *A. thaliana* mesophyll cells.[17] External magnetic forces also influenced the intracellular uptake of functionalized magnetic SWCNTs into canola (*B. napus*) and carrot (*D. carota*) cells.[18] But there is negligible proof for the intake of SWCNTs by intact plants. SWCNTs were found sticking to the external surface of the root and none of them entered into the tissues of young seedlings of six species of plant in a study conducted by Cañas et al.[19] Similarly, in another study using maize, MWCNTs was present only on the root surface and intercellular areas of root cortex.[20] Several studies concluded that high concentration of SWCNTs is essential for the effective penetration.

Seed germination stimulated by SWCNTs treatment has been reported for salvia (*Salvia macrosiphon*), tall fescue (*Festuca arundinacea*), and pepper (*Capsicum annuum*). Concentration of SWCNTs taken and type of plants played a crucial role in germination. For example, highest germination rate was achieved when 30 mg/L SWCNTs was taken for salvia, but for pepper, it was 10 mg/L.[21] A similar concentration did not at all affect the germination in maize seeds.[20]

6.3.1.1.2 *Effect of MWCNTs in Germination*

Even though the diameter and length of MWCNTs are greater than that of SWCNTs, plant uptake and translocation has also been found successful in the case of MWCNTs. It can penetrate into the rigid seed coats and seedlings by creating new pores.[10] There were studies on the detection of MWCNTs with a diameter in the range of 15–40 nm in the seeds of soybean (*G. max*), barley (*H. vulgare*), maize (*Z.mays*),[22] roots of wheat (*T. aestivum*)[23] and red spinach (*A. tricolor*) seedlings.[24] In a study done on wheat by Wild and Jones, they found that MWCNTs having a diameter of 110–170 nm were able to penetrate the wall of epidermal cells and could translocate about 4 μm into the cytoplasm of its root hairs.[25] MWCNTs have the ability to translocate from root to shoot through the process of transpiration[26,27] and reach various parts of plant including vegetable shoot organs, flowers,[28] and even in xylem and phloem cells.[29]

There is a wider exposure of MWCNTs over SWCNTs for seed germination which is mainly due to the lesser toxicity of MWCNTs. Accelerated speed of germination was achieved for several crops, such as tomato (*S. lycopersicum*),[10,30] rice (*O. sativa*),[31] barley (*Hordeum vulgare*), maize (*Z. mays*) soybean (*G. max*),[13,22,] mustard seeds (*B. juncea*).[13] Beneficial effects of MWCNTs are observed only in certain plant species which is attributed to the genotypic differences associated with each plant types. Researchers have proved that different techniques of application of CNTs did not influenced the germination rate but the chemical composition of medium used for germination tests played a crucial role. The growth medium can be distilled water, agar with mineral nutrients, Murashige-Skoog medium with bioactive compounds, such as minerals, vitamins, amino acids, and hormones.[32] The positive influence on germination is also reflected in the studies done on six crop types, such as radish, lettuce, cucumber, rye, maize, and rapeseed (*Brassica napus)*.[15] Treatment of MWCNTs exhibited high germination rate in species including tomato, rice (*Oryza sativa*), urd bean (*Vigna mungo*), Indian mustard (*Brassica juncea*), and hybrid Bt cotton (*Gossypium hirsutum*).[33–35] The effect of MWCNTs differs from that of SWCNTs in several aspects but an increase in germination rate was achieved with increased water uptake in the case of CNTs-treated seeds than the control seeds.

The sedimentation of MWCNTs on the surface of seeds, such as corn, barley, and soybean witnessed the activation in the germination process.[36] An increase in concentration of MWCNTs led to reduced corn seedling germination but showed considerable increase with low MWCNT concentration.

6.3.1.2 SEEDLING AND PLANT GROWTH

Carbon nanotubes have the potential to enhance the growth of plants through nutrient uptake and thereby increase the crop yield. It is the capability of CNTs to influence the morphological and physiological properties of plant cells that are responsible for regulating the growth of plant species. The enhanced growth of plant parts is attributed to the uptake, translocation, and accumulation of CNTs that plays a major role in modeling various functions of plants. CNTs are involved not only in vegetative growth but also in reproductive induction in plants. There were reports on the production of twice the amount of fruits in CNT applied plants than the control plants due to the activation of reproductive system of plants by CNTs.[28]

6.3.1.2.1 *Effect of SWCNTs in Plant Growth*

There have been several reports on the effect of SWCNTs on the growth of young seedlings in different mediums. A wide range of plant species including fig plants (*Ficus carica*), tomato (*S. lycopersicum*), and maize (*Zea mays*) seedlings showed stimulatory growth effects.[20,42,43] But there were remarkable variability in the plant responses toward CNTs which may arise due to the genotypic differences or sizes of the plant species. The authors found that small-seeded species with lower surface-to-volume ratio are more sensitive compared with large-seeded species due to their lower interaction with CNTs. The process is concentration-dependent in certain cases and it is necessary to find the apt concentration otherwise the result will become paradoxical. This was proved by one of the experiments in which a concentration of 10–30 mg/L of SWCNTs resulted in an increase in the seedling biomass but at 40 mg/L of SWCNTs, the result was negative for salvia (*S.macrosiphon*), tall fescue (*F. arundinacea*) and pepper (*C. annuum*).[21] Treatment of seeds of *Hyoscyamus niger* with SWCNTs activated the defense system and enhanced the water uptake thereby tolerating the drought level at moderate conditions[64] (Figure 6.3). Results demonstrate that treatment with low concentrations of SWCNTs can induce tolerance in seedlings against low to moderate levels of drought through enhancing the water uptake and activating the plant defense system.

In some cases, contradictory effects were also reported in which the functionalized and nonfunctionalized SWCNTs in short-term application gave different results on different plant species. SWCNTs with no functionalization stimulated root growth in cucumber (*C. sativus*) and onion (*A. cepa*) but it showed inhibitory effect in tomato (*S. lycopersicum*). At the same time, SWCNTs functionalized with poly-3-aminobenzenesulfonic acid showed no effect on cabbage (*B. oleracea*) and carrot (*D. carota*) but inhibited the root elongation in lettuce (*L. sativa*)[19] and germination rate of maize (*Z. mays*) seeds.[20] Cano and coworkers in a comparitive study evaluated the effect of pure SWCNTs, OH-functionalized and

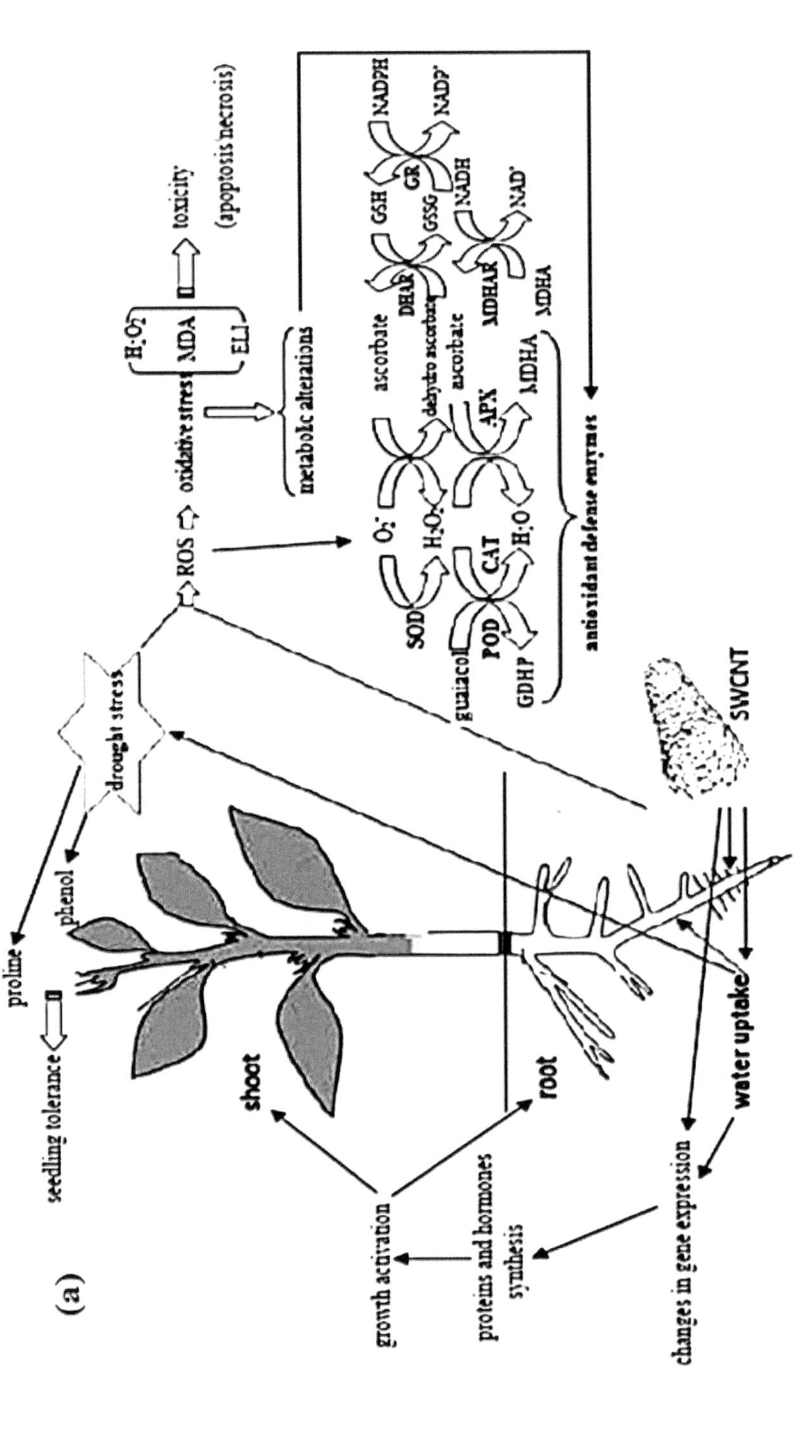

FIGURE 6.3 Schematic representation of possible mechanisms of growth activation and drought resistance achieved by SWCNTs exposure.

Source: Reprinted with permission from Ref. 64. © 2017 Elsevier.

surfactant-stabilized SWCNTs on the growth of corn seeds. They found that the nonfunctionalized CNTs were superior in enhancing the root length elongation than the surfactant-stabilized and OH-functionalized CNTs.[47]

6.3.1.2.2 *Effect of MWCNTs in Plant Growth*

A wide range of plant species such as mustard (*B. juncea*), soybean (*G. max*), wheat (*Triticum aestivum*),[23] black lentil (*P. mungo*),[33] maize (*Z. mays*),[22] tomato (*S. lycopersicum*)[10,43,44] showed positive effect on root and shoot growth in the presence of MWCNTs. It is the rapid uptake of MWCNTs followed by translocation at different parts of the plant that are responsible for the improved growth. The process was concentration-dependent with favorable effects at lower levels and inhibitory effects at higher levels.[33,45] Uniform distribution of MWCNTs in the form of dispersion resulted in better yield than large agglomerates of similar MWCNTs.[44] Retention of water, improvement of biomass, fruit yield, flowering, and enhancement of medicinal properties are some of the interesting characteristics of MWCNTs.[46] Seedling development and early enhancement in growth were also achieved in red spinach, rice, lettuce, and cucumber with the application of MWCNTs.[24,48,49] Small seeds were found to be more sensitive toward MWCNTs like that of SWCNTs.

Major studies related to the effects of MWCNTs have been done using hydroponics or other media as growth substrates instead of using soil culture. CNTs have the capability of inducing defense responses in plants based on the intensity of stimulus thereby acting as stress factors. There has been evidence for the rapid enhancement in the root elongation of alfalfa and wheat seedlings upon MWCNT exposure without remarkable translocation and uptake but only through adsorption on the root surface.[40] Inspite of its wide range of advances, there have been reports for inhibitory effects also[24,45,50,51] (see Table 6.2).

6.3.2 *REDUCTION OF APPLIED AGROCHEMICALS*

There is a steep rise in the use of agrochemicals in modern agriculture in order to resist the attack of phytopathogens, plant diseases, and to increase the productivity of crops. And the result is the increased resistance of pathogens, pollution due to pesticide accumulation, and health risks to all

TABLE 6.2 Effect of Different Types of Carbon Nanotubes on Plant Growth.

Plant species	Carbon nanotubes (CNTs)	Effect on plant growth	Ref
Alfalfa wheat	CNTs	Enhanced elongation in roots	[40]
Tomato	CNTs	Increase in the production of fruits and flowers	[28]
Chickpea	CNTs	Increase in growth rate	[52]
Tomato	CNTs	Increase in seedling growth	[34]
Arabidopsis	CNTs	Reduction in photosynthesis, carbon gain, chlorophyll, fluorescence, and transpiration	[53]
Common gram	CNTs	Increase in growth rate	[52]
Cucumber	SWCNT	Enhanced elongation in roots	[19]
Onion	fSWCNT		
Tomato	SWCNTs	Increase in fresh and dry plant biomass	[43]
Rice	SWCNTs	Enhanced uptake of water and production of healthier seedlings	[31]
Blackberry	SWCNTs	Enhanced elongation in roots	[59]
Maize	SWCNTs	Enhanced elongation in roots	[20]
Onion	Poly-3-amino	Enhanced elongation in roots	[19]
cucumber	Benzene sulfonic acid functionalized SWCNTs		
Hyoscyamus niger	SWCNTs	Enhanced resistance against drought stress at low concentration by increasing water uptake and activating defense system of plants	[64]
Lettuce	Poly-3-amino	Inhibited elongation in roots	[19]
	Benzene sulfonic acid functionalized SWCNTs		

TABLE 6.2 *(Continued)*

Plant species	Carbon nanotubes (CNTs)	Effect on plant growth	Ref
Cabbage	SWCNT, fSWCNT	No effect	[19]
Carrot			
Tomato	SWCNT, fSWCNT	Inhibited elongation in roots	[19]
Lettuce			
Arabidopsis	SWCNTs	Inducing programmed cell death	[60]
Rice			
Broccoli	MWCNTs	Enhanced growth rate	[63]
Maize	MWCNTs	Increase in growth rate	[45]
Rice	MWCNTs	Improvement in seed germination, uptake of water, and production of healthier seedlings	[31]
Red spinach	MWCNTs	Exhibited growth inhibition, cell death (−)	[24]
Wheat	MWCNTs	Enhancement in root growth	[23]
Tobacco	MWCNTs	Increase in growth rate	[61]
Tomato	MWCNTs	Increase in height and number of flowers	[28]
Castor	MWCNTs	Improvement in growth rate	[62]
Rye	MWCNTs	Enhanced root length	[15]
Lettuce	MWCNTs	Reduction in root length	[15]
Barley	MWCNTs	Increased length of leaves and biomass in maize, increased root length in soybean	[24]
Maize			
Soybean			
Satureja	MWCNTs	Increased growth of callus	[57]

TABLE 6.2 *(Continued)*

Plant species	Carbon nanotubes (CNTs)	Effect on plant growth	Ref
Wheat	MWCNTs	Elongation of root and shoot	[37]
Maize			
Peanut			
Garlic			
Tomato	MWCNTs	Increase in stem length and biomass	[27]
Tomato	MWCNTs	Increase in total biomass and up regulation of stress-related genes	
Wheat	MWCNTs	No effect	[27]
Rapeseed			
Arabidopsis	MWCNTs	Decrease in cell dry weight, cell chlorophyll content cell viabilities, and superoxide dismutase activities	[58]
Onion	MWCNTs	DNA damage	[54]
Onion	MWCNTs	DNA damage	[48]
Cucumber, rice, lettuce	MWCNTs	Decrease in shoot length and biomass, Increase in membrane leakage	[49]
Red spinach			
Zucchini	MWCNTs	Reduction in biomass	[41]
Rice	MWCNTs	Decrease in cell density	[56]
Rice	MWCNTs	Decrease in cell viability and increased ROS	[55]
Maize	MWCNTs	Stimulates growth of maize but inhibits growth of soybean uptake and translocation	[29]
Soybean			

living beings.[65,66] The stability and high surface-to-volume ratio of carbon nanotubes have made them efficient in carrying nutrients in the agricultural sector. Agrochemicals usually include fertilizers which are substances added to soil to increase fertility and pesticides which are substances used for destroying pests that are harmful to cultivated plants. Both are used with a single aim that is to increase the productivity without causing any damage to the plant species.

CNTs are regarded as smart nanofertilizers as they can release water and nutrients as per the requirement thereby reducing the amount of synthetic chemicals. The target-specific approach of CNTs provides nutrition to the cell, hence preventing soil and water pollution, and can be considered as sustainable agents. Conventional agricultural practices of fertilizer application result in the loss of nutrients by leaching whereas CNT-based controlled release method completely eliminates the loss by leaching and there will be no risk of facing effects of overdose.

6.3.2.1 CNTs AS NANOFERTILIZERS

Nanofertilizers are products that are made of nanoparticles with a purpose of improving nutrient efficiency in crops.[67,68] They can boost the efficiency of nutrient uptake in plants thereby increasing the crop yield and decreasing the harmful effects of traditional fertilizers.[69] The main reason for using nanofertilizers is the ability of precisely timely release of nutrients according to the biological demands of plant and environmental stimulations associated with the ecosystem. They could also regulate the dispersion and solubility of micronutrients, reduce the rate of loss of fertilizers into the soil through leaching, release fertilizers on-site target, and minimize the contamination that occur globally.[70]

Carbon nanotubes are employed as complement of the nutrients required for crop plants. There have been several reports and patents associated with the application of carbon nanotubes as nanofertilizers. MWCNTs were used as a carrier to enhance the nutrient uptake of plant cells which was possible due to the improved absorption of nitrogen that leads to plant growth upon successful intake of CNTs.[71,72] In another work done by Yatim et al., urea functionalized MWCNTs (UF-MWNTs) were used as fertilizer to provide nutrition to rice plant and were found as efficient nutrient delivery systems.[14] Several spectroscopic and microscopic analyses were done to confirm the bonding in UF-MWNTs. Their results concluded the significance of functionalization and quantity of MWCNTs employed for the development of novel area of MWCNTs-based fertilizers. CNTs also have the ability to carry macro and microelements including nitrogen, phosphorous, and potassium and regulate their slow release to plants upon requirement.[73] An investigation on the effect of foliar application of CNTs loaded with NPK fertilizer on the French bean seedling growth, absorption and translocation at various concentration lead to the finding that low-dose concentrations suited well for achieving better crop yield.[74]

6.3.2.2 CNTs AS NANOPESTICIDES

Pesticides are an unavoidable commodity in agricultural practices since it is very crucial to protect the plant from causing diseases and make it resistant to many environmental aspects in order to get good yield and reduce the crop losses.[76] Depending on the target objects, pesticide can be a herbicide, insecticide, or fungicide but the main objective of every pesticide is to either completely wipe out pests or make the plant resistant. Hence, the synthesis of this pesticide using the nanotechnology not only helps in the slow and efficient release but also decreases the soil runoff rate thereby reduces soil pollution to a great extent.[75] Carbon nanotubes due to their structural and chemical properties are regarded as excellent candidates for the synthesis of nanopesticides.

Carbon nanotubes were employed as nanopesticides through different ways due to their antimicrobial and antifungal properties. Gorczyca et al. investigated the effect of MWCNTs on conidia of a fungus *Paecilomyces fumosoroseus* that was entomopathogenic in nature and found that the spore production of mycelium is strongly inhibited on application of CNTs.[77] Strong antifungal activity was exhibited in a polyethylene–carbon nanotube composite that was synthesized by solution casting from xylene boiling.[78] In a work done by Wang et al., the antifungal activities of MWCNTs with surface modifications against *Fusarium graminearum* was examined. They found that the surface-enhanced MWCNTs inhibited the spore germination remarkably.[79] In another work, different carbon nanomaterials including SWCNTs, MWCNTs, fullerene, graphene oxide (GO), reduced graphene oxide (rGO), and activated carbon were investigated against two plant fungi, *Fusarium graminearum* and *Fusarium poae* and found that SWCNTs showed best result as an antifungal agent after MWCNTs, GO and rGO. And no significant activity was shown by fullerene and activated carbon.[80]

There are also reports on encapsulation of pesticides on carbon nanotubes showing superior antifungal activity. MWCNTs coated with polycitric acid acted as a carrier for dithiocarbamate fungicides which was optimized at a pH range of 6–8 and stirring time 30–80 min.[81]

Better inhibition effects were indicated on the plant microns like *Ralstonia solanacearum*, *Fusarium graminearum*, and *F. oxysporum* by CNTs.[81–84] MWCNTs have a remarkable antimicrobial activity and along with metallic nanomaterials including silver or zinc, can control the growth of phytopathogenic fungi *Aspergillus fumigates* and *A. ochraceus*.[85] SWCNTs showed higher antimicrobial properties compared with multiwalled carbon nanotubes considering the toxicological point of view.[86] CNTs are also employed through delivery of DNA or RNA that can induce several resistance mechanisms to defend the plant from disease causing pathogens.[87,88] In another work, the effect of CNTs on the soil microbial community was investigated and found that the abundance of *Bacteroidetes* and *Firmicutes* increased whereas *Proteobacteria* and *Verrucomicrobia* showed a decrease in abundance with increase in CNT concentration. Even though there are several reports on the application of CNTs as nanopesticides, there are only a few works of investigating the actual mechanism.

6.3.3 IMPROVED FARMING

Intelligent application of nanotechnology in agriculture led to the innovation of "smart agriculture" which not only enhances the crop yield through different paths but also monitors the pollution level, sense the disease causing pathogens and recovers the salt-affected soils.[89] This could definitely improve farming in all fields especially in places where there is lesser area available for farming and helps the farmers to get maximum yield. Carbon-based nanomaterials are excellent candidates for sensors out of which carbon nanotubes are predominant due their unique physicochemical properties. There are many researches in literature related to carbon nanotubes in sensing applications. CNTs inserted inside the plants or insects can work as a distant sensing system in order to detect harmful gases. This digital device based on CNT could hence detect plant diseases.[90] It highlights the potential of CNTs in examining the effects on plant health and improving farming.

Pesticides are necessary for improving crop yield but their intense use had resulted in their accumulation in the fresh and edible plant parts which poses a serious risk to all living beings.[91,92] Furthermore, this could create great shakes in the global trade marketing and affect farming across the world.[93] One of the prime step to address this issue is to detect these pesticides using a simple, cheap, and selective sensing platform. Carbon nanotubes are emerging nanosensors that are being enrolled for the detection of pesticide residues and other harmful biomolecules.

6.3.3.1 MONITORING OF PESTICIDES

Carbon nanotubes due to their fascinating properties, such as rigidity, high mechanical strength, lightness and conductivity, have made it a potential nanosensor compared with traditional sensors. Ultrasensitive detection of analytes is possible because of the one-dimensional properties in which the surface atoms could detect even minor changes in the chemical environment thereby changing the electrical properties. Out of different types of sensors, electrochemical sensor-based CNTs have gained much more attention. Detection of various target analytes has been done using different electrode architectures which include pristine CNTs and its combination with other modifiers. These can be divided into seven different CNT-based electrochemical sensors including (1) CNT sensors, (2) Phthalocyanine/CNT, (3) Molecularly Imprinted Polymers/CNT, (4) Ionic Liquid/CNT (5) Metallic Nanoparticles/CNT, (6) β-Cyclodextrin/CNT, and (7) Fullerene and Quantum Dots/CNT sensors, which are discussed in detail in the further section. Electrochemical sensors based on CNTs have remarkable potential in analyzing very minuscule amount of pesticide residues on the plant surface which in future is going to have a great significance in the safety aspects of agriculture.

6.3.3.1.1 Carbon Nanotube Sensors

Carbon nanotube-based electrochemical sensors are reported widely in the literature. Among the different electrode architectures based on carbon nanotubes as bare electrodes, the two types that are mostly reported include carbon nanotube paste electrode (CNPE) and glassy carbon electrodes as support. CNPE consists of a mixture of carbon nanotubes and organic liquid such as mineral oil or silicon oil in a diameter and depth defined glass or plastic tube.[94,95] And glassy carbon electrode (GCE) contains a surface that is rigid and flat with a honeycomb aspect.

An insecticide known as methiocarb was detected with a detection limit of 2.0×10^{-6}mol/L using a MWCNT paste electrode and its application in soil and river samples was very satisfactory.[96] A combination of square-wave stripping voltammetry and CNT sensors was comparatively better due to the considerable increase obtained for the recorded signals. MWCNT paste electrode was able to determine the insecticide cyromazine by square-wave adsorptive stripping voltammetry with a remarkable limit of detection. Glassy carbon electrode modified with functionalized MWCNTs detected carbendazim with a wide linear range of concentration using differential pulse adsorptive stripping voltammetry.[97] There are similar reports using the CNT-modified glassy carbon electrode for the detection of pesticides.[98,99] The major problem faced by the researchers is the detection of pesticides in low concentrations. Another combination of CNT with graphene has allowed significant performance in detecting pesticides.[100] Sensitive detection of herbicides known as diuron and fenuron was done by modifying the glassy carbon electrode surface with graphene oxide and MWCNTs and it showed outstanding electrocatalytic activity and enhanced peak currents.[101]

6.3.3.1.2 Phthalocyanine/CNT Sensors

Phthalocyanines are compounds with excellent capabilities, such as electrical properties and accommodation of different metal ions including Ni^{2+}, Mg^{2+}, Fe^{3+}, and Cu^{2+}. There are several studies reported on the use of metal phthalocyanines and porphyrins along with MWCNTs for the detection of pesticides.[102–104] Electrocatalytic detection of amitrole carried out by Siswana et al. explains the successful application of graphite electrode modified with MWCNTs and iron (II) tetra-aminophthalocyanine (MPc) with a detection limit of 0.5 nmol/L.[105] The proposed electrode does not show any interference from other pesticides. Simultaneous detection of two pesticides, amitrole and diuron was possible by using GCE modified with iron (II) tetra-aminophthalocyanine – SWCNT (Figure 6.4).[102] All the chemical species involved helped in enhancing the electrocatalytic oxidation of amitrole and diuron.

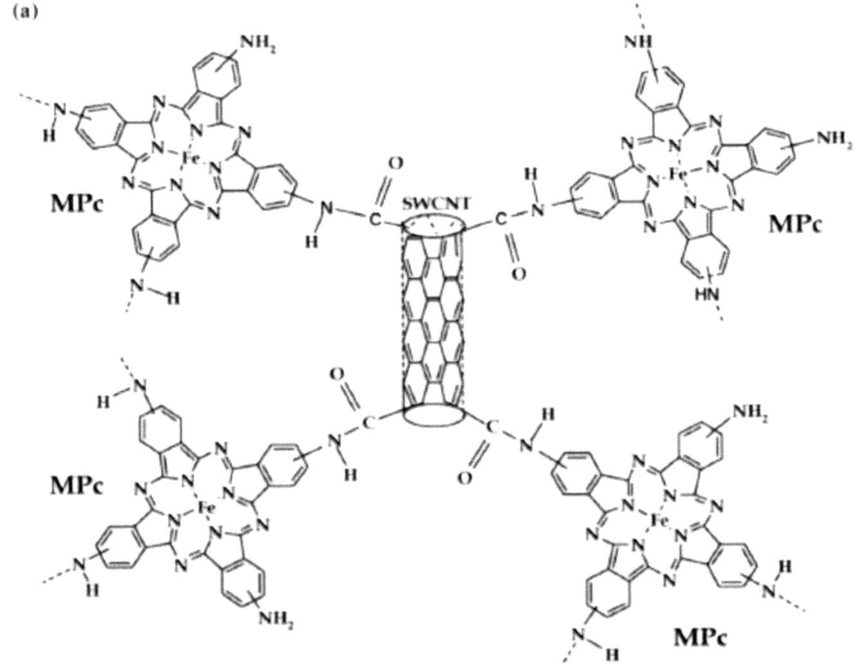

FIGURE 6.4 Schematic representation of dendrimers formed between a SWCNT and iron (II) tetra-aminophthalocyanine.

Source: Reprinted with permission from Ref. 102. © 2010 Elsevier.

6.3.3.1.3 *β-Cyclodextrin/CNT Sensors*

Cyclodextrin (CD) molecules are defined as cyclic oligosaccharides, composed of six to twelve α-D-glucopyranose monomers which are connected by one and four carbon atoms.[106] β-cyclodextrin is a type of CD molecule consisting of seven D-glucopyranose units with a hydrophilic exterior and a hydrophobic interior cavity.[107] Rahemi et al. has contributed a series of works based on pesticide sensors using β-Cyclodextrin/CNT combinations.[107–109] They have identified the properties of CD, such as molecular recognition and selectivity in pre-concentration of analytes and used it for the construction electrochemical sensors.[110–112] Their first work was concentrated on the sensing of a herbicide known as MCPA (4-chloro-2-methylphenoxyacetic acid) using a sensor composed of glassy carbon electrode modified by MWCNT, β-cyclodextrin, and polyaniline film.[107] The same electrochemical sensor architecture was used with cyclic voltammetry for the detection of another herbicide bentazone[108] and MCPA and its metabolite 4-chloro-2-methylphenol[109] and found successful with excellent accuracy.

6.3.3.1.4 Molecularly Imprinted Polymers/CNT Sensors

Molecularly imprinted polymers (MIP) are those materials that have the capability of biomimetic recognition of analyte species similar to antigen–antibody or enzyme substrate interaction. It involves selective detection of certain pesticides with accuracy. A sensor based on MIP and MWCNT on glassy carbon electrode was designed for the sensing of 6-benzyl-aminopurine (BAP) by Gan et al.[113] It showed enhanced sensitivity and selectivity and found to be very cost-effective for the raid detection of BAP in vegetable and fruit samples.

6.3.3.1.5 Metallic Nanoparticles/CNT Sensors

High electrocatalytic activity and surface area are some of the advantages of metallic nanoparticles that make it an excellent sensor along with carbon nanotubes. The combined effect of both the metallic nanoparticles and carbon nanotubes has been exploited in the recent years for the selective detection of pesticides. An increase in the electroactive surface area that is responsible for the high electroanalytical performance is responsible for their ability to detect pesticides with lower limit of detection. Zhang et al. reported the use of gold and palladium nanoparticles along with MWCNTs for the modification of glassy carbon electrode for the detection of parathion in spiked water and vegetable samples.[114] In a similar work, parathion was detected by using GCE modified with MWCNTs and palladium nanoparticles with a limit of detection 1.7×10^{-1} mol/L.

6.3.3.1.6 Ionic Liquid/CNT Sensors

Ionic liquids are those compounds defined by unique features, such as high conductivity, high stability, high viscosity, low toxicity, and nonvolatility. In liquid state, it exists as a salt with positively and negatively charged ions. When an electrode surface is modified with ionic liquid and MWCNT, it resulted in synergisms that lead to the development of numerous analytical techniques for the detection of pesticides.[115,116] Another fungicide applied to seeds known as pyrimethanil (PMT) was determined using a GCE modified with MWCNTs and 1-butyl-3-methylimidazolium hexafluorophosphate.[117] The benefit of adding ionic liquid to MWCNTs is that it helps in increasing the effective surface area by unfolding the highly entangled MWCNTs thereby exhibiting higher current response (Figure 6.5). The sensor exhibited high reproducibility and stability with negligible interference and was found effective in PMT determination in real samples.

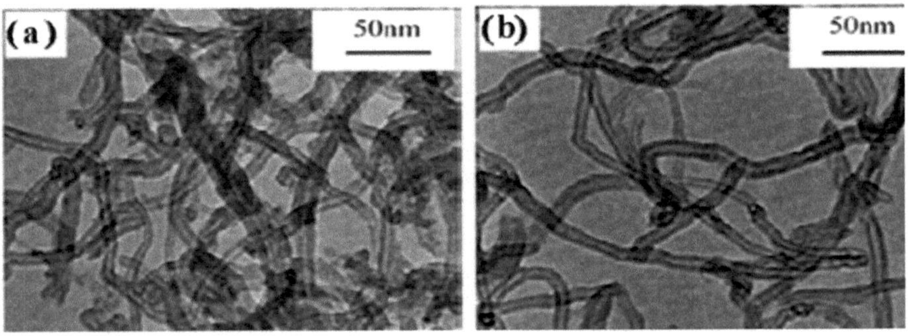

FIGURE 6.5 TEM images of (a) MWCNTs and (b) MWCNTs–Ionic liquid.
Source: Reprinted with permission from Ref. 117. © 2015 Elsevier.

6.3.3.1.7 Fullerene and Quantum Dots/CNT Sensors

Enhanced performance was obtained when fullerene functionalized carbon nanotubes were explored for the design of electrochemical sensors for the detection of pesticides. Two major works reported on the application of fullerene functionalized CNTs were used for the detection of carbendazim and vinclozolin.[118,119] In the work which describes the detection of endocrine disruptor vinclozolin, the design of sensor involved GCE surface modified with fullerene functionalized MWCNT in a particular ratio (2:1 w/w) followed by treatment in a 0.1 mol/L tetrabutylammonium hexafluorophosphate (TBAPF6) solution prepared in acetonitrile (Figure 6.6).[118] Another work for the determination of carbendazim also involved the use of glassy carbon electrode as the bare electrode with modification.[119] Both the sensors were successful in detecting pesticides in the real samples with efficiency. There are also reports on quantum dot (QD) decorated MWCNTs for the determination of pesticides. Feng et al. reported the design of sensor based on ZnSe QD decorated MWCNTs for the determination of pentachlorophenol by using differential pulse voltammetric technique.[120]

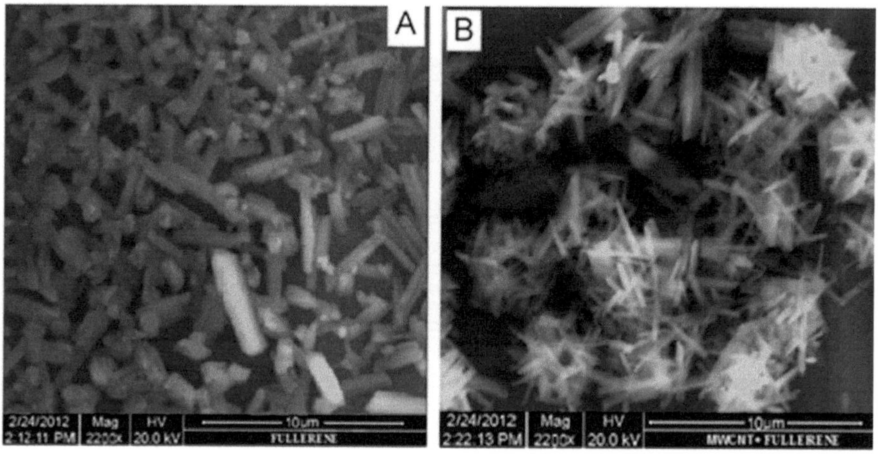

FIGURE 6.6 Scanning electron micrograph of (A) C60/GCE, and (B) C60–MWCNTs/GCE.

Source: Reprinted with permission from Ref. 118. © 2012 Elsevier.

6.4 FOOD-RELATED APPLICATIONS OF CNTs

One of the most important consumable of all living beings is food along with water. Chemical sensing has been used for monitoring food management and production. Nanotechnology in food science plays a vital role in food safety, preservation, packaging, monitoring of food ripening, quality assurance and soon. According to the reports about 1.3 billion tons of food, that is approximately one-third of the food produced for consumption are wasted or lost due to the inadequate steps taken for food safety and storage.[126] Economically affordable sensing platforms could definitely avoid this situation. Carbon nanotubes with its small size, simplicity and low consumption of power can act as a sensor in different applications of food sector.

TABLE 6.3 Carbon Nanotube-Based Electrochemical Sensors for the Detection of Pesticides.

Electrochemical sensor	Pesticide detected	Linear range	Limit of detection	Ref
MIP-MWNT@SnS2/GCE	6-benzylamino purine	1.0×10^{-10}–1.0×10^{-2}	5.0×10^{-11}	[113]
Fullerene/MWCNT/Nafion/GCE	Carbendazim	2.0×10^{-8}–3.5×10^{-7}	1.7×10^{-8}	[119]
GO–MWCNTs/GCE	Diuron and fenuron	9.0×10^{-6}–3.8×10^{-4} and 9.0×10^{-7}–4.7×10^{-5}	1.5×10^{-6} and 3.5×10^{-7}	[101]
MIP-MWCNT/Cu	Lindane	1.0×10^{-10}–1.0×10^{-3}	1.0×10^{-10}	[121]
PANI-_-CD/fMWCNT/GCE	MCPA and 4-chloro-2-methylphenol	1.0×10^{-5}–5.0×10^{-5}	1.1×10^{-6} and 1.9×10^{-6}	[109]
QDs-MWCNT/GCE	Pentachlorophenol	8.0×10^{-8}–4.0×10^{-6}	2.0×10^{-9}	[120]
IL-MWCNT/GCE	Pyrimethanil	1.0×10^{-7}–1.0×10^{-4}	1.6×10^{-8}	[117]
PANI-_-CD/fMWCNT/GCE	Bentazone	1.0×10^{-5}–8.0×10^{-5}	1.6×10^{-6}	[108]
GO–MWNTs/GC	Carbendazim	1.0×10^{-8}–4.0×10^{-6}	5.0×10^{-9}	[100]
FePy-MWCNT/CPE	2,4-dichlorophenoxyacetic acid	9.9×10^{-6}–1.4×10^{-4}	2.1×10^{-6}	[104]
CNT/PE	Methiocarb	6.7×10^{-6}–2.6×10^{-4}	2.0×10^{-6}	[96]
MWCNT/GCE	Carbendazim	2.6×10^{-5}–3.1×10^{-6}	5.5×10^{-8}	[98]
MWCNTPE	Cyromazine	2.5×10^{-6}–5.0×10^{-4}	7.2×10^{-7}	[122]
P3MT/MWCNT/GCE	Cypermethrin	3.6×10^{-8}–6.0×10^{-7}	3.6×10^{-12}	[123]
P3MT/MWCNT/GCE	Deltamethrin	3.8×10^{-8}–4.9×10^{-7}	3.8×10^{-12}	[123]
P3MT/MWCNT/GCE	Dicofol	1.4×10^{-7}–6.7×10^{-7}	1.4×10^{-11}	[123]
P3MT/MWCNT/GCE	Fenvalerate	1.4×10^{-7}–5.9×10^{-7}	1.4×10^{-11}	[123]
P3MT/MWCNT/GCE	Isoproturon	3.4×10^{-7}–1.2×10^{-7}	3.4×10^{-11}	[123]
PANI-_-CD/fMWCNT/GCE	MCPA	1.0×10^{-5}–1.0×10^{-4}	9.9×10^{-7}	[107]
C60–MWCNTs/GCE	Vinclozolin	2.5×10^{-6}–8.8×10^{-6}	9.1×10^{-8}	[118]

TABLE 6.3 *(Continued)*

Electrochemical sensor	Pesticide detected	Linear range	Limit of detection	Ref
P3MT/MWCNT/GCE	Voltage	3.4×10^{-7}–1.2×10^{-6}	3.4×10^{-11}	[123]
FeTAPc-SWCNT/GCE	Amitrole and diuron	5.0×10^{-5}–1.0×10^{-4}	$2.1 \times 10^{-7}$and 2.6×10^{-7}	[102]
FMWCNTs/GCE	Carbendazim	5.2×10^{-11}–2.6×10^{-4}	5.2×10^{-11}	[97]
Pd/MWCNTs/GC	Methyl parathion	3.4×10^{-7}–4.8×10^{-5}	1.7×10^{-7}	[124]
MWCNT/CoPc/GC	Carbaryl	3.3×10^{-7}–6.6×10^{-6}	5.5×10^{-9}	[103]
MWNT-PMRE	Carbendazim	2.0×10^{-7}–1.0×10^{-5}	9.0×10^{-9}	[125]
Au/CNTs/GCE	Methyl parathion	3.8×10^{-7}–5.3×10^{-5}	1.9×10^{-7}	[114]
MWCNT/GCE	Triclosan	1.7×10^{-7}–6.0×10^{-6}	5.7×10^{-8}	[99]
BMIMPF6–SWNT/GCE	Methyl parathion	2.0×10^{-9}–4.0×10^{-6}	1.0×10^{-9}	[115]
FeTAPc-MWCNT/BPPGE	Amitrole	–	5.0×10^{-10}	[105]

Au:	gold
β-CD:	β-cyclodextrin
BMIMPF6:	1-butyl-3-methylimidazolium hexafluorophosphate
BPPGE:	basal plane pyrolytic graphite electrode
CoPc:	cobalt phthalocyanine
C60:	fullerene
FePy:	5,10,15,20-tetrakis(pentafluorophenyl)-21H,23H-porphyrin iron(III)
FeTAPc:	iron(II) tetra-aminophthalocyanine
FMWCNTs:	functionalized multiwalled carbon nanotubes
GO:	graphene oxide
IL:	ionic liquid
MWCNTPE:	multiwalled carbon nanotubes paste electrode
PANI:	polyaniline
PMRE:	polymeric methyl red film
P3MT:	poly(3-methyl thiophene)

6.4.1 FOOD QUALITY ASSURANCE

6.4.1.1 DETECTION OF TASTE AND SMELL

Carbon nanotube-based sensors are able to determine taste and smell of various food types and can be considered as electronic tongues and electronic noses.[127,128] The design of sensor is such that it could distinguish between different smell and taste just like humans. A collaboration of the four hundred olfactory receptors in humans are responsible for the sense of smell.[129] The neural signal that is generated due to the binding of odorant to receptor is the key point that the researchers try to incorporate to the CNT sensors. Olfactory receptors from humans,[130–136] insects,[140] canines[141] and rodents[137–139] have been used for the development of carbon nanotube sensors for the detection of taste and smell. Initially, the olfactory receptors were coated on network of SWCNTs. A reduction in the mobile charge carrier density indicates binding of the odorant since it undergoes a shift from inactive to active state upon binding. This sensor was able to detect amyl butyrate at picomolar range of concentrations.[142] Further modification by the same group introduced nanovesicles decorated olfactory receptors on solid support consisting of CNT channels.[143] The binding of odorant to receptor induces a field effect on CNT due to the accumulation of positive charges inside the vesicle which is triggered by the binding effect. CNT sensors were also successful in quantifying several tastes including bitter,[132,134,138] sweet,[131] umami,-140 salty[137] and sour[139] tastes. Furthermore, CNTs were able to make a distinction between different qualities of food. CNT sensors introduced by Wei et al. were able to distinguish between Chinese rice wines based on the age.[144] In another work, MWCNT/polymer composite was designed such that it could differentiate between alcohol vapors based on the signals generated.[145,146]

6.4.1.2 FRUIT RIPENESS

Managing and monitoring the optimal time for harvesting and preserving freshness in storage and transportation is essential to ensure better quality of fruits in markets. Ethylene which is generally known as fruit ripening hormone regulates color, texture, aroma, and flavor of fruit. The concentration of ethylene in atmosphere either it is produced naturally or artificially can control ripening and senescing of fruits.[147] Carbon nanotube-based sensors can detect ethylene by experimental[148–150] and computational methods.[151] Ethylene at room temperature was detected using sensors composed of MWCNTs decorated with tin oxide nanoparticles.[149] Selectivity was a major problem faced by the researches at the early stages since there is possibility of interference from other gases. This was overcome when the SWCNTs were coordinated with a tris(pyrazolyl)-borate copper(I) complex and the test was found practical in a wide range of fruits.[148]

6.4.2 SAFETY OF FOOD

6.4.2.1 PREVENTING SPOILAGE OF FOOD

Ensuring food safety is a social concern since it can prevent consumption of harmful products and also reduce unnecessary wastage of food. CNT-based sensors can detect biomarkers that indicates the spoilage of food. Biomarkers can be volatile biogenic compounds, such as amines, ammonia, hexanal which indicates spoilage in meat,[151] fish,[152] and milk[153] respectively. More focus is given to small biomarkers than complex ones. CNT-based chemiresistive sensors with olfactory receptors have been used to identify spoilage in oysters,[154,155] meat[133] and other seafoods.[156] A clear differentiation between samples stored at room temperature and at 4°C in the refrigerator can be obtained from these sensors. CNT sensors can reduce food-related illness which usually occurs due to the consumption of spoiled food.

6.4.2.2 ENSURING INTEGRITY IN PACKAGED FOOD

The quality of packaged food can be monitored by checking the integrity of food packaging. Packaging in an inert atmosphere is necessary for maintaining the quality of the product in order to prevent oxidative degradation.[157,158] Intactness of the packaged material can be identified by detecting the oxygen inside. There have been several reports for the application of pristine CNTs and functionalized CNTs for the design of oxygen-sensing devices. Even though the actual mechanism of interaction between the molecules of oxygen and carbon nanotubes is not clear but it is assumed initially that p-doping is induced in CNTs due to the physisorption of oxygen. Doping effect was confirmed both experimentally and computationally.[159–162] But further investigation by Derycke et al. led to the conclusion that Schottky barriers of carbon nanotube and metal contacts is modified upon adsorption of oxygen rather than doping effect.[163] It may be the dissimilarities in the quality of carbon nanotubes that lead to disagreeing experimental reports. In addition to this drawback, the slow desorption of oxygen from the nanotubes also occurs on CNT-based sensors.

Irreversible adsorption of oxygen is useful in areas where its presence is detrimental, such as in packaging of food items so that it could indicate the increasing level of oxygen content.[164] A wireless oxygen dosimeter was designed by Zhu et al. which is composed FeII–eII–poly(4-vinylpyridine)–SWCNT composites with a passive radiofrequency identification (RFID) tag. SWCNTs is dispersed on Poly(4-vinylpyridine) and FeII species is bound through pyridyl ligands. The transfer of electron density from FeII to CNTs eliminates hole carriers and increases the resistivity. But the exposure of oxygen oxidizes FeII to FeIII and increases the conductivity. This can be detected by a smart phone through wireless RFID tags thereby indicating the presence of oxygen inside the food packaging.

6.4.3 DETECTION OF FOOD-BORNE PATHOGENS

A study conducted on food-borne diseases by World Health Organization (WHO) in the United States reports that about 5000 cases were led to death and 32,500 required hospitalizations among the 76 million cases that occurred annually.[165] Most of the diarrheal infections occurred due to pathogens like *Escherichia coli, Listeria monocytogenes, Salmonella*, and *Campylobacter*.[166] The causes of infections can be the consumption of contaminated food or contact with infected persons.[167] Many deaths that occur due to such infections can be prevented upon real time recognition of contaminated food items. Several CNT-based sensors are reported for the identification of pathogenic microorganisms that causes infections. In most of the works published, CNTs are functionalized with antibodies that have specific interaction with the pathogen. A paper-based electrochemical sensor was developed for the selective detection of *Staphylococcus aureus* (*S. aureus*).[168] The electrode was coated with antibody functionalized CNTs which showed an increase in peak current with binding of *S. aureus* in milk within a response time of 30 minutes.

A bacterial protein toxin called cholera toxin (CT) secreted by the bacterium *Vibrio cholera* was detected by the electrochemical sensor composed of poly(3,4-ethylenedioxythiophene)/antibody (PEDOT/Ab) composite on a MWCNT electrode.[170] In another work an antibody functionalized MWCNT/sodium alginate composite was designed as an electrochemical sensor for the detection of *Shigella flexneri.*[169] There have been reports on the rapid detection of pathogens using functionalized nanojunctions.[171,172]

6.5 NEGATIVE IMPACTS OF CNTs IN AGRICULTURAL AND FOOD SECTOR APPLICATIONS

Carbon nanotubes have a wide range of applications in the development of a green environment.[184–187] Apart from the beneficial aspects of carbon nanotubes, there are some associated negative impacts also that should be given special focus. In agricultural applications, carbon nanotubes have enormous applications due to their extraordinary properties. At the same time, it also imparts some harmful effects on plant growth and soil microbes. There have been reports on decrease in plant growth production of reactive oxygen species upon exposure to MWCNTs at high concentrations. Changes in the morphology of leaf and root along with necrotic lesions in cells also occur in some cases.[24] A decrease in cell dry weight, chlorophyll content, viability and activity of superoxide dismutase in the cell suspension of *Arabidopsis thaliana* was observed in the presence of MWCNTs.[173] The reason for programmed cell death in *Arabidopsis thaliana* was due to the oxidative stress induced by the adverse effect of SWCNTs on leaves and protoplasts.[60] The impact of carbon nanotubes varies with plants and types of CNTs employed. Types of carbon nanotubes include whether single-walled or multiwalled, functionalized or nonfunctionalized. Inhibition of root elongation was observed in lettuce with functionalized CNTs whereas inhibition occurred with nonfunctionalized CNTs in tomato.[19] The controversial effect of CNTs is still an untangled matter of debate and it is of great significance to standardize the methods for a better result and comparison.

The cytotoxic effects of carbon nanotubes must be given great importance even though they have has numerous advantages in food industry. They can cause serious threats to human health due to their small size since there is high risk of penetrating into the cell causing health-related problems in living beings.[174,175] Application of carbon nanotubes in food sector is mainly on assuring food quality, safety and on detection of food-borne pathogens. So there is less possibility of direct contact with CNT, and it indicates that there is no space for doubt regarding safety issues related to food sector applications.

6.6 CONCLUSIONS AND FUTURE OUTLOOKS

In this chapter, we summarized the application of carbon nanotubes in agricultural and food-related applications. Carbon nanotubes with their unique properties are an inevitable part for the development of greener environment. The growing population and less availability of land suitable for agriculture are the future challenges that we need to address with adequate care and attention. The use of nanotechnology in agriculture has led to the evolution of smart agriculture with various environment-friendly key features. We attempt to focus on the influence of CNTs in the agri-food sector. The part played by CNTs in improving crop yield, farming, and reducing the applied agrochemicals is given major attention with special focus on nanofertlizers, nanopesticides, and CNT-based sensors in pesticide monitoring. The effect of SWCNTs and MWCNTs in seed germination and plant growth is also discussed in detail. It was found that the effect of CNTs varies from plant to plant and also relies on the quality of CNTs. The wide use of carbon nanotubes as a favorable nanomaterial with high stability shows their potential of emerging as sustainable next-generation fertilizers. It helps to maintain an ecological balance by reducing the excess use of pesticides and fertilizers. But there are gaps still existing in the literature which need to be filled for the commercialization of these materials. Optimization of the concentration of CNTs for specific crop, mechanism of interaction between CNTs and plant systems, degradation patterns of CNTs in environment, exact knowledge of the long-term impact on human health and environment are the future endeavors to be accomplished in the field.

In food-related applications, role of CNTs in assuring quality and safety of food and detection of food-borne pathogens are described along with their ability to prevent spoilage of food and ensuring the integrity in food packaging by acting as an oxygen sensor. The unique properties of CNTs led to the design of sensors that could detect taste and smell thereby ensuring the quality of food. All these research outcomes remain in words until more trustable reports regarding the safety of using CNTs in agro-food sector come into sight. We hope that this chapter has provided the readers a deeper insight into the developments of CNTs in agro-food sector toward a greener environment, the existing gaps and the need for new horizons of advanced research in this area.

KEYWORDS

- **carbon nanotubes**
- **agriculture**
- **nanopesticides**
- **nanofertilizers**
- **food safety**
- **sensors**
- **environment**

REFERENCES

1. Zhu, J.; Holmen, A.; Chen, D. Carbon Nanomaterials in Catalysis: Proton Affinity, Chemical and Electronic Properties, and Their Catalytic Consequences. *ChemCat Chem.* **2013,** *5*, 378–401.
2. Karami, M.; Bahabadi, M. A.; Delfani, S.; Ghozatloo, A. A New Application of Carbon Nanotubes Nanofluid as Working Fluid of Low-Temperature Direct Absorption Solar Collector. *Sol. Energy Mater. Sol. Cells.* **2014,** *121*, 114–118.
3. Esfe, M. H.; Saedodin, S.; Yan, W.-M.; Afrand, M.; Sina, N. Study on Thermal Conductivity of Water-based Nanofluids with Hybrid Suspensions of CNTs/Al_2O_3 Nanoparticles. *J. Therm. Anal. Calorim.* **2016,** *124*, 455–460.
4. Mauter, M. S.; Elimelech, M. Environmental Applications of Carbon-based Nanomaterials. *Environ. Sci. Technol.* **2008,** *42*, 5843–5859.
5. Thirumal, V.; Pandurangan, A.; Jayavel, R.; Krishnamoorthi, S.; Ilangovan, R. Synthesis of Nitrogen Doped Coiled Double Walled Carbon Nanotubes by Chemical Vapor Deposition Method for Supercapacitor Applications. *Curr. Appl. Phys.* **2016,** *16*, 816–825.
6. Sinha, N.; Ma, J.; Yeow, J. T. Carbon Nanotube-based Sensors. *J. Nanosci. Nanotechnol.* **2006,** *6*, 573–590.
7. Jacobs, C. B.; Peairs, M. J.; Venton, B. J. Carbon Nanotube Based Electrochemical Sensors for Biomolecules. *Anal. Chim. Acta.* **2010,** *662*, 105–127.
8. Masenelli Varlot, K.; McRae, E.; Dupont- Pavlovsky, N. Comparative Adsorption of Simple Molecules on Carbon Nanotubes Dependence of the Adsorption Properties on the Nanotube Morphology. *Appl. Surf. Sci.* **2002,** *196*, 209–215.

9. Milne, W. I.; Teo K. B. K.; Amaratunga, G. A. J.; Legagneux, P.; Ganglof, L.; Schnell, J. P.; Semet, V.; Binh, V. T.; Groening, O. Carbon Nanotubes as Field Emission Sources. *J. Mater. Chem.* **2004,** *14*, 933–943.
10. Khodakovskaya, M.; Dervishi, E.; Mahmood, M.; Xu, Y.; Li, Z.; Watanabe, F.; Biris, A. S. Carbon Nanotubes Are Able to Penetrate Plant Seed Coat and Dramatically Affect Seed Germination and Plant Growth. *ACS Nano*. **2009,** *3*, 3221–3227.
11. Liu, Q.; Chen, B.; Wang, Q.; Shi, X.; Xiao, Z.; Lin, J.; Fang, X. Carbon Nanotubes as Molecular Transporter for Walled Plant Cell. *Nano Lett.* **2009,** *9*, 1007–1010.
12. Serag, M. F.; Kaji, N.; Tokeshi, M.; Baba, Y. Introducing Carbon Nanotubes into Living Walled Plant Cells through Cellulose Induced Nanoholes. *RSC Adv*. **2012,** *2*, 398–400.
13. Mondal, A.; Basu, R.; Das, S.; Nandy, P. Beneficial Role of Carbon Nanotubes on Mustard Plant Growth: An Agricultural Prospect. *J. Nanopart. Res.* **2011,** *13*, 4519–4528.
14. Yatim, N. M.; Shaaban, A.; Dimin, M. F.; Yusof, F. Statistical Evaluation of the Production of Urea Fertilizer Multiwalled Carbon Nanotubes Using Plackett Burman Experimental Design. *Proc. Soc. Behav. Sci.* **2015,** *195*, 315–323.
15. Lin, D.; Xing, B. Phytotoxicity of Nanoparticles: Inhibition of Seed Germination and Root Growth. *Environ. Pollut.* **2007,** *150* (2), 243–250.
16. Liu, Q.; Chen, B.; Wang, Q.; Shi, X.; Xiao, Z.; Lin, J.; Fang, X. Carbon Nanotubes as Molecular Transporters for Walled Plant Cells. *Nano Lett.* **2009,** *9*, 1007–1010.
17. Yuan, H.; Hu, S.; Huang, P.; Song, H.; Wang, K.; Ruan, J.; He, R.; Cui, D. Singlewalled Carbon Nanotubes Exhibit Dual-phase Regulation to Exposed Arabidopsis Mesophyll Cells. *Nanoscale Res. Lett.* **2010,** *297*, 787.
18. Hao, Y.; Yang, X.; Shi, Y.; Xing, J.; Marowitch, J.; Chen, J.; Chen, J. FITC Delivery into Plant Cells Using Magnetic Single-walled Carbon Nanotubes. *J. Nanosci. Nanotechnol.* **2012,** *12*, 6287–6293.
19. Cañas, J. E.; Long, M.; Nations, S.; Vadan, R.; Da,i L.; Luo, M.; Ambikapathi, R.; Lee, E. H.; Olszyk, D. Effects of Functionalized and Non Functionalized Single Walled Carbon Nanotubes on Root Elongation of Select Crop Species. *Environ. Toxicol. Chem.* **2008,** *27*, 1922–1931.
20. Yan, S.; Zhao, L.; Li, H.; Zhang, Q.; Tan, J.; Huang, M.; He, S.; Li, L. Single-walled carbon nanotubes selectively influence maize root tissue development accompanied by the change in the related gene expression. *J. Hazard. Mater.* **2013,** *246–247*, 110–118.
21. Pourkhaloee, A.; Haghighi, M.; Saharkhiz, M. J.; Jouzi, H.; Doroodmand, M. M. Carbon Nanotubes Can Promote Seed Germination via Seed Coat Penetration. *Seed Technol.* **2011,** *33*, 155–169.
22. Lahiani, M. H.; Dervishi, E.; Chen, J.; Nima, Z.; Gaume, A.; Biris, A. S.; Khodakovskaya, M. V. Impact of Carbon Nanotube Exposure to Seeds of Valuable Crops. *ACS Appl. Mater. Interf.* **2013,** *5*, 7965–7973.
23. Wang X.; Han, H.; Liu, X.; Gu, X.; Chen, K.; Lu, D. Multi-walled Carbon Nanotubes Can Enhance Root Elongation of Wheat (Triticum aestivum) Plants. *J. Nanopart. Res.* **2012,** *14*, 841.
24. Begum, P.; Fugetsu, B. Phytotoxicity of Multi-walled Carbon Nanotubes on Red Spinach (Amaranthus tricolor L) and the Role of Ascorbic Acid as an Antioxidant. *J. Hazard. Mater.* **2012,** *243*, 212–222.

25. Wild, E.; Jones, K. C. Novel Method for the Direct Visualization of In Vivo Nanomaterials and Chemical Interactions in Plants. *Environ. Sci. Technol.* **2009,** *43*, 5290–5294.
26. Chen, G.; Qiu, J.; Liu, Y.; Jiang, R.; Cai, S.; Liu, Y.; Zhu, F.; Zeng, F.; Luan, T.; Ouyang, G. Carbon Nanotubes Act as Contaminant Carriers and Translocate within Plants. *Sci. Rep.* **2015,** *5*, 15682.
27. Larue, C.; Pinault, M.; Czarny, B.; Georgin, D.; Jaillard, D.; Bendiab, N.; Mayne-L'Hermite. M.; Taran, F.; Dive, V.; Carrière, M. Quantitative Evaluation of Multi-walled Carbon Nanotube Uptake in Wheat and Rapeseed. *J. Hazard. Mater.* **2012,** *227*, 155–163.
28. Khodakovskaya, M. V.; Kim, B.; Kim, J. N.; Alimohammadi, M.; Dervishi, E.; Mustafa, T.; Cernigla, C. E. Carbon Nanotubes as Plant Growth Regulators: Effects on Tomato Growth, Reproductive System, and Soil Microbial Community. *Small.* **2013,** *9*, 115–123.
29. Zhai, G.; Gutowski, S. M.; Walters, K. S.; Yan, B.; Schnoor, J. L. Charge, Size, and Cellular Selectivity for Multiwall Carbon Nanotubes by Maize and Soybean. *Environ. Sci. Technol.* **2015,** *49*, 7380–7390.
30. Biris, A. S.; Khodakovskaya, M. V. Method of Using Carbon Nanotubes to Affect Seed Germination and Plant Growth. Patent US 20120233725, July 19, 2016.
31. Nair, R.; Mohamed, M. S.; Gao, W.; Maekawa, T.; Yoshida, Y.; Ajayan, P. M.; Kumar, D. S. Effect of Carbon Nanomaterials on the Germination and Growth Of Rice Plants. *J. Nanosci. Nanotechnol.* **2012,** *12*, 2212–2220.
32. Murashige, T.; Skoog, F. A Revised Medium for Rapid Growth and BioAssays with Tobacco Tissue Cultures. *Physiol. Plant.* **1962,** *15*, 473–497.
33. Ghodake, G.; Seo, Y. D.; Park, D.; Lee, D. S. Phytotoxicity of Carbon Nanotubes Assessed by *Brassica juncea* and *Phaseolus mungo. J. Nanoelectron. Optoelectron.* **2010,** *5* (2), 157–160.
34. Morla, S.; Ramachandra Rao, C. S. V.; Chakrapani, R. Factors Affecting Seed Germination and Seedling Growth of Tomato Plants Cultured In Vitro Conditions. *J. Chem. Biol. Phys. Sci. B.* **2011,** *1*, 328.
35. Nalwade, A. R.; Neharkar, S. B. Carbon Nanotubes Enhance the Growth and Yield of Hybrid Bt Cotton Var. ACH-177-2. *Int. J. Adv. Sci. Technol. Res.* **2013,** *3*, 840.
36. Lahiani, M. H.; Chen, J.; Irin, F.; Puretzky, A. A.; Green, M. J.; Khodakovskaya, M. V. Interaction of Carbon Nanohorns with Plants: Uptake and Biological Effects. *Carbon.* **2013,** *81*, 607–619.
37. Srivastava, A.; Rao, D. P. Enhancement of Seed Germination and Plant Growth of Wheat, Maize, Penut and Garlic Using Multiwalled Carbon Nanotubes. *Eur. Chem. Bull.* **2014,** *3*, 502–504.
38. Jiang, Y.; Hua, Z.; Zhao, Y.; Liu, Q.; Wang, F.; Zhang, Q. The Effect of Carbon Nanotubes on Rice Seed Germination and Root Growth. In *Proceedings of the 2012 International Conference on Applied Biotechnology (ICAB2012)*. Springer: Berlin, 2012; pp 1207–1212.
39. Haghighi, M.; Teixeira da Silva, J. A. The Effect of Carbon Nanotubes on the Seed Germination and Seedling Growth of Four Vegetable Species. *J. Crop Sci. Biotechnol.* **2014,** *17*, 201–208.

40. Miralles, P.; Johnson, E.; Church, T. L.; Harris, A. T. Multiwalled Carbon Nanotubes in Alfalfa and Wheat: Toxicology and Uptake. *J. R. Soc. Interf.* **2012,** *9*, 3514–3527.
41. Stampoulis, D.; Sinha, S. K.; White, J. C. Assay-dependent Phytotoxicity of Nanoparticles to Plants. *Environ. Sci. Technol.* **2009,** *43*, 9473–9479.
42. Flores, D.; Chaves, J. S.; Chacón, R.; Schmidt, A. A Novel Technique Using SWCNTs to Enhanced Development and Root Growth of Fig Plants (*Ficus carica*). In *Nanotechnology 2013. Bio Sensors, Instruments, Medical, Environment and Energy Technical Proceedings of the 2013 NSTI Nanotechnology Conference and Expo*; Nano Science and Technology Institute: Boston, 2013; pp 167–170.
43. Khodakovskaya, M. V.; de Silva, K.; Nedosekin, D. A.; Dervishi, E.; Biris, A. S.; Shashkov, E. V.; Galanzha, E. I.; Zharov, V. P. Complex Genetic, Photothermal, and Photoacoustic Analysis of Nanoparticle-plant Interactions. *Proc. Natl. Acad. Sci.* **2011,** *108*, 1028–1033.
44. Villagarcia, H.; Dervishi, E.; de Silva, K.; Biris, A. S.; Khodakovskaya, M. V. Surface Chemistry of Carbon Nanotubes Impacts the Growth and Expression of Water Channel Protein in Tomato Plants. *Small.* **2012,** *8*, 2328–2334.
45. Tiwari, D. K.; Dasgupta-Schubert, N.; Cendejas, L. M.; Villegas, J.; CarretMontoya, L.; García, S. E. Interfacing Carbon Nanotubes (CNT) with Plants: Enhancement of Growth, Water and Ionic Nutrient Uptake in Maize (*Zea mays*) and Implications for Nanoagriculture. *Appl. Nanosci.* **2014,** *4*, 577–591.
46. Husen, A.; Siddiqi, K. S. Carbon and Fullerene Nanomaterials in Plants System. *J. Nanobiotechnol.* **2014,** *12*, 1–10.
47. Cano, A. M.; Kohl, K.; Deleon, S.; Payton, P.; Irin, F.; Saed, M.; Shah, S. A.; Green, M. J.; Canas-Carrel, J. E. Determination of Uptake, Accumulation, and Stress Effects in Corn (Zea mays L.) Grown in Single-Wall Carbon Nanotube Contaminated Soil. *Chemosphere* **2016,** *152*, 117–122.
48. Ghosh, M.; Bhadra, S.; Adegoke, A.; Bandyopadhyay, M.; Mukherjee, A. MWCNT Uptake in *Allium cepa* Root Cells Induces Cytotoxic and Genotoxic Responses and Results in DNA Hyper-Methylation. *Mutat. Res.* **2015,** *774*, 49–58.
49. Begum, P.; Ikhtiari, R.; Fugetsu, B.; Matsuoka, M.; Akasaka, T.; Watari, F. Phytotoxicity of Multi-Walled Carbon Nanotubes Assessed by Selected Plant Species in the Seedling Stage. *Appl. Surf. Sci.* **2012,** *262*, 120–124.
50. Ikhtiar, R.; Begum, P.; Watari, F.; Fugetsu B. Toxic Effect of Multiwalled Carbon Nanotubes on Lettuce (*Lactuca sativa*). *Nano Biomed.* **2013,** *5*, 18–24.
51. Begum, P.; Ikhtiari, R.; Fugetsu, B. Potential Impact of Multi-walled Carbon Nanotubes Exposure to the Seedling Stage of Selected Plant Species. *Nanomaterials* **2014,** *4* (2), 203–221.
52. Tripathi, B. P.; Shahi, V. K. Organic–inorganic Nanocomposite Polymer Electrolyte Membranes for Fuel Cell Applications. *Prog. Polym. Sci.* **2011,** *36* (7), 945–979.
53. Voleti, R.; Wait, D. A. Effect of Carbon Nanotubes on Plant Growth and Gas Exchange Using *Arabidopsis thaliana*. In *Technical Proceedings of the 2014 NSTI Nanotechnology Conference and Expo*. NSTI Nanotech: Cambridge, **2014**; p 3.
54. Ghosh, M.; Chakraborty, A.; Bandyopadhyay, M.; Mukherjee, A. Multi-walled Carbon Nanotubes (MWCNT): Induction of DNA Damage in Plant and Mammalian Cells. *J. Hazard. Mater.* **2011,** *197*, 327–336.

55. Tan, X.; Lin, C.; Fugetsu, B. Studies on Toxicity of Multi-walled Carbon Nanotubes on Suspension Rice Cells. *Carbon.* **2009,** *47,* 3479–3487.
56. Tan, X.; Fugetsu, B. Multi-walled Carbon Nanotubes Interact with Cultured Rice Cells: Evidence of a Self-defense Response. *J. Biomed. Nanotechnol.* **2007,** *3,* 285–288.
57. Ghorbanpour, M.; Hadian, J. Multi-walled Carbon Nanotubes Stimulate Callus Induction, Secondary Metabolites Biosynthesis and Antioxidant Capacity in Medicinal Plant *Satureja khuzestanica* Grown In Vitro. *Carbon.* **2015,** *94,* 749–759.
58. Lin, C.; Fugetsu, B.; Su, Y.; Watari, F. Studies on Toxicity of Multi-walled Carbon Nanotubes on Arabidopsis T87 Suspension Cells. *J. Hazard. Mater.* **2009,** *170,* 578–583.
59. Flores, D.; Chacón, R.; Alvarado, L.; Schmidt, A.; Alvarado, C.; Chaves, J. Effect of Using Two Different Types of Carbon Nanotubes for Blackberry (*Rubus adenotrichos*) In Vitro Plant Rooting, Growth and Histology. *Am. J. Plant Sci.* **2014,** *5,* 3510–3518.
60. Shen, C. X.; Zhang, Q. F.; Li, J.; Bi, F. C.; Yao, N. Induction of Programmed Cell Death in Arabidopsis and Rice by Single-Wall Carbon Nanotubes. *Am. J. Bot.* **2010,** *97,* 1602–1609.
61. Khodakovskaya, M. V.; de Silva, K.; Biris, A. S.; Dervishi, E.; Villagarcia, H. Carbon Nanotubes Induce Growth Enhancement of Tobacco Cells. *Am. Chem. Soc. Nano.* **2012,** *6* (3), 2128–2135.
62. Fathi, Z.; Nejad, R-A. K.; Mahmoodzadeh, H.; Satari, T. S. Investigating of a Wide Range of Concentrations of Multi-Walled Carbon Nanotubes on Germination and Growth of Castor Seeds(*Ricinus communis* L.). *J. Plant Prot. Res.* **2017,** *57,* 228–236.
63. Martínez-Ballesta, M. C.; Zapata, L.; Chalbi, N.; Carvajal, M. Multiwalled Carbon Nanotubes Enter Broccoli Cells Enhancing Growth and Water Uptake of Plants Exposed to Salinity. *J. Nanobiotechnol.* **2016,** *14,* 42–56.
64. Hatami, M.; Hadian, J.; Ghorbanpour, M. Mechanisms underlying toxicity and stimulatory role of single walled carbon nanotubes in Hyoscyamus niger during drought stress simulated by polyethylene glycol. *J. Hazard. Mater.* **2017,** *324,* 306–320.
65. Rai, M. C.; Mattoso, L.; Duran, N. *Nanotechnologies in Food and Agriculture*; Springer: New York, London, 2015; p 347.
66. Solanki, P.; Bhargava, A.; Chhipa, H.; Jain, N.; Panwar, J. Nano-fertilizers and Their Smart Delivery System. In *Nanotechnologies in Food and Agriculture*; *Springer*, 2015; pp 81–101.
67. Bala, R.; Kalia, A.; Dhaliwal, S. S. Evaluation of Efficacy of ZnO Nanoparticles as Remedial Zinc Nanofertilizer for Rice. *J. Soil Sci. Plant Nutr.* **2019,** *19,* 379–389.
68. Kalia, A.; Kaur, H. Nanofertilizers: An Innovation towards New Generation Fertilizers for Improved Nutrient Use Efficacy (NUE) and Environmental Sustainability. In *Emerging Trends in Nano Bio Medicine*; Bhoop, B., Katare, O., Souto, E., Eds.; Taylor & Francis: Boca Raton, FL, 2019; pp 45–61.
69. Abobatta, W. F. Overview of Nano-fertilizers. *Asian J. Ethnopharmacol. Med. Foods.* **2018,** *4,* 17–20.
70. Mastronardi, E.; Tsae, P.; Zhang, X.; Monreal, C.; DeRosa, M. C. Strategic Role of Nanotechnology in Fertilizers: Potential and Limitations. In *Nanotechnologies in Food and Agriculture*; Springer, 2015; pp 25–67.
71. Torney, F.; Trewyn, B.; Lin, V. S. Y.; Wang, K. Mesoporous Silica Nanoparticle Deliver DNA and Chemicals Into Plant. *Nat. Nanotechnol.* **2007,** *2,* 295–300.

72. Serag, M. F.; Kaji, N.; Gaillard, C.; Okamoto, Y.; Terasaka, K.; Jabasini, M.; Tokeshi, M.; Mizukami, H.; Bianco, A.; Baba, Y. Trafficking and Subcellular Localization of Multiwalled Carbon Nanotubes in Plant Cells. *ACS Nano*. **2011,** *5*, 493–499.
73. Hasaneen, M. N. A.; Abdel-Aziz, H. M. M.; Omer, A. M. Characterization of Carbon Nanotubes Loaded with Nitrogen, Phosphorus and Potassium Fertilizers. *Am. J. Nano Res. Appl.* **2017,** *5* (2), 12–18.
74. Hasaneen, M. N. A.; Abdel-aziz, H. M. M. Effect of Foliar Application of Engineered Nanomaterials: Carbon Nanotubes NPK and Chitosan Nanoparticles NPK Fertilizer on the Growth of French Bean Plant. *Biochem. Biotechnol. Res*. **2016,** *4*, 68–76.
75. Chinnamuthu, C. R.; Murugesa Boopathi, P. Nanotechnology and Agroecosystem. *Madras Agric. J.* **2009,** *96* (1–6), 17–31.
76. Borgatta, J.; Ma, C.; Hudson-Smith, N.; Elme, W.; Plaza-Pérez, C. D.; De La Torre-Roche, R.; Hamers, R. J. Copper Based Nanomaterials Suppress Root Fungal Disease in Watermelon (*Citrullus lanatus*): Role of Particle Morphology, Composition and Dissolution Behaviour. *ACS Sustain. Chem. Eng.* **2018,** *6* (11), 14847–14856.
77. Gorczyca, A.; Kasprowicz, M. J.; Lemek, T. Physiological Effect of Multi–walled Carbon Nanotubes MWCNTs on Conidia of the Entomopathogenic Fungus, *Paecilomyces fumosoroseus* Deuteromycotina, Hyphomycetes. *J. Environ. Sci. Health A.* **2009,** *44* (14), 1592–1597.
78. Asgari, P.; Moradi, O.; Tajeddin, B. The Effect of Nanocomposite Packaging Carbon Nanotube Based on Organoleptic and Fungal Growth of Mazafati Brand Dates. *Int. Nano. Lett.* **2014,** *4*, 1–5.
79. Wang, X.; Zhou, Z.; Chen, F. Surface Modification of Carbon Nanotubes with an Enhanced Antifungal Activity for the Control of Plant Fungal Pathogen. *Materials* **2017,** 10, 1375.
80. Wang. X.; Liu, X.; Chen, J.; Han, H.; Yuan, Z. Evaluation and Mechanism of Antifungal Effects of Carbon Nanomaterials in Controlling Plant Fungal Pathogen. *Carbon* **2014,** *68*, 798–806.
81. Wang, L.; Fortner, J. D.; Hou, L.; Zhang, C.; Kan, A. T.; Tomson, M. B.; Chen, W. Contaminant-mobilizing Capability of Fullerene Nanoparticles (nC60): Effect of Solvent-Exchange Process innC60 Formation. *Environ. Toxicol. Chem.* **2013,** *32*, 329–336.
82. Wang, H.; Ma, H.; Zheng, W.; An, D.; Na, C. Multifunctional and Recollectable Carbon Nanotube Ponytails for Water Purification. *ACS Appl. Mater. Interf.* **2014,** *6*, 9426–9434.
83. Wang, T.; Zhao, D.; Guo, X.; Correa, J.; Riehl, B. L.; Heineman, W. R. Carbon Nanotube-Loaded Nafion Film Electrochemical Sensor for Metal Ions: Europium. *Anal. Chem.* **2014,** *86*, 4354–4361.
84. Wang, X.; Liu, X.; Chen, J.; Han, H.; Yuan, Z. Evaluation and Mechanism of Antifungal Effects of Carbon Nanomaterials in Controlling Plant Fungal Pathogen. *Carbon* **2014,** *68*, 798–806.
85. Fosso-Kankeu, E.; De Klerk, C. M.; Botha, T. A.; Waanders, F.; Phoku, J.; Pandey, S. The Antifungal Activities of Multi-Walled Carbon Nanotubes Decorated with Silver, Copper and Zinc Oxide Particles. In *International Conference on Advances in Science, Engineering, Technology and Natural Resources (ICASETNR-16)*, Parys, South Africa, November 24–25, 2016.

86. Oyelami, A. O.; Semple, K. T. Impact of Carbon Nanomaterials on Microbial Activity in Soil. *SoilBiol Biochem.* **2015,** *86,* 172–180.
87. Brandelli, A. Nanobiotechnology Strategies for Delivery of Antimicrobials in Agriculture and Food. In *Nanotechnologies in Food and Agriculture*; Rai, M.; Ribeiro, C.; Mattoso, L.; Duran, N., Eds.; Springer: New York, 2015; p 119.
88. Aouada, F. A.; De Moura, M. R. Nanotechnology Applied in Agriculture: Controlled Release of Agrochemicals. In: *Nanotechnologies in Food and Agriculture*; Rai, M.; Ribeiro, C.; Mattoso, L.; Duran, N., Eds.; Springer: New York, 2015; p 103.
89. Kaushal, M.; Wani, S. P. Nanosensors: Frontiers in Precision Agriculture. In *Nanotechnology*; Springer: Singapore, 2017; p 279.
90. Lee, K.; Park, J.; Lee, M. S.; Kim, J.; Hyun, B. G.; Kang, D. J.; Na, K.; Lee, C. Y.; Bien, F.; Park, J. U. In–Situ Synthesis of Carbon Nanotube–Graphite Electronic Devices and Their Integrations onto Surfaces of Live Plants and Insects. *Nano Lett.* **2014,** *14* (5), 2647–2654.
91. Damalas, C. A.; Eleftherohorinos, I. G. Pesticide Exposure Safety Issues, and Risk Assessment Indicators. *Int. J. Environ. Res. Public Health.* **2011,** *8,* 1402–1419.
92. Sangeetha, J.; Thangadurai, D.; Hospet, R.; Purushotham, P.; Manowade, K. R.; Mujeeb, M. A.; Mundaragi, A. C.; Jogaiah, S.; David, M.; Thimmappa, S. C.; Prasad, R.; Harish, E. R. Production of Bionanomaterials from Agricultural Wastes. In *Nanotechnology*; Prasad, R.; Kumar, M.; Kumar, V., Eds.; Springer Nature: Singapore Pvt Ltd, 2017; p 33.
93. Gholipour, Y.; Erra-Balsells, R.; Nonami, H. Integrative Analysis of Physiological Phenotype of Plant Cells by Turgor Measurement and Metabolomics. *Eng. Lett.* **2012,** *20* (4), EL01.
94. Švancara, I.; Vytřas, K.; Kalcher, K.; Walcarius, A.; Wang, J. Carbon Paste Electrodes in Facts, Numbers, and Notes: A Review on the Occasion of the 50-years Jubilee of Carbon Paste in Electrochemistry and Electroanalysis. *Electroanalysis.* **2009,** *21,* 7–28.
95. Apetrei, C.; Apetrei, I. M.; Saja, J. A. D.; Rodriguez-Mendez, M. L. Carbon Paste Electrodes made from Different Carbonaceous Materials: Application in the Study of Antioxidants. *Sensors* **2011,** *11,* 1328–1344.
96. Inam, R.; Bilgin, C. Square Wave Voltammetric Determination of Methiocarb Insecticide Based on Multiwall Carbon Nanotube Paste Electrode. *J. Appl. Electrochem.* **2013,** *43,* 425–432.
97. Sundari, P. L. A.; Palaniappan, S. P.; Manisankar, P. Enhanced Sensing of Carbendazim, a Fungicide on Functionalized Multiwalled Carbon Nanotube Modified Glassy Carbon Electrode and Its Determination in Real Samples. *Anal. Lett.* **2010,** *43,* 1457–1470.
98. Ribeiro, W. F.; Selva, T. M. G.; Lopes, I. C.; Coelho, E. C. S.; Lemos, S. G.; de Abreu, F. C.; do Nascimento, V. B.; de Araújo, M. C. U. Electroanalytical Determination of Carbendazim by Square Wave Adsorptive Stripping Voltammetry with a Multiwalled Carbon Nanotubes Modified Electrode. *Anal. Methods.* **2011,** *3,* 1202–1206.
99. Yang, J.; Wang, P.; Zhang, X.; Wu, K. Electrochemical Sensor for Rapid Detection of Triclosan Using a Multiwall Carbon Nanotube Film. *J. Agric. Food Chem.* **2009,** *57,* 9403–9407.
100. Luo, S.; Wu, Y.; Gou, H. A Voltammetric Sensor Based on Go–MWNTS Hybrid Nanomaterial-Modified Electrode for Determination of Carbendazim in Soil and Water Samples. *Ionics.* **2013,** *19,* 673–680.

101. Mani, V.; Devasenathipathy, R.; Chen, S.-M.; Wu, T.-Y.; Kohilarani, K. High-Performance Electrochemical Amperometric Sensors for the Sensitive Determination of Phenyl Urea Herbicides Diuron and Fenuron. *Ionics.* **2015,** *21*, 2675–2683.
102. Mugadza, T.; Nyokong, T. Electrocatalytic Oxidation of Amitrole and Diuron on Iron (II) Tetraaminophthalocyanine Singlewalled Carbon Nanotube Dendrimer. *Electrochim. Acta.* **2010,** *55*, 2606–2613.
103. Moraes, F. C.; Mascaro, L. H.; Machado, S. A.; Brett, C. M. Direct Electrochemical Determination of Carbaryl Using a Multi-Walled Carbon Nanotube/Cobalt Phthalocyanine Modified Electrode. *Talanta.* **2009,** *79*, 1406–1411.
104. Wong, A.; Sotomayor, M. D. P. T. Biomimetic Sensor Based on 5,10,15,20-Tetrakis (pentafluorophenyl)-21H,23Hporphyriniron (III) Chloride and MWCNT for Selective Detection of 2,4-d. *Sens. Actuat. B.* **2013,** *181*, 332–339.
105. Siswana, M.; Ozoemena, K.; Nyokong, T. Electrocatalytic Detection of Amitrole on the Multi-Walled Carbon Nanotube—Iron (II) Tetra-Aminophthalocyanine Platform. *Sensors.* **2008,** *8*, 5096.
106. Sambasevam, K.; Mohamad, S.; Sarih, N.; Ismail, N. Synthesis and Characterization of the Inclusion Complex of -Cyclodextrin and Azomethine. *Int. J. Mol. Sci.* **2013,** *14*, 3671–3682.
107. Rahemi, V.; Vandamme, J. J.; Garrido, J. M. P. J.; Borges, F.; Brett, C. M. A.; Garrido, E. M. P. J. Enhanced Host–Guest Electrochemical Recognition of Herbicide MCPA Using a -Cyclodextrin Carbon Nanotube Sensor. *Talanta.* **2012,** *99*, 288–293.
108. Rahemi, V.; Garrido, J. M. P. J.; Borges, F.; Brett, C. M. A.; Garrido, E. M. P. J. Electrochemical Determination of the Herbicide Bentazone Using a Carbon Nanotube _-Cyclodextrin Modified Electrode. *Electroanalysis.* **2013,** *25*, 2360–2366.
109. Rahemi, V.; Garrido, J. M. P. J.; Borges, F.; Brett, C. M. A.; Garrido, E. M. P. J. Electrochemical Sensor for Simultaneous Determination of Herbicide MCPA and Its Metabolite 4-chloro-2-methylphenol. Application to Photodegradation Environmental Monitoring. *Environ. Sci. Pollut. Res.* **2015,** *22*, 4491–4499.
110. Ferancová, A.; Labuda, J. Cyclodextrins as Electrode Modifiers. *J. Anal. Chem.* **2001,** *370*, 1–10.
111. Liu, Z.; Zhang, A.; Guo, Y.; Dong, C. Electrochemical Sensor for Ultrasensitive Determination of Isoquercitrin and Baicalin Based on DM-_Cyclodextrin Functionalized Graphene Nanosheets. *Biosens. Bioelectron.* **2014,** *58*, 242–248.
112. Wang, X.; Li, X.; Luo, C.; Sun, M.; Li, L.; Duan, H. Ultrasensitive Molecularly Imprinted Electrochemical Sensor Based on Magnetism Graphene Oxide/_-cyclodextrin/au nanoparticles composites for chrysoidine analysis. *Electrochim. Acta.* **2014,** *130*, 519–525.
113. Gan, T.; Lv, Z.; Sun, Y.; Shi, Z.; Sun, J.; Zhao, A. Highly Sensitive and Molecular Selective Electrochemical Sensing of 6-Benzylaminopurine with Multiwall Carbon Nanotube@ SNS2-assisted Signal Amplification. *J. Appl. Electrochem.* **2016,** *46*, 389–401.
114. Zhang, Y.; Kang, T.-F.; Wan, Y.-W.; Chen, S.-Y. Gold Nanoparticles-carbon Nanotubes Modified Sensor for Electrochemical Determination of Organophosphate Pesticides. *Microchim. Acta.* **2009,** *165*, 307–311.
115. Fan, S.; Xiao, F.; Liu, L.; Zhao, F.; Zeng, B. Sensitive Voltammetric Response of Methyl Parathion on Single-Walled Carbon Nanotube Paste Coated Electrodes Using Ionic Liquid as Binder. *Sens. Actuat. B.* **2008,** *132*, 34–39.

116. Luo, M.; Liu, D.; Zhao, L.; Han, J.; Liang, Y.; Wang, P.; Zhou, Z. A Novel Magnetic Ionic Liquid Modified Carbon Nanotube for the Simultaneous Determination of Aryloxyphenoxy-Propionate Herbicides and Their Metabolites in Water. *Anal. Chim. Acta*. **2014,** *852*, 88–96.
117. Yang, J.; Wang, Q.; Zhang, M.; Zhang, S.; Zhang, L. An Electrochemical Fungicide Pyrimethanil Sensor Based on Carbon Nanotubes/Ionic-Liquid Construction Modified Electrode. *Food Chem.* **2015,** *187*, 1–6.
118. Rather, J. A.; De Wael, K. C60-Functionalized MWCNT Based Sensor for Sensitive Detection of Endocrine Disruptor Vinclozolin in Solubilised System and Wastewater. *Sens. Actuat. B*. **2012,** *171–172*, 907–915.
119. Teadoum, D. N.; Noumbo, S. K.; Arnaud, K. T.; Ranil, T. T.; Ze Mvondo, A. D.; Tonle, I. K. Square Wave Voltammetric Determination of Residues of Carbendazim Using a Fullerene/Multiwalled Carbon Nanotubes/Nafion/Coated Glassy Carbon Electrode. *Int. J. Electrochem.* **2016,** 7839708.
120. Feng, S.; Yang, R.; Ding, X.; Li, J.; Guo, C.; Qu, L. Sensitive Electrochemical Sensor for the Determination of Pentachlorophenol in Fish Meat Based on ZnSe Quantum Dots Decorated Multiwall Carbon Nanotubes Nanocomposite. *Ionics*. **2015,** *21*, 3257–3266.
121. Anirudhan, T. S.; Alexander, S. Design and Fabrication of Molecularly Imprinted Polymer-Based Potentiometric Sensor from the Surface Modified Multiwalled Carbon Nanotube for the Determination of Lindane (-Hexachlorocyclohexane), an Organochlorine Pesticide. *Biosens. Bioelectron.* **2015,** *64*, 586–593.
122. Mercan, H.; ˙Inam, R.; Aboul-Enein, H. Y. Square Wave Adsorptive Stripping Voltammetric Determination of Cyromazine Insecticide with Multi-Walled Carbon Nanotube Paste Electrode. *Anal. Lett.* **2011,** *44*, 1392–1404.
123. Sundari, P. A.; Manisankar, P. Development of Nano Poly(3-methyl thiophene)/ Multiwalled Carbon Nanotubes Sensor for the Efficient Detection of Some Pesticides. *J. Braz. Chem. Soc.* **2011,** *22*, 746–755.
124. Huang, B.; Zhang, W.-D.; Chen, C.-H.; Yu, Y.-X. Electrochemical Determination of Methyl Parathion at APD/MWCNTS-Modified Electrode. *Microchim. Acta.* **2010,** *171*, 57–62.
125. Li, J.; Chi, Y. Determination of Carbendazim with Multiwalled Carbon Nanotubes-Polymeric Methyl Red Film Modified Electrode. *Pestic. Biochem. Physiol.* **2009,** *93*, 101–104.
126. Food Losses and Waste in the Context of Sustainable Food Systems. A Report by the High Level Panel of Experts on Food Security and Nutrition of the Committee on World Food Security. *HLPE Rep.* **2014,** *1–6*, 1–116.
127. Son, M.; Lee, J. Y.; Ko, H. J.; Park, T. H. Bioelectronic Nose: An Emerging Tool for Odor Standardization. *Trends Biotechnol.* **2017,** *35*, 301–307.
128. Son, M.; Park, T. H. The Bioelectronic Nose and Tongue Using Olfactory and Taste Receptors: Analytical Tools for Food Quality and Safety Assessment. *Biotechnol. Adv.* **2018,** *36*, 371–379.
129. Wasilewski, T.; Gębicki, J.; Kamysz, W. Advances in Olfaction-Inspired Biomaterials Applied to Bioelectronic Noses. *Sens. Actuat. B*. **2018,** *257*, 511–537.
130. Son, M.; Cho, D. G.; Lim, J. H.; Park, J.; Hong, S.; Ko, H. J.; Park, T. H. Real-Time Monitoring of Geosmin and 2-Methylisoborneol, Representative Odor Compounds

in Water Pollution Using Bioelectronic Nose with Human-like Performance. *Biosens. Bioelectron.* **2015,** *74*, 199–206.

131. Song, H. S.; Jin, H. J.; Ahn, S. R.; Kim, D.; Lee, S. H.; Kim, U. K.; Simons, C. T.; Hong, S.; Park, T. H. Bioelectronic Tongue Using Heterodimeric Human Taste Receptor for the Discrimination of Sweeteners with Human-like Performance. *ACS Nano.* **2014,** *8*, 9781–9789.
132. Kim, T. H.; Song, H. S.; Jin, H. J.; Lee, S. H.; Namgung, S.; Kim, U.; Park, T. H.; Hong, S. Bioelectronic Super-Taster Device Based on Taste Receptor-Carbon Nanotube Hybrid Structures. *LabChip.* **2011,** *11*, 2262–2267.
133. Son, M.; Kim, D.; Ko, H. J.; Hong, S.; Park, T. H. A Portable and Multiplexed Bioelectronic Sensor Using Human Olfactory and Taste Receptors. *Biosens. Bioelectron.* **2017,** *87*, 901–907.
134. Song, H. S.; Kwon, O. S.; Lee, S. H.; Park, S. J.; Kim, U. K.; Jang, J.; Park, T. H. Human Taste Receptor-Functionalized Field Effect Transistor as a Human-like Nanobioelectronic Tongue. *NanoLett.* **2013,** *13*, 172–178.
135. Lee, S. H.; Jin, H. J.; Song, H. S.; Hong, S.; Park, T. H. Bioelectronic Nose with High Sensitivity and Selectivity Using Chemically Functionalized Carbon Nanotube Combined with Human Olfactory Receptor. *J. Biotechnol.* **2012,** *157*, 467–472.
136. Lee, S. H.; Kwon, O. S.; Song, H. S.; Park, S. J.; Sung, J. H.; Jang, J.; Park, T. H. Mimicking the Human Smell Sensing Mechanism with an Artificial Nose Platform. *Biomaterials.* **2012,** *33*, 1722–1729.
137. Liu, Q.; Zhang, F.; Zhang, D.; Hu, N.; Wang, H.; Jimmy Hsia, K.; Wang, P. Bioelectronic Tongue of Taste Buds on Microelectrode Array for Salt Sensing. *Biosens. Bioelectron.* **2013,** *40*, 115–120.
138. Qin, Z.; Zhang, B.; Hu, L.; Zhuang, L.; Hu, N.; Wang, P. A Novel Bioelectronic Tongue in Vivo for Highly Sensitive Bitterness Detection with Brain-Machine Interface. *Biosens. Bioelectron.* **2016,** *78*, 374–380.
139. Zhang, W.; Chen, P.; Zhou, L.; Qin, Z.; Gao, K.; Yao, J.; Li, C.; Wang, P. A Biomimetic Bioelectronic Tongue: A Switch for On and Off- Response of Acid Sensations. *Biosens. Bioelectron.* **2017,** *92*, 523–528.
140. Lee, M.; Jung, J. W.; Kim, D.; Ahn, Y. J.; Hong, S.; Kwon, H. W. Discrimination of Umami Tastants Using Floating Electrode-Based Bioelectronic Tongue Mimicking Insect Taste Systems. *ACSNano.* **2015,** *9*, 11728–11736.
141. Park, J.; Lim, J. H.; Jin, H. J.; Namgung, S.; Lee, S. H.; Park, T. H.; Hong, S. A Bioelectronic Sensor Based on Canine Olfactory Nanovesicle–carbon Nanotube Hybrid Structures for the Fast Assessment of Food Quality. *Analyst.* **2012,** *137*, 3249–3254.
142. Kim, T. H.; Lee, S. H.; Lee, J.; Song, H. S.; Oh, E. H.; Park, T. H.; Hong, S. Single-Carbon-Atomic-Resolution Detection of Odorant Molecules Using a Human Olfactory Receptor-Based Bioelectronic Nose. *Adv. Mater.* **2009,** *21*, 91–94.
143. Jin, H. J.; Lee, S. H.; Kim, T. H.; Park, J.; Song, H. S.; Park, T. H.; Hong, S. Nanovesicle-Based Bioelectronic Nose Platform Mimicking Human Olfactory Signal Transduction. *Biosens. Bioelectron.* **2012,** *35*, 335–341.
144. Wei, Z.; Zhang, W.; Wang, J. Nickel and Copper Foam Electrodes Modified with Graphene or Carbon Nanotubes for Electrochemical Identification of Chinese Rice Wines. *Microchim. Acta.* **2017,** *184*, 3441–3451.

145. Wang, L. C.; Tang, K. T.; Chiu, S. W.; Yang, S. R.; Kuo, C. T. A Bio-Inspired Two-Layer Multiple-Walled Carbon Nanotube-Polymer Composite Sensor Array and a Bio-Inspired Fast-Adaptive Readout Circuit for a Portable Electronic Nose. *Biosens. Bioelectron.* **2011,** *26*, 4301–4307.
146. Chiu, S. W.; Wu, H. C.; Chou, T. I.; Chen, H.; Tang, K. T. A Miniature Electronic Nose System Based on an MWNT-Polymer Microsensor Array and a Low-Power Signal-Processing Chip Chemosensors and Chemoreception. *Anal. Bioanal. Chem.* **2014,** *406*, 3985–3994.
147. Theologis, A. One Rotten Apple Spoils the Whole Bushel: The Role of Ethylene in Fruit Ripening. *Cell.* **1992,** *70*, 181–184.
148. Esser, B.; Schnorr, J. M.; Swager, T. M. Selective Detection of Ethylene Gas Using Carbon Nanotube-Based Devices: Utility in Determination of Fruit Ripeness. *Angew. Chem. Int. Ed.* **2012,** *51*, 5752–5756.
149. Leghrib, R.; Llobet, E.; Pavelko, R.; Vasiliev, A. A.; Felten, A.; Pireaux, J. J. Gas Sensing Properties of MWCNTs Decorated with Gold or Tin Oxide Nanoparticles. *Procedia Chem.* **2009,** *1*, 168–171.
150. Zhang, Z.; Huang, Y.; Ding, W.; Li, G. Multilayer Interparticle Linking Hybrid MOF-199 for Non invasive Enrichment and Analysis of Plant Hormone Ethylene. *Anal. Chem.* **2014,** *86*, 3533–3540.
151. Ruiz-Capillas, C.; Jiménez-Colmenero, F. Biogenic Amines in Meat and Meat Products. *Crit. Rev. Food Sci. Nutr.* **2005,** *44*, 489–599.
152. Naila, A.; Flint, S.; Fletcher, G.; Bremer, P.; Meerdink, G. Control of Biogenic Amines in Food-Existing and Emerging Approaches. *J. Food Sci.* **2010,** *75*, R139–R150.
153. Mehta, R. S.; Bassette, R. Organoleptic, Chemical and Microbiological Changes in Ultra-High-Temperature Sterilized Milk Stored at Room Temperature. *J. Food Prot.* **1978,** *41*, 806–810.
154. Lee, S. H.; Lim, J. H.; Park, J.; Hong, S.; Park, T. H. Bioelectronic Nose Combined with a Microfluidic System for the Detection of Gaseous Trimethylamine. *Biosens. Bioelectron.* **2015,** *71*, 179–185.
155. Lee, K. M.; Son, M.; Kang, J. H.; Kim, D.; Hong, S.; Park, T. H.; Chun, H. S.; Choi, S. S. A Triangle Study of Human, Instrument and Bioelectronic Nose for Non-Destructive Sensing of Seafood Freshness. *Sci. Rep.* **2018,** *8*, 547.
156. Lim, J. H.; Park, J.; Ahn, J. H.; Jin, H. J.; Hong, S.; Park, T. H. A Peptide Receptor-Based Bioelectronic Nose for the Real-Time Determination of Seafood Quality. *Biosens. Bioelectron.* **2013,** *39*, 244–249.
157. Brody, A. L.; Strupinsky, E. R.; Kline, L. R. *Active Packaging for Food Applications*; CRC Press: Boca Raton, FL, 2001.
158. Mahajan, R.; Templeton, A.; Harman, A.; Reed, R. A.; Chern, R. T. The Effect of Inert Atmospheric Packaging on Oxidative Degradation in Formulated Granules. *Pharm. Res.* **2005,** *22*, 128–140.
159. Kang, D.; Park, N.; Ko, J.; Bae, E.; Park, W. Oxygen-Induced p-Type Doping of a Long Individual Single-Walled Carbon Nanotube. *Nanotechnology.* **2005,** *16*, 1048–1052.
160. Rajavel, K.; Lalitha, M.; Radhakrishnan, J. K.; Senthilkumar, L.; Rajendra Kumar, R. T. Multiwalled Carbon Nanotube Oxygen Sensor: Enhanced Oxygen Sensitivity

at Room Temperature and Mechanism of Sensing. *ACS Appl. Mater. Interf.* **2015,** *7*, 23857–23865.

161. Sumanasekera, G.; Adu, C.; Fang, S.; Eklund, P. Effects of Gas Adsorption and Collisions on Electrical Transport in Single-Walled Carbon Nanotubes. *Phys. Rev. Lett.* **2000,** *85*, 1096–1099.
162. Tchernatinsky, A.; Desai, S.; Sumanasekera, G. U.; Jayanthi, C. S.; Wu, S. Y.; Nagabhirava, B.; Alphenaar, B. Adsorption of Oxygen Molecules on Individual Single-Wall Carbon Nanotubes. J. *Appl. Phys*. **2006,** *99*, 034306.
163. Derycke, V.; Martel, R.; Appenzeller, J.; Avouris, P. Controlling Doping and Carrier Injection in Carbon Nanotube Transistors. *Appl. Phys. Lett.* **2002,** *80*, 2773–2775
164. Zhu, R.; Desroches, M.; Yoon, B.; Swager, T. M. Wireless Oxygen Sensors Enabled by Fe(II)-Polymer Wrapped Carbon Nanotubes. *ACS Sensors.* **2017,** *2*, 1044–1050.
165. *Food Safety and Food borne Illness*; World Health Organization: Geneva, 2007; Fact sheet no. 237.
166. Velusamy, V.; Arshak, K.; Korostynska, O.; Oliwa, K.; Adley, C. An Overview of Food borne Pathogen Detection: In the Perspective of Biosensors. *Biotechnol. Adv.* **2010,** *28*, 232–254.
167. Frenzen, P. D.; Drake, A.; Angulo, F. J. Economic Cost of Illness Due to Escherichia Coli O157 Infections in the United States. *J. Food Prot.* **2005,** *68*, 2623–2630.
168. Bhardwaj, J.; Devarakonda, S.; Kumar, S.; Jang, J. Development of a Paper-Based Electrochemical Immunosensor Using an Antibody-Single Walled Carbon Nanotubes Bio-Conjugate Modified Electrode for Label-Free Detection of Food borne Pathogens. *Sens. Actuat. B.* **2017,** *253*, 115–123.
169. Zhao, G.; Zhan, X.; Dou, W. A Disposable Immunosensor for Shigella Flexneri Based on Multiwalled Carbon Nanotube/Sodium Alginate Composite Electrode. *Anal. Biochem.* **2011,** *408*, 53–58.
170. Viswanathan, S.; Wu, L. C.; Huang, M. R.; Ho, J. A. A. Electrochemical Immunosensor for Cholera Toxin Using Liposomes and Poly(3,4-Ethylenedioxythiophene)-Coated Carbon Nanotubes. *Anal. Chem.* **2006,** *78*, 1115–1121.
171. Yamada, K.; Kim, C.-T.; Kim, J.-H.; Chung, J.-H.; Lee, H. G.; Jun, S. Single Walled Carbon Nanotube-Based Junction Biosensor for Detection of Escherichia Coli. *PLoS One*. **2014,** *9*, 105767.
172. Yamada, K.; Choi, W.; Lee, I.; Cho, B. K.; Jun, S. Rapid Detection of Multiple Food borne Pathogens Using a Nanoparticle-Functionalized Multi-Junction Biosensor. *Biosens. Bioelectron.* **2016,** *77*, 137–143.
173. Lin, B.; Fugetsu, Y.; Su, F. Watari, Studies on Toxicity of Multi-Walled Carbon Nanotubes on Arabidopsis T87 Suspension Cells. *J. Hazard. Mater*. **2009,** *170*, 578–583.
174. Gonzalez, N.; Johnston, L. Safety of Engineered Nanomaterials. *Chem. Int.* **2018,** *40*, 28–29.
175. Yuan, X.; Zhang, X.; Sun, L.; Wei, Y.; Wei, X. Cellular Toxicity and Immunological Effects of Carbon-Based Nanomaterials. Part. *Fibre Toxicol.* **2019,** *16*, 18.
176. Parisi, C.; Vigani, M.; Rodríguez-Cerezo, E. Proceedings of a Workshop on Nanotechnology for the Agricultural Sector: From Research to the Field. JRC-IPTS(Seville), November 21–22, 2013; Luxembourg, 2014.

177. Khot L. R.; Sankaran, S.; Maja, J. M.; Ehsani, R.; Schuster, E. W. Applications of Nanomaterials in Agricultural Production and Crop Protection: A Review. *Crop Prot.* **2012,** 35, 64–70.
178. Hong, J.; Peralta-Videa, J. R.; Gardea-Torresdey, J. L. Nanomaterials in Agricultural Production: Benefits and Possible Threats? In *Sustainable Nanotechnology and the Environment: Advances and Achievements*; Shamim, N., Sharma, V. K., Ed., Vol. 1124; American Chemical Society: Washington, DC, 2013; p 73.
179. Gogos, A.; Knauer, K.; Bucheli, T. D. Nanomaterials in Plant Protection and Fertilization: Current State, Foreseen Applications, and Research Priorities. *J. Agric. Food Chem.* **2012,** *60,* 9781–9792.
180. Suresh Kumar, R. S.; Shiny, P. J.; Anjali, C. H.; Jerobin, J.; Goshen, K. M.; Magdassi, S.; Mukherjee, A.; Chandrasekaran, N. Distinctive Effects of Nanosized Permethrin in the Environment. *Environ. Sci. Pollut. Res. Int.* **2013,** *20,* 2593–2602.
181. Sarlak, N.; Taherifar, A.; Salehi, F. Synthesis of Nanopesticides by Encapsulating Pesticide Nanoparticles Using Functionalized Carbon Nanotubes and Application of New Nanocomposite for Plant Disease Treatment. *J. Agric. Food Chem.* **2014,** *62,* 4833–4838.
182. Pereira, A.; Grillo, R.; Mello, N. F.; Rosa, A. H.; Fraceto, L. F. Application of Poly(Epsilon-Caprolactone) Nanoparticles Containing Atrazine Herbicide as an Alternative Technique to Control Weeds and Reduce Damage to the Environment. *J. Hazard. Mater.* **2014,** *268,* 207–215.
183. Auernhammer, H. Precision Farming—The Environmental Challenge. *Comput. Electron. Agric.* **2001,** *30,* 31–43.
184. Sebastian, M.; Mathew, B. Multiwalled Carbon Nanotube Based Ion Imprinted Polymer as Sensor and Sorbent for Environmental Hazardous Cobalt Ion. *J. Macromol. Sci.* **2018,** *55,* 455–465.
185. Aravind, A.; Mathew, B. An Electrochemical Sensor and Sorbent Based on Mutiwalled Carbon Nanotube Supported Ion Imprinting Technique for Ni(II) Ion from Electroplating and Steel Industries. *SN Appl. Sci.* **2018,** *1,* 23.
186. Sebastian, M.; Mathew, B. Carbon Nanotube-Based Ion Imprinted Polymer as Electrochemical Sensor and Sorbent for Zn(II) Ion from Paint Industry Wastewater. *Int. J. Polym. Anal. Charact.* **2017,** *23,* 18–28.
187. Sebastian, M.; Mathew, B. Ion Imprinting Approach for the Fabrication of an Electrochemical Sensor and Sorbent for Lead Ions in Real Samples Using Modified Multiwalled Carbon Nanotubes. *J. Mater. Sci.* **2017,** *53,* 3557–3572.

CHAPTER 7

Green Energy Applications of Carbon Nanotubes

APARNA RAJ[1] and RIJU K. THOMAS[2,*]

[1]*School of Pure and Applied Physics, Mahatma Gandhi University, Kottayam, Kerala, India*

[2]*Bharata Mata College, Thrikkakara, Ernakulam, Kerala, India*

[*]*Corresponding author. E-mail: rijukthomas@gmail.com*

ABSTRACT

Carbon nanotubes (CNTs) are a historic class of technological materials, whose discovery makes breakthroughs in numerous applications. No previous material showed the combinations of supreme electrical, thermal, structural, and mechanical properties. A lot of research has been done to reveal the unique properties of CNTs. Currently, research focused to enhance the properties of CNTs corresponding to the desired applications. The idiomatic combination of properties of carbon nanotubes makes them attractive for abroad range of applications, which includes sensors, drug delivery, environmental monitoring, electrochemical sensors, biomedical applications, actuators, displays, bio-imaging, fabrics and fibers, energy transformation and storage, super capacitors, photonics, waste water treatment, green energy applications, etc. The energy crisis arises from the rapid increase of population and environmental pollution arises from the conventional energy sources demands the development of green energy sources. Due to the fusion of special characteristics, CNTs are extremely suitable materials for various steps of different green energy production. Here, we are presenting a general readership about the green energy applications of carbon nanotubes.

7.1 INTRODUCTION

Carbon is one of the limited elements cognized since antiquity. The reason which enables this element as an important one to all known life is its plenitude of organic compounds, capability to form polymers even in normal temperatures, and its superabundance in nature. The carbon atoms themself bind conjointly to form a variety of allotropes of carbon.[1] When compared to other metals, carbon has the largest number of identified allotropes. The commonly known allotropes are graphite, charcoal, diamond, and C_{60} buckminsterfullerene.[2] According to the allotropic form, the physical properties vary differently even if the basic material is carbon atom.[3] If we compare graphite and diamond it is very easy to understand the above statement. The two forms had extremely contrary properties and behavior. The lattice arrangement makes the diamond a poor electrical conductor. But graphite is an excellent electrical conductor. The basic structural units of graphite are single-layered 2D hexagonal lattice called graphene. Graphite is a highly stable material but very flexible and soft but not elastic. Diamond is highly stable but hardest material. Before unearthing the fullerenes in 1985, the only known allotropes were graphite and diamond.

The trio, Sir Harold W. Kroto, Richard E. Smalley, and Robert F. Curl, Jr., shared the Nobel Prize for the dynamite discovery of fullerenes. The accidental discovery befalls when graphite rods were vaporizing in helium atmosphere. They found out a soccer ball-like cage structure having 60 carbon atoms joined together by single and double bonds. This astonishing structure was named as C_{60} buckminsterfullerene after an American architect R. Buckminster Fuller, the great engineer who developed the only practical sort of building that has no restriction of dimensions (geodesic dome). The empirical formulae of a fullerene can be written as C_n, where n is the number of carbon atoms.[4] Later in 1991, Iijima Sumio found an unfamiliar structure consisting of beaucoup concentric "graphene" cylinders of closed ends. They are carbon nanotubes (CNTs). They are tiny tubes having length of micrometer range and diameter of nanometer range. The first identified structure was multi-walled CNT (MWCNT). But in 1996, single-walled CNT (SWCNTs) were produced. Carbon nanotubes have wide varieties of interesting physical and chemical properties.

The significant discovery of the third allotrope of carbon, that is, the fullerenes has opened an entirely new chapter of nanoscience and nanotechnology.

7.2 INTRODUCTION TO CARBON NANOTUBES

CNTs, a member of fullerene family are one of the most unique inventions in the last decade of 20th century. It is an idiomatic nanostructure of cylindrical shape. CNTs conceptually exhibit the essential features of a quantum wire.[5] The CNTs are composed of rolled up graphene sheets in particular directions, that is, graphene are the building blocks of CNTs. So the one dimension CNTs has an SP^2 orbital hybridization exactly as graphene.

One of the major factors which govern the properties of CNTs is the alteration of their tubular structure which is caused by the cylindrical twist of graphene. Depending on the rolling angle (chirality and chiral angle) of graphene, CNTs can exist in three different configurations. They are chiral, armchair, and zigzag configuration.[6] With the help of graphene sheet, we can define chiral vector and chiral angle. As we know graphene lattice is composed of hexagonal rings of carbon atoms. Figure 7.1 explains the possible rolling directions of graphene sheet. The chiral vector C can be delineated in respect of the unit vectors of hexagonal (a_1 and a_2) and translational indices (n and m). The chiral vector can be expressed as, $C = na_1 + ma_2$. The degree of wresting of the tube is arbitrating by chiral angle "α." It is the angle between a_1 and C vectors so that the angle α has deviates in $0° \leq |\alpha| \leq 30°$ range. The two extreme cases 0 and 30 degree correspond to zigzag and armchair configuration and because of the mirror plane they both are achiral structure. The other case, $0° < |\theta| < 30°$ corresponds exactly to chiral.[7]

Since graphite is the basic element, CNTs pick up many characteristics inherently. Just like graphite, they have good electrical conductivity, extremely stable, and prominently corrosion resistant. Other than the immanent properties, they have wide varieties of idiosyncratic fingerprints. The first and most important one is the high aspect ratio of at least 1000. Comparing with the length, carbon nanotubes are ultrathin. Assimilated with steel, the highest tensile material, mechanical tensile strength of CNTs is very high.[6] Density of the structure is 1.3–1.4 g/cm^3 only. This makes them a light-weight material but strong. The thermal conductivity is higher than diamond, the second allotrope of carbon which has a thermal conductivity value of 2000 watt per meter per kelvin. This value is at least five times higher than the best thermal conductor copper (pure). A special feature that makes them extremely solitary as good electron field emitters is the diminutive surface area of their tip which is very much near to the theoretical limit. Also the hollow interior of CNTs can be suitable for many applications since nanomaterials loaded in the interior can be shielded from surroundings. These unsurpassed or individual peculiarities make them a fascinating material for the future.[9–11] For example, CNTs can be used in therapeutic drugs for neurological disease treatment. CNTs will become the key for neural disease therapy in near future. Figure 7.2 explains the applications of CNTs in vitro and in vivo for neurological disease treatments.

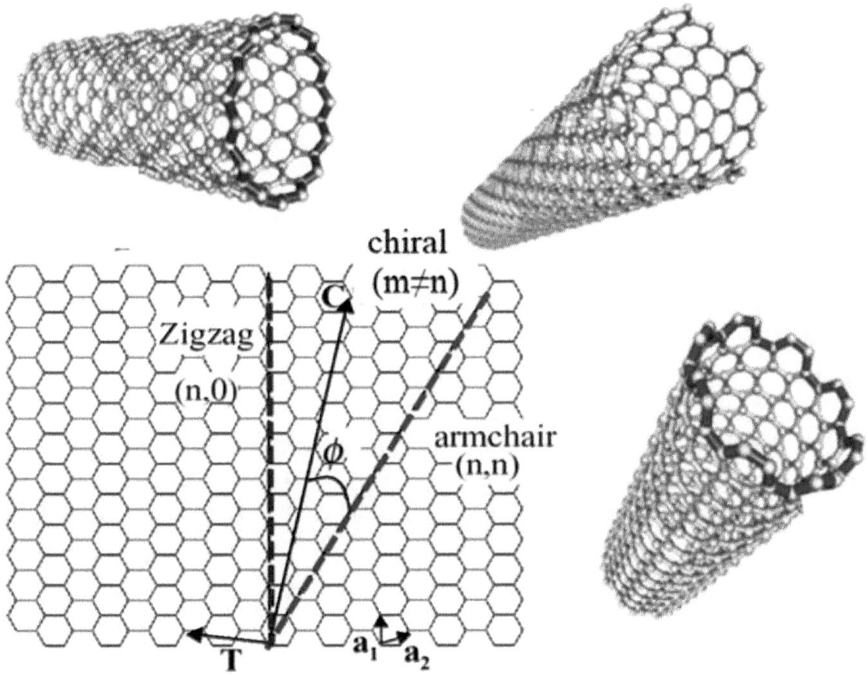

FIGURE 7.1 Different possible rolling directions of graphene that result in SWCNTs with different chirality.

Source: Reproduced with permission from Ref. [8]. https://creativecommons.org/licenses/by/4.0/

The factors suitable for the CNTs in numerous applications are:

- Longer inner space comparing with diameter makes them fit for entrapment
- Needle-like shape

- Light weight but strong and highly elastic
- Biocompatible
- Environmentally safe
- Both ends of the tube are open
- Not capable of inducing any immune response in human body
- It is possible to dissimilarly modify the inner and outer surface

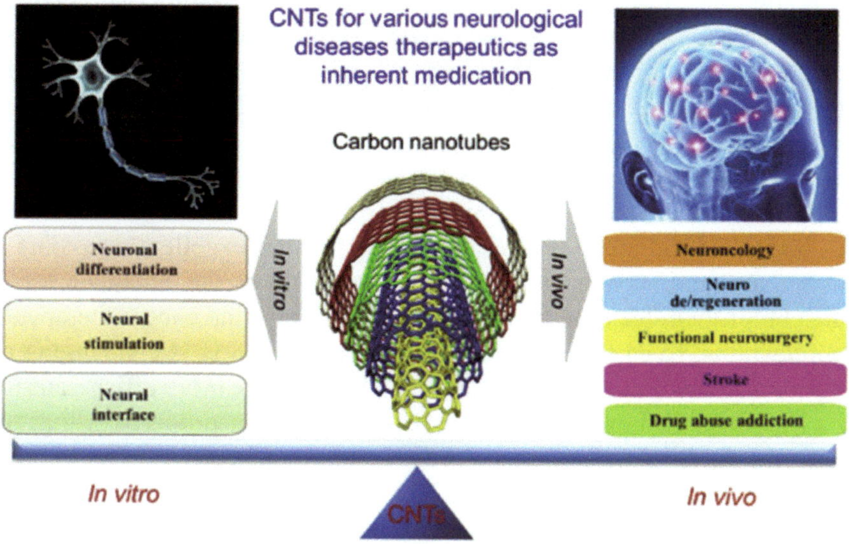

FIGURE 7.2 CNTs for various neurological diseases therapeutics as inherent medication. *Source:* Reprinted with permission from Ref. [12]. © 2020 Elsevier.

7.2.1 DIFFERENT TYPES OF CARBON NANOTUBES

Based on the number of graphene sheets, carbon nanotubes categorized into two:

1. Single-walled carbon nanotubes (SWCNTs)
2. Multiple-walled carbon nanotubes (MWCNTs)

Single-walled carbon nanotubes (SWCNTs):

As the name suggests, SWCNTs consist of a single layer of graphene plane which is enfolded to form a cylinder. The ends of the cylinder are enclosed by hemi fullerenes of felicitous diameter. This depiction of

SWCNTs will exemplify its bipartite nature, that is, they have display the properties of both micro and nanomaterials which can be appreciated as macromolecules or as nanomaterials. Whatever the growth procedure, the inclusionary morphology of isolated SWCNTs is unequalled, that is, a hollow tiny tube whose diameter can be varied from 0.4 to 5 nm and length diverse as nanometers to a few centimeters. The inordinate values of length and diameter correspond to peculiar cases. The diameter of the nanotube has a dependence on the synthesis temperature. The diameter will increase with higher growth temperature. The structure of SWCNTs may be achiral or chiral. The electronic property of SWCNTs originates from a single graphene sheet. So single-walled CNTs of armchair configuration acquit-like metals. But the chiral and zigzag configurations will act like semiconductors and metals accordingly with the curvature of graphene sheet. Many of the unique properties like ultra-high surface area as large as 1300 m^2/g of SWCNTs make it as an excellent candidate for many applications especially medical applications. The adequate surface area helps them for bioconjugation, drug delivery, and cancer therapeutics applications. Even though bulk synthesis of SWCNTs is challenging and purity is low, for medical applications, SWCNTs are more efficacious than multi-walled carbon nanotubes. Another important property is that they will not accumulate in human body. Functionalization of SWCNTs will result in defects of the structure. But the characterizations are comparatively easy. Even then they are good nano candidates for biomedical fields and not as much suitable for other applications.[13,14]

Multiple-walled carbon nanotubes (MWCNTs):

Multiple-walled carbon nanotubes (MWCNTs) are built from many graphene sheets encircle a hollow core with the distance between graphene sheets is nearly 0.3 nm. Or simply, MWCNT is made up of "x" no of SWCNTs (with $\times > 1$) having varying diameters. It is like an annular cylinder having inner radius between 1 and 3 and outer radius less than 100 nm. MWCNTs are complex structures and they are less flexible so it will be difficult to twist this tubular structure. It is comparatively easy for the bulk synthesis of the high purity MWCNTs than SWCNTs since MWCNTs can be produced without using catalyst. But the perfect characterization and synthesis of the improved versions will be a difficult task. A multi-scale analysis procedure is needed in order to accurately analyze the quality and determine the quantity of MWCNTs.[15] Different from SWCNTs they will accumulate in human body. So the medical applications are limited. However, this unique nanomaterial has exceptional electrical, thermal, and mechanical properties. So they have a predominant role in industrial as well as other developing fields. Along with the pure form they also can be decorated with other nano-structures and bonded by physical interactions for numerous applications like energy applications, magnetic data storage, chemical and biosensors, electronic devices, etc.

7.2.2 METHODS OF CARBON NANOTUBE SYNTHESIS

The very common methods for the synthesis of CNTs are:

1. Plasma-based synthesis method or arc-discharge method
2. Laser ablation method
3. Thermal synthesis process
4. Chemical vapor deposition (CVD)
5. Plasma-enhanced CVD (PECVD)

Different techniques have been evolved for the small-scale production of CNTs.[17] Arc-discharge method and laser ablation methods are the methods used in earlier days. They are high-temperature methods. But later, CVD method was introduced. The schematic diagram of the three methods is illustrated in Figure 7.3. This is a low-temperature method comparing with the other two. Now, these three are the common methods.[17,18] The basic elements in these methods are carbon source, catalyst, and adequate energy. The main attribute of the arc-discharge and laser ablation methods is the use of the large quantity of energy to induce the arrangements of carbon atoms into CNTs. The temperature is also very high. So these methods are not relevant for mass production. Also, the continuously substitute of the carbon target source, high production cost and vacuum conditions also make them inappropriate for large-scale production.

In the arc-discharge method, pure graphite is used as the source and the suitable condition is low-pressure inert gas. The typical yield is 30%–90% which includes single-walled tubes with diameters of 0.5–1.4 nm and multi-walled tube with an inner diameter of 1–3 nm and outer diameter nearly 10 nm. In this method, SWCNTs can be easily produced with only minor structural defects and MWCNTs can be produced without using a catalyst. The main disadvantage of the arc-discharge method is that they produce length-wise small tubes in a random direction. Laser ablation is

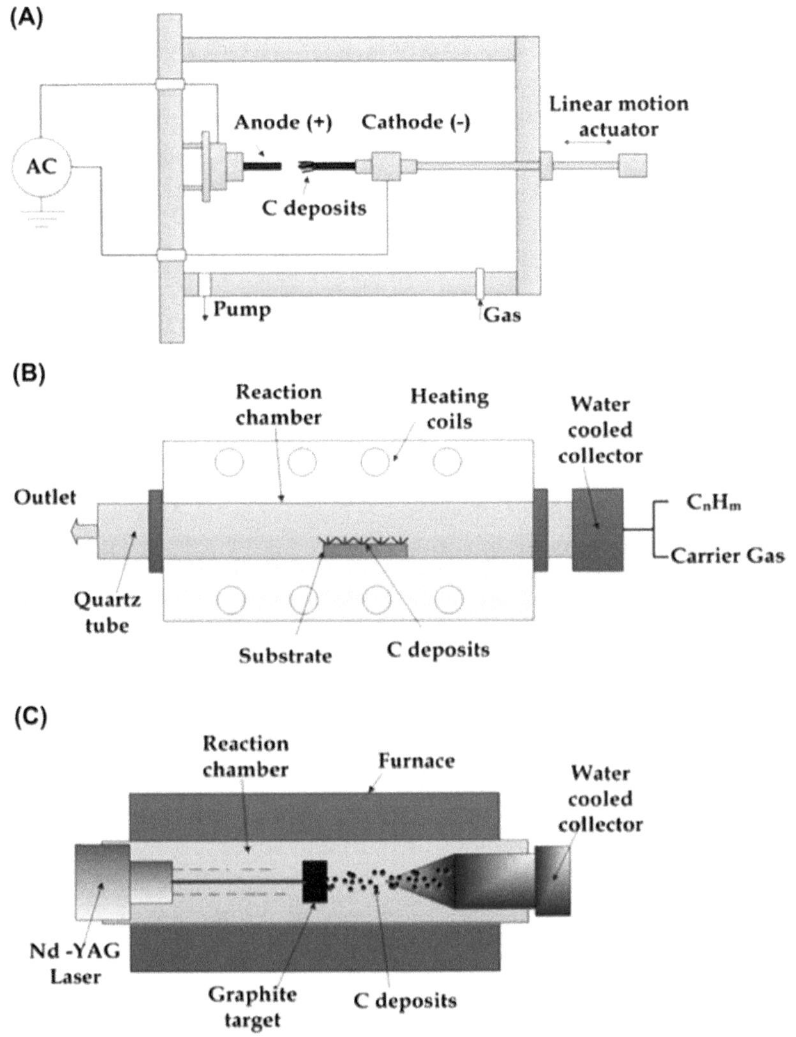

FIGURE 7.3 Methods used for CNT synthesis. (a) Arc- discharge, (b) chemical vapor deposition, and (c) laser ablation.

Source: Reprinted with permission from Ref. [16]. © 2017 Elsevier.

an improved method than arc-discharge method but also a costly technique which needs an extravagant laser source. Carbon source here also graphite but the condition for the synthesis is nitrogen gas atmosphere. This method helps in the production of SWCNTs of length up to 20 μm with good quality. The typical yield of the method is up to 70%. By comparing the three common methods, CVD is the best method for mass production. In CVD method, a carbon compound decomposes to produce carbon atoms in the presence of catalyst (e.g., metallic nanoparticles) at temperature ranges from 500°C to 1000°C. These carbon atoms arrange to form CNTs. The yield of this low-cost method sometimes increased up to 100% which includes lengthy single-walled and multi-walled CNTs. Through this method, it is simple to produce diameter-controlled SWCNTs in pure form. Recently, PECVD and thermal synthesis methods are also used for CNT synthesis (Figure 7.4).[19]

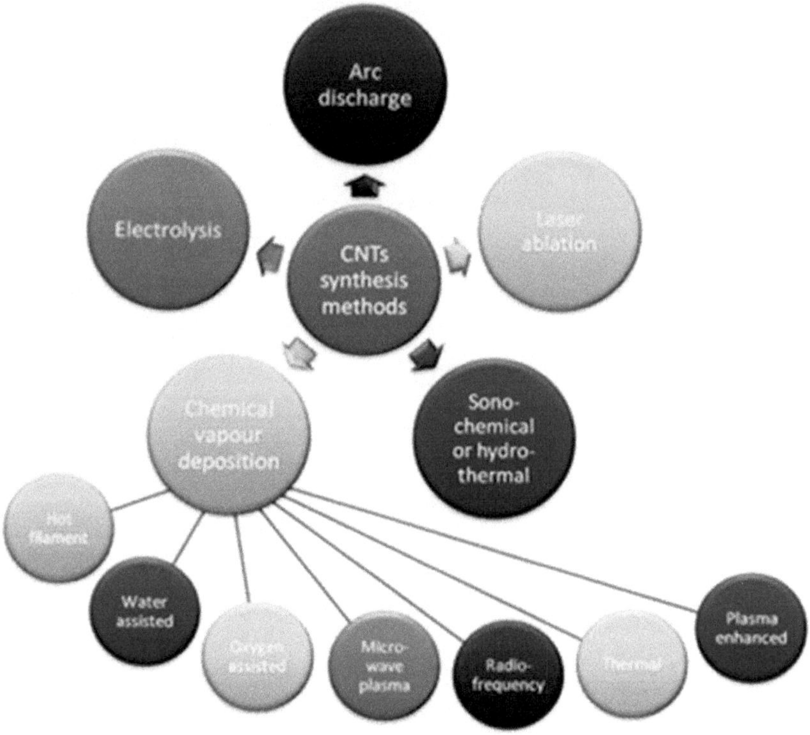

FIGURE 7.4 Schematic diagram for CNTs synthesis methods.

Source: Reprinted with permission from Ref. [17]. © 2011 Royal Society of Chemistry

The different-sized CNTs produced using various methods that can be concluded using the above-schematic diagram. The choice of growth methods will determine the properties of CNTs.

7.3 APPLICATIONS OF CNT

The sole properties of CNTs make them extremely attractive for a variety of applications, which include environmental monitoring, electrochemical sensors, biomedical applications, drug delivery, bio-imaging, fabrics and fibers, energy conversion and storage, super capacitors, wastewater treatment, green energy applications, etc.[20–23] In earlier days, CNTs have been used in electronics, optics, plastics, and other materials of nanotechnology fields. Later, CNTs are introduced in other fields such as biomedical and green energy applications. Carbon nanotubes marked a new epoch in a wide variety of disciplines (Figure 7.5).[24–26]

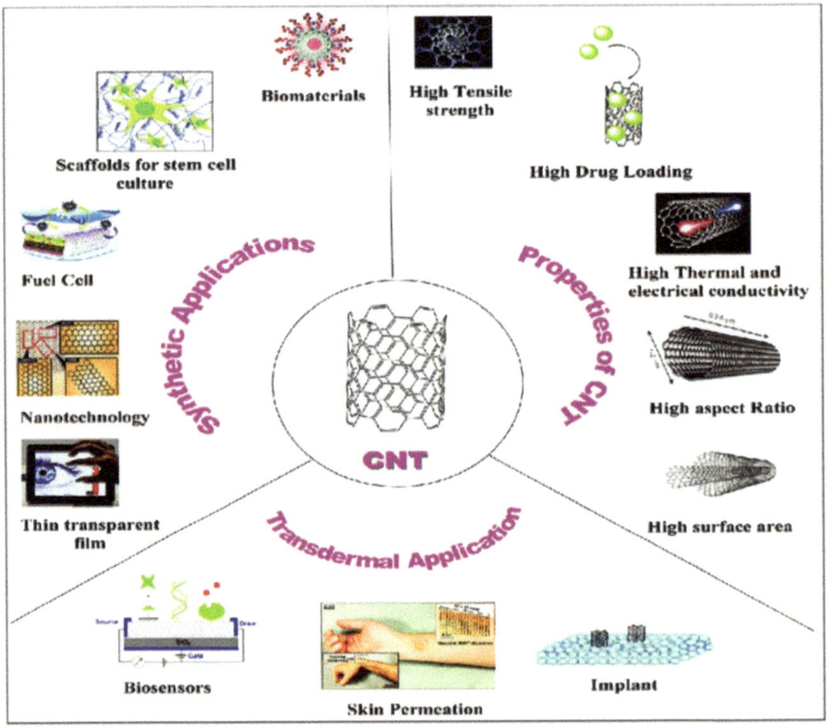

FIGURE 7.5 The overview of properties of CNTs with synthetic and transdermal applications.
Source: Adapted with permission from Ref. [27].

Biomedicine applications include particularly in fields like drug delivery, biosensors, supporting framework in tissue engineering, diagnosis, biomedical imaging, and gene delivery to organs or cells.[28–33] CNT is a biocompatible material and the toxicity level on human body is very low since the human body contains a good amount of carbon. Due to the hydrophobic surfaces, CNTs are insoluble in most of the solvents. In order to solubilize CNT, surface modification or functionalization is needed. Biomedical applications of CNTs after successful functionalization are shown in Figure 7.6. Generally, functionalized CNTs are used in drug delivery applications. The CNT–drug combination enters into the body by orally or by injection and then directed on the target by an external stimulus. They will carry the drug molecule and vaccine through cellular and nuclear membrane without any toxic effects. Comparing with traditional preparation of drug, the drug fixed inside the functionalized CNT and the surface will be effective and harmless. Currently, transdermal patches are also beneficial for the drug delivery, for example, administration of bio-actives due to their effective clinical output. Other different important properties along with transdermal and synthetic applications are established as shown in Figure 7.5. In cancer therapy, CNT acts as a carrier vehicle for drugs. This will overcome the disadvantages like limited cellular penetration, toxicity, harmful effect of anticancer drug on healthy tissues and organs, etc. CNT can be also effectively used in antitumor immunotherapy, tissue regeneration and artificial implants, infection therapy, biosensor vehicle, neurodegenerative diseases, DNA therapy, etc.

The strong value of electrical conductivity and extraordinary sharpness of tip makes the CNTs as an incredible field emitter. Since the sharpness results in a smaller radius of curvature, the electric field will concentrate on the tip. Due to this reason, the field emission will increase and emit even at extremely small voltages. The low-power electrical devices utilize this property of CNTs. The produced current will be highly stable. Currently, these field emitters are used in different applications such as flat panel displays and electron microscope sources since this electron source is far better than the conventional sources. The electrical conductivity along with the geometry makes the CNT as perfect switches in molecular electronics. The molecular electronics is a nanofield in which the dimension of electronic circuits is reduced to nanoscale. The applications of CNTs in electronics such as in advanced computing are due to the thermal properties. The anisotropic thermal conductivity of CNTs is enabling many applications where heat needs to move from one place to another. The intrinsic properties of CNTs are extremely suitable for the materials which can be used as electrodes in capacitors and batteries. Also they have been used in applications like energy storage, catalysis, fuel cells, etc. For energy and catalysis applications, the opportunities of CNTs can be enhanced with the modification of CNTs with nitrogen doping along with suitable functionalization using metals, quantum dots, biomolecules, or polymers. Figure 7.7 represents the schematic presentation of the nitrogen-doped CNTs.

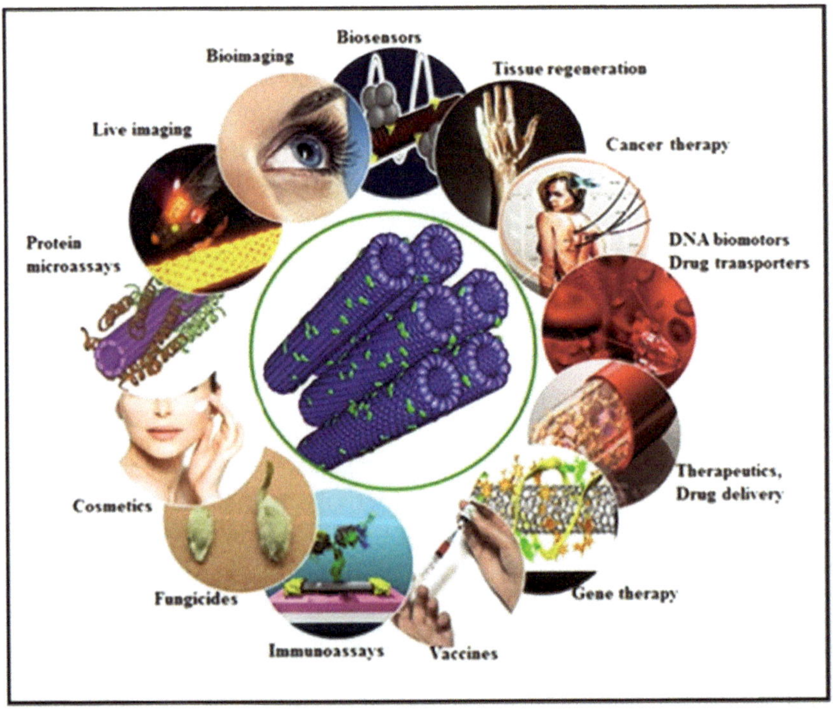

FIGURE 7.6 Biomedical applications of CNTs after functionalized successfully.
Source: Reprinted with permission from Ref. [34]. © 2017 Elsevier.

7.4 GREEN ENERGY APPLICATIONS OF CNT

The rapidly growing energy consumption is one of the major problems facing the whole world. The current energy crisis cannot be eliminated or rectified by conventional energy sources. However, this preeminent energy usage demands more power generation than the usual capacity. Energy from the burning fossil fuels is a reason for environmental pollution and the quality degradation. In order to address these consequences, green energy sources can be used for the power production. Green energy is the part of renewable energy that furnishes the highest environmental advantages.[36,37,39] Or simply green energy is the energy from natural sources. The electricity from solar, Earth's heat, tide, plants, microorganisms, qualified biomass, running water, wind, and biogas is known as green power. Water and solar are the most effective and efficacious green energy sources. Research and progression in green energy have detonated in the past decades resulting in numerous hopeful techniques that can consolidate our dependence on conventional energy sources like coal, oil, and natural gas since they hold the majority of the energy market. More precisely, nanotechnology and CNTs play an important role in the development of energy sources from nature to satisfy our energy demands and environmental concern by bringing about a major or fundamental change in the applications like energy conversion and storage, green engineering of eco-friendly materials as well as environmental monitoring.

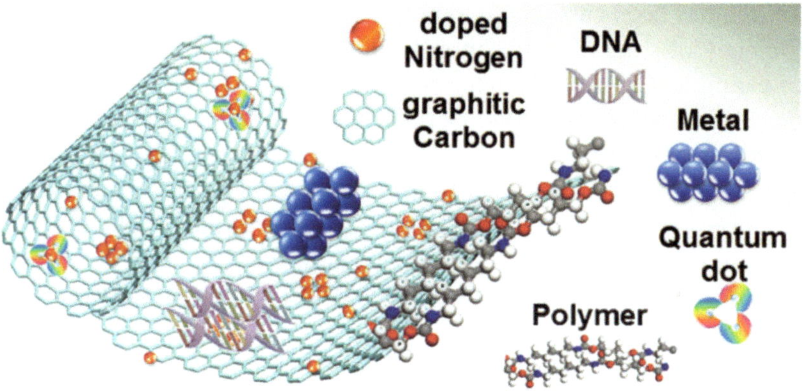

FIGURE 7.7 Nitrogen-doped CNTs with various functional materials.

Source: Reprinted with permission from Ref. [35]. © 2014 Royal Society of Chemistry.

7.4.1 SOLAR ENERGY

Sun is the most ideal source for producing energy since it is the pollution free and tenable energy source due to its profusion and more evenly distribution in nature than any other renewable energy sources. Solar energy consists of both heat and light. So the sun's energy can be used for power production by using both light and heat in different technologies like artificial photosynthesis, solar cell, and solar thermal technology. Solar thermal technology utilizes heat from the sun to make water to steam. A photovoltaic cell otherwise known as solar cell is an electrical device that converts the light energy directly into electricity by the photovoltaic process. Solar cell is the most used technique for the energy extraction from sun. Photovoltaic (PV) devices (solar cells) can be categorized into first, second, and third generation cells based on how advanced they are. The primary generation is the conventional cells made of crystalline silicon. These are the commercially preponderant cells due to the high conversion efficiency. However, the production of Si-based cells is expensive. Second-generation cells are thin-film technologies-based solar cells. Some of the examples are amorphous silicon, copper indium gallium selenide (CIGS) and gallium arsenide (GaAs). The third-generation solar cells include organic/polymer solar cells, dye-sensitized solar cells (DSSCs), perovskite solar cells, quantum dot solar cells, etc.[39,40] The third-generation cells are the most advanced ones and they are the most promising photovoltaic devices due to their low manufacturing cost and high-efficiency value. There is continuous research has been carried out to put into execute the high photoconversion efficiency of solar cells by introducing carbon nanotubes into the device. For example, CNT/silicon heterojunction is used for photovoltaic applications, such as photodetectors. Because along with the electronic properties, CNTs thin films have optical response over a long frequency range from UV to IR radiation. The below figures represent SWCNT/Si heterojunction photodetector with electrical contacts. These figures include schematic diagram, optical photograph, SEM image of SWCNT belt and broadband response, and on/off ratios of the SWCNT/Si heterojunction as a function of reverse bias (Figure 7.8).

Earlier, the active area, that is, a metallic frame in the center of silicon wafer limits the industrial applications of carbon nanotube–silicon solar cell technology. But later, CNTs/Si solar cell is designed with a front- and back-junction which enables for the whole wafer to be used as the active region. The cells are fabricated by using intermixed Nafion layer which is an antireflection and passivation layer. Also act as a nanotube dopant. This design enables the fabrication of solar cell with area of the active region up to 16 cm^2. Jianhui Chen et al. made front- and back-junction devices (solar cells) with record power conversion efficiencies of 15.2% and 18.9% for 1 and 3 cm^2 active areas, respectively. The entire concept can be explained by Figure 7.9.

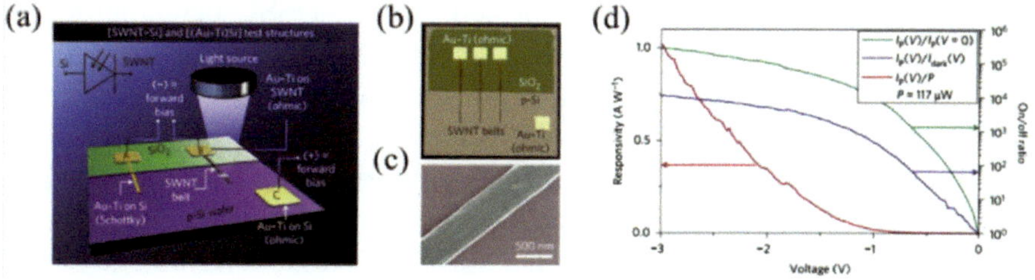

FIGURE 7.8 (a) Schematic and (b) optical photograph of a SWCNT/Si heterojunction structure with electrical contacts. (c) SEM image of a submicron-width SWCNT belt. (d) Broadband response and on/off ratios of the SWCNT/Si heterojunction as a function of reverse bias.

Source: Reprinted with permission from Ref. [41]. © 2019 Elsevier.

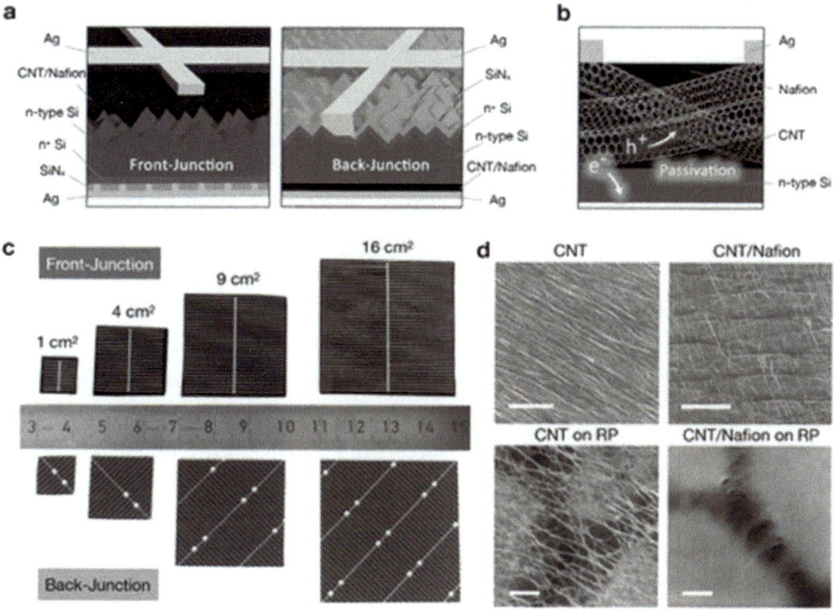

FIGURE 7.9 Schematic of (a) cross-sectional of the front- and back-junction CNT/Si solar cell fabrication and (b) hybrid PCSC operation: holes are collected by nanotubes and traverse to a Ag contact, which for a front-junction cell is a finger array (as depicted) and for a back-junction cell is a full contact electrode. Electrons are collected by the silicon and holes by the nanotubes. Nafion will passivize the exposed regions of silicon. (c) top-contact design for the front- and back-junction design for solar cells with an active area up to 16 cm². (d) SEM images of CNT films on the random pyramid (RP) textured and polished silicon with and without Nafion.

Source: Reprinted with permission from Ref. [44]. https://creativecommons.org/licenses/by/4.0/

Recently, a new generation solar cell, that is, DSSC perennial to gain attention since its low production expenditure and environmental supportability provides a well-founded base for modern applications.[40] Dye-sensitized solar cells are belonging to thin-film solar cells sets. However, the conventional rigid DSSCs are generally made using fluorine-doped tin oxide glass, which generates constraints in practical application. During a recent period of time, CNTs have been used in DS solar cells to take advantage of their lower electrical resistance, porous structure, mechanical strength, high aspect ratio, light weight, flexibility, and amazing electrocatalytic activity. Solar cell consists of counter electrode (CE) and working electrode (WE). The counter electrode plays a big role in the photoconversion efficiency of solar cells since it is the major factor which controls the electron reception or electron flow from the outer current loop. So it is understood that counter electrode and hence its material type of DSSCs show clearly the reason of different photovoltaic performance efficiency. The majority of the recent research found that carbon materials like CNTs are suitable to DSSCs. Latest studies tell that platinized CNTs are most efficient in DSSCs as CEs.[42,43] The incorporation of carbon nanotubes has started even before 2000. After that several results have published on highly flexible 3D DSSCs. In DSSCs, light is absorbed by dye molecules and then incorporated to semiconductor oxides in the system.

The basis electrochemical processes in the DS solar cells are charge separation and electron flow and these processes can be enhanced by the nanostructured surface of CNTs. So obviously, CNTs may be used as working or counter electrode, or even as the conductive materials. A conventional DSSC had the porous nanocrystalline TiO_2 layer as working electrode. The incident photon strikes the dye molecule and is absorbed on the surface of TiO_2. The photon-generated charge will carry out through the transparent conducting oxide (TCO) glass from TiO_2.[45,46] The TCO glasses are very expensive and it plays a significant part more than 50% of the total cost of the solar cell. To reduce the cost of manufacturing and increase the conversion efficiency carbon-based materials are introduced. Considerable conversion efficiency can be accomplished by inserting the SWCNT films to construct solar cells.[47] In comparison with the typical-conjugated polymers which absorb the UV and visible radiations, the SWCNTs intense absorption capacity toward visible and near-infrared (NIR) radiation. Also in contrast to other thin film photovoltaic materials and donor polymers, CNTs have upper hands due to high photostability, exciton lifetime, and carrier mobility. TiO_2 structures incorporated with CNTs have a significant role in charge collection and transport toward the conductive support. SWCNTs network in working electrode improved the electrocatalytic functions, photoconversion efficiency as well as the current generation. The solar cell with working electrode made by depositing SWCNT films and TiO_2 on carbon fiber electrode is nearly doubly efficient than the cells having TiO_2 of same structure without carbon nanotubes. Ghartavol et al. reported a fabrication of DSSC in which conventional nanocrystalline TiO_2 furnished with one-dimensional CNT–TiO_2 core–shell structures (CTH) acted as photoanodes. The Figure 7.10 represents the schematic diagram of the device and the electron transport mechanism.

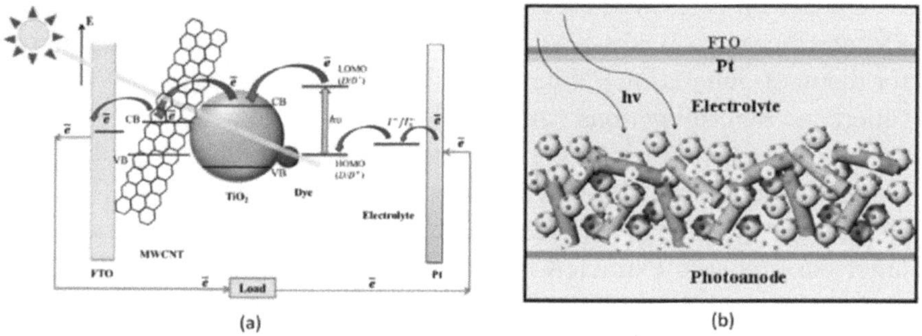

FIGURE 7.10 Schematic diagram of (a) the electron transport mechanism in DSSC and (b) the device made of mixtures of TiO_2 NPs and the CTH.

Source: Reprinted with permission from Ref. [48]. © 2002 Royal Society of Chemistry.

Some of the recent research showed that the incident photon conversion efficiency of solar cells increased more than two times. So it is well understood that the role of SWCNTs in charge collection in electrodes is very much significant. Based on the donor-acceptor nano constructions on the working electrode there are striking solar cell were reported in which the CNTs were employed as the charge acceptor substance because the perfect charge transport pathway in the photoactive layer will be a one-dimensional nanostructure like CNTs. For reducing the unwanted charge recombination and thereby enhancing the cell performance of solar cells, charge is transferred from the excited state of the donor to the semiconductor nanoparticle's conduction band through CNTs as the charge acceptor. Even though the highly pure SWCNTs help in the production of low-cost, highly efficient photovoltaic material, there are some factors may limit the significant advances for SWCNT incorporated PV cells. For example, let us say in order to decrease the complexity of design, many researchers have a selected small range of SWCNT chirality although this selection may ultimately bind the power conversion efficiency (PCE) values. Mao gang Gong and team reported a solar cell structure that increases the maximum photocurrent using polychiral SWCNTs while holding secure high photovoltage and it leads to a record a high efficiency SWCNT–fullerene solar cells with the best photoconversion efficiency of 3.1%.[49] Moreover, these cells show strong absorption in the NIR portion of the solar spectrum that is currently impossible by many important thin-film PV devices.

Blackburn compared the basic underpinnings of two different models; PV interfaces involved s-SWCNTs which help to establish the design rules for the next-generation PV devices.[59] The two interfaces are s-SWCNT—fullerene heterojunctions and s-SWCNTs—photo-excited perovskite active Layer ($MaPbI_3$). In the first case, charges are generated by photo-induced electron transfer (PET) or photo-induced hole transfer (PHT) and exciton dissociation at the interface drives solar energy conversion. In the latter case, charge extracted from the interface between s-SWCNTs and a photo-excited perovskite active layer. The s-SWCNT interfacial layer is used to extract the holes so that fast PHT is observed from $MAPbI_3$ to s-SWCNTs, while recombination is slow. Both structures are compared and studied separately. The understanding from this study can help in the rational design of solar energy harvesting systems containing s-SWCNTs for achieving the ultimate goal of efficient and inexpensive photovoltaic and solar fuels.[59] Figure 7.11 represents the schematic diagram of the comparison study done by Blackburn.

Carbon-based materials are magnificent in the field of photovoltaic considering their ability to provide long-term stable perovskite solar cells. Carbon-based perovskite solar cells (C-PSCs) have gained much industrial heed and been proved to be an excellent replacement for hole-transporting materials (HTMs) based PSCs.[51] They are cheap and readily available and show anticorrosive nature.[52] Under ambient atmosphere, non-encapsulated devices are reported to retain their initial photovoltaic performance for over 1000 h.[53] However, recombination of electrons at the perovskite/carbon interface and the low conductivity of carbon electrode yields a poor conversion efficiency, thus hinders the commercialization of C-PSCs.[54] As a solution to the poor conversion efficiency of conventional carbon-based

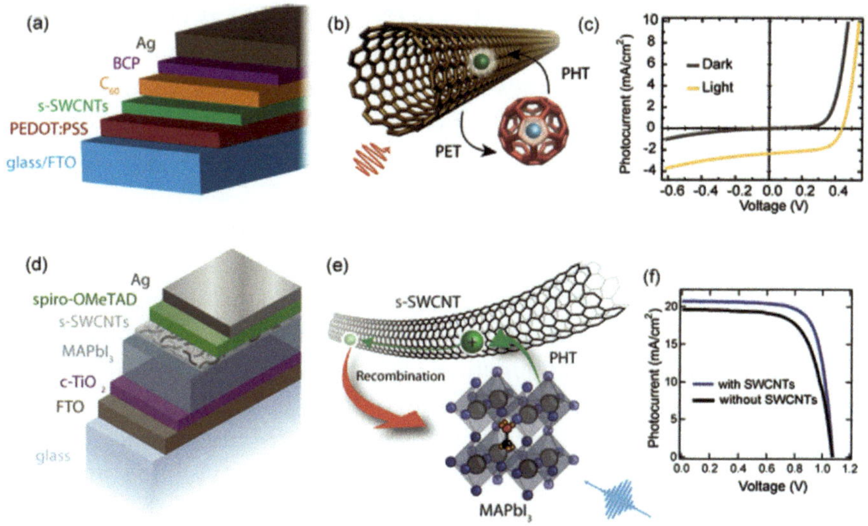

FIGURE 7.11 (a) Schematic of the SWCNT–C60 solar cell. (b) Charges are generated by (PET) or (PHT) at the SWCNT–C60 interface. (c) J–V curve of a representative SWCNT–C60 solar cell. (d) Schematic of a $MaPbI_3$ solar cell, using s-SWCNT interfacial layer to extract holes. (e) Fast PHT is observed from $MAPbI_3$ to s-SWCNTs. (f) J–V curve of a representative $MAPbI_3$ solar cell with and without s-SWCNT hole extraction layer.
Source: Reprinted with permission from Ref. [50]. © 2017 Americal Chemical Society.

solar cells, CNTs are being used owing to their excellent chemical and thermal stability, high surface area, and exceptional electrical conductivity.[55,56] Wu et al. experimented on the C-PSCs and reported a commendable outcome.[53] In order to reduce the carrier recombination, they inserted an organic CuSCN as HTM between perovskite and carbon electrode. The introduction of a highly conducting CNT network as a charge collection electrode resulted in the high performance of C-PSC. The morphology of the CuSCN film was controlled by low-temperature spin-coating process. Systematic solution-based deposition technique is found to increase the device performance dramatically. By taking the advantage of CNT electrode instead of conventional carbon black electrode, they achieved an improved V_{oc} of 1.01 V with a conversion efficiency of 17.58% (measured in backward scan), the highest reported PCE in carbon-based perovskite solar cells. Inorganic interface of CuSCN/CNT paid impressive device stability retaining around 80% of initial photovoltaic performance under the illumination of one sun condition for 1000 h. Mehmood et al. reported the use of nickel impregnated MWCNTs (Ni/MWCNT) as counter electrode material replacing the expensive platinum electrode.[57] Cyclic voltammetry and tafel polarization analyses were done to estimate the catalytic activity of Ni/MWCNT catalyst. The fabricated DSSC using tubular geometrical Ni/MWCNT demonstrated a power conversion efficiency of 9.72% which is greater than that of Pt-based DSSC (8.85%). Catalytic efficiency is found strongly dependent on the percentage of Ni loading. The new class of two-dimensional inorganic compounds, MXene has found applications ranging from energy storage to medicine. A mixed electrode structure was employed by Mi et al. incorporating one-dimensional CNTs and 2D Ti_3C_2-MXene nanosheets into commercial carbon paste.[58] Network structure and multi-dimensional charge transfer path of these mixed electrodes provided an efficiency of 7.09%. This efficiency can be attributed to the appearance of CNTs which boosted the carrier transport by providing a high-speed transport circuit for the carriers.

7.4.2 MICROBIAL FUEL CELL (MCF)

For the increasing demand for environmental-friendly technologies, there has been a progressive development in microbiology and biotechnology. These technologies make a major contribution to satisfying the worldwide need for green energy sources. There are many research teams working on the photosynthetic microbial systems. Photosynthetic microorganisms are capable of capturing solar energy, which is a cheap, existing in large amounts, and depleted-utilized energy source. Captivity of at least a small fraction of the accessible solar energy can make a large contribution toward global energy needs. Photosynthetic microorganisms can be employed for power solar-driven microbial fuel cells (MFCs). Nowadays in industries, the metabolic capabilities of photosynthetic microorganisms are redirected to produce high-energy chemicals. Many of the microorganisms have enormous potential to grow in different environments and accordingly have a clever metabolic stratagem to extract energy from the accessible nutrients and procreate a variety of by-products which are capable of being put to use. Bioelectricity production is the power production by organisms due to their metabolism resulting in the production of electricity.[59]

The power generated by the microbial cell biomass can be extracted using MFCs.[61] Microbial Fuel Cell consists of a cathode chamber, anode chamber (microorganisms will be present in the anode chamber), and protein exchange membrane. In contrast to the other fuel cells, this one uses biocatalyst, which is composed of algae or bacteria. Microbial Fuel Cell has the ability to produce electrical power from biodegradable raw materials and organic waste. But the commercial use of this method is confined mainly by the cost of construction materials. In MFCs, electrochemically active microorganisms live in the self-produced polymeric matrix to a great extent and it is attached to the electrode surface. Biofilm has a complex nature and is called a biofilm and it may study from different aspects like environmental impacts, growth, and metabolic activity in terms of formation, etc. Backbone of bacteria–electrode interaction is the bacterial biofilms. Figure 7.12 represents schematic diagram of the basic MFCs. The electrochemically active microorganisms oxidize the organic compounds and the produced electrons will transferred to the electrode by direct contact with the cell outer membrane or indirect electron transfer with the assistance of redox mediators. The electrons then transfer to the cathode, where electron acceptors (e.g., O_2) are reduced upon catalysis process. The efficiency of MFCs can be improved by the use of CNTs which can be used to increase the conductivity. Pure CNTs or functionalized CNTs can be incorporated into electrodes. Huan Wei and co-workers reported amine-terminated ionic liquid functionalized CNTs for enhancing the quality and quantity of the interfacial electron transfer of *Shewanella putrefaciens* (*S. putrefaciens*) anode in MFCs.[62] The Figure 7.13 schematically represents the two-chambered MFCs in which *Shewanella* algae were grown anaerobically and using an amine-capped MWCNTs-based nanocomposite. This anode achieved a power density at least three times than the anode with the unmodified CNT. Nguyen and team observed the use of mesoporous polysulfone matrix coated with SWCNT as the anode material of MFC which improved the direct electron energy transfer of bacterial community, and enhanced the anode performance.[63] CNTs can also be conjugated with natural polymers for MFC anode preparation like CNT hydrogels-coated substrate which is a low cost and highly porous candidate.[61]

7.4.3 HYDROGEN ENERGY

Except solar energy, hydrogen is a clean energy source when comparing with conventional fuel. Hydrogen fuel has been taken into account as a good approach for the developing technologies of green energy.[65,66] Microbial electrolysis cells (MECs) are the new technology for hydrogen production and it is developed on the basis of MFCs. The hydrogen will produce through microbial electrolysis from organic waste. The organic materials will be oxidized by microorganisms and produced carbon dioxide. These processes will also result in the release of electrons from the oxidative reactions to the anode and transferring protons into the solution. With an additional supply of voltage to the cathode will result in a proton–electron interaction along with the production of hydrogen. The CNTs can be incorporated with biocathode in the MEC. In particular, polymer MWCNT composite can be employed as the composition of a biocathode for biohydrogen yield in a single-chamber MEC.[67,68]

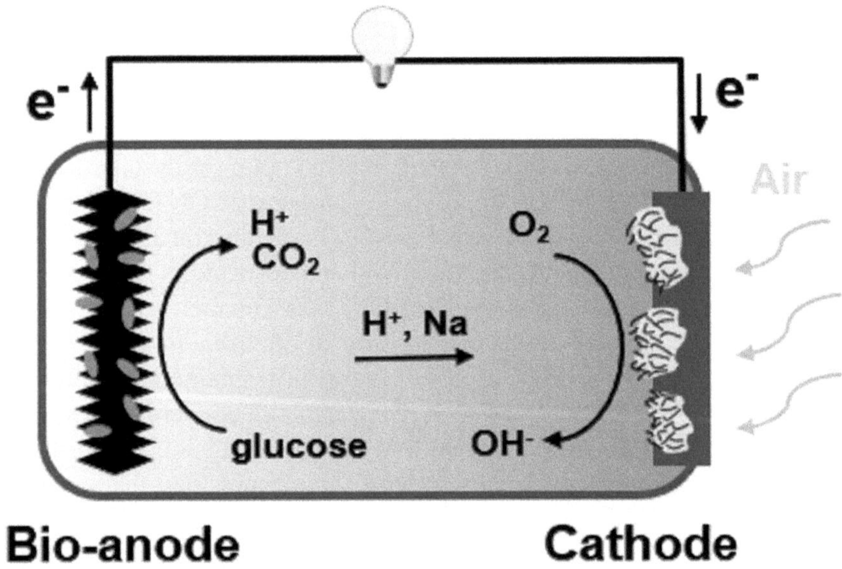

FIGURE 7.12 Schematic setup of MFCs.

Source: Reprinted with permission Ref. [60]. © 2019 Elsevier.

Hydrogen energy system may have the potential to replace the conventional energy sources in the future. One excellent importance of hydrogen fuel lies in the reduction of environmental pollution. Figure 7.14 explains a solar-microbial fuel device based on graphene/carbon nanotube biofilm for enhancing the hydrogen production.[69] The most important application of hydrogen fuel will be in automobile industries. In order to make this target, the development of hydrogen storage facility is important. The latest finding of the reversible and large hydrogen storage capacity of

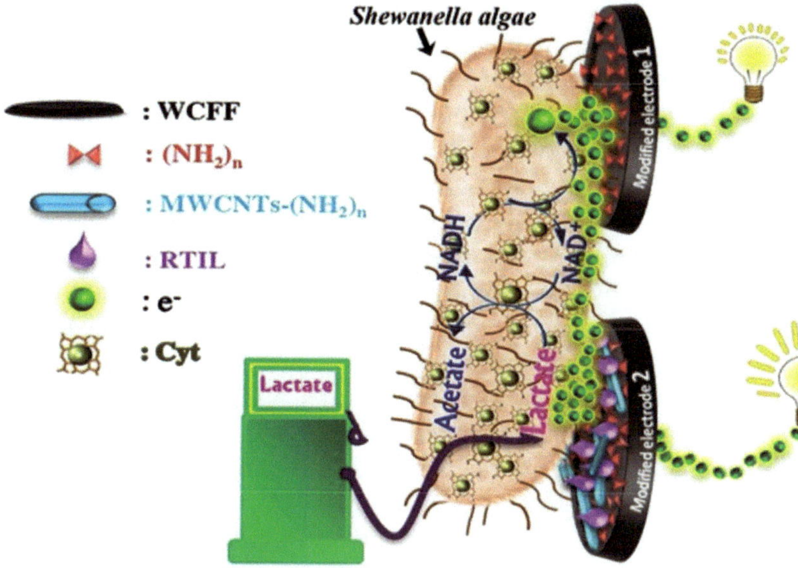

FIGURE 7.13 Schematic representation of an efficient *Shewanella algae* two-chambered microbial fuel cell using a CNT–RTIL-based nanocomposite.

Source: Reprinted with permission from Ref. [64]. © 2013 Royal Society of Chemistry.

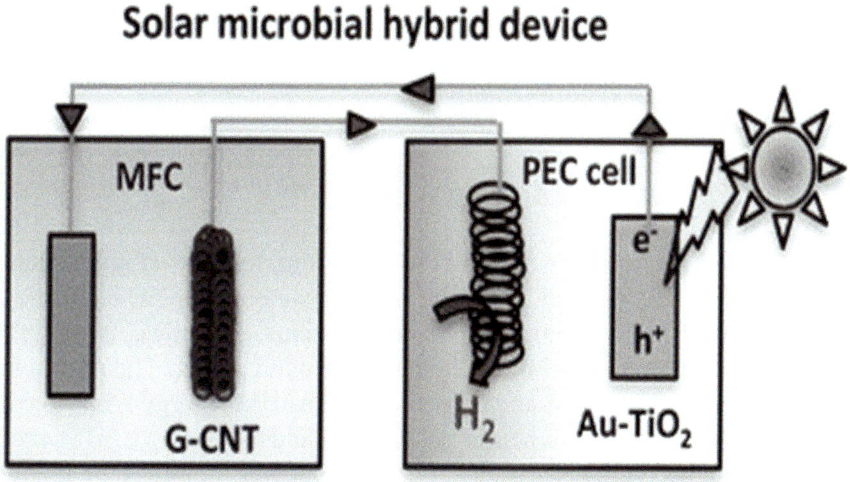

FIGURE 7.14 Solar-microbial fuel device based on graphene/carbon nanotube biofilm for enhanced production of hydrogen.

Source: Reprinted with permission from Ref. [69]. © 2016 Royal Society of Chemistry.

CNTs makes such a system very promising. CNTs are the potential storage substance due to their small density, distribution of nano-sized pores, high-surface area, affordability, high-chemical stability, and recycling characteristics. Depending upon the nature of the array and orientation of CNTs, the quantity of hydrogen absorption varies. Also some theoretical studies showed that the outer surface of CNT is more preferable for hydrogen absorption than inner core. Due to the high-bulk density value, SWCNTs are more preferable for volumetric storage. Ambient conditions for the storage are low temperature (as liquid nitrogen temperature) and high pressure. Otherwise the hydrogen storage will be less than 1 wt.%. Also the hydrogen physisorption alone was not sufficient to meet the expectation. Researchers were used hybrid CNT/metal compounds to initiate chemisorption. The hydrogen storage behavior in SWCNTs may be the chemisorption or physisorption. The hydrogen storage capacity can be reinforced by transition or alkali metal-doped CNTs, with s–p–d hybridization through spill-over mechanism. To enhance the chemisorption of hydrogen in CNTs, structural defects also play an important role to another potential approach as those substitutes to the metal doping methods. The structural defects in CNTs will increase the pore volume and surface area and in turns rise the interaction between hydrogen and CNTs, and thereby increases the adsorption of hydrogen. Theoretically is sure that CNTs are the most suitable materials for the hydrogen storage technology.[70] But other than theoretical proofs, more efforts are needed to achieve the results experimentally.

7.4.4 HYDROPOWER AND WIND ENERGY

Hydropower and wind energy are the most commonly used green energy. Hydropower is produced from the Earth's water cycle which includes rainfall and force of water that stored in dams and tides. Wind in the atmosphere of Earth can be used to generate wind power. The mechanical energy produced is stored and converted to electrical energy when needed. Conversion of energy depends on the demands of energy. The energy storage can be typically divided into hydropower storage and flywheel energy storage. Energy storage using proper material with high mechanical properties is very important. CNTs have high tensile strength, high young's modulus, and high breaking strain. Since the CNTs are the strongest material ever known, they have shown promising potential for the storage of mechanical energy. Energy density of the flywheel energy system will determine by the rotating speed of the flywheel. So obviously flywheel materials should be of high strength to resist successfully the centrifugal force at large rotating speed. To enrich the efficiency of FES systems, CNTs can be employed in flywheel due to the extraordinary mechanical properties like tensile-strain energy density, which has an enormous value (1125 Wh/kg) comparing with steel springs and electrochemical batteries.[71]

7.5 FUTURE PERSPECTIVE

The revolution in science and technology brings the humans to a high standard of life. The constantly increasing the global population and rapidly improving the quality of life demand a large amount of energy. Currently, the human society reliances upon the preliminary energy sources like natural gas, coal, and oil which will be depleted in a very near future. The environmental pollution by these sources also became a worldwide concern. So the need of alternative energy sources with less pollution is important. Green energy sources are the perfect solution for these problems. Solar energy, wind energy, hydroenergy, hydrogen fuel, and bio-energy are necessary to grow rapidly to fill the energy gap. Also the high-performance conversion technologies, efficacious production, and effectual storage and transportation are the factors to achieve for wide applications. This can be accomplished by the introduction of CNTs in various energy applications. Carbon nanotubes have the great potentials in energy and environmental applications with their conspicuous electronic, structural, and mechanical properties. The incorporation of CNTs into the energy conversion technologies and energy storage system has yielded good results to address the energy challenge for the future. Further researches have to be done for improvising the performance of CNTs in order to achieve toward commercialization and practical applications of the perfect clean and sustainable energy sources.

KEYWORDS

- **carbon nanotubes**
- **green energy**
- **solar energy**
- **solar cell**
- **microbial fuel cell**
- **hydrogen energy**

REFERENCES

1. Nasir, S. Carbon-Based Nanomaterials/Allotropes: A Glimpse of Their Synthesis, Properties and Some Applications. **2018**, 1–24. https://doi.org/10.3390/ma11020295.
2. Klupp, G. Fullerenes $ C 60 : Buckminsterfullerene. **2016**, No. May 2015, 1–6. https://doi.org/10.1016/B978-0-12-803581-8.02299-2.
3. Science, C. M. Graphene: Synthesis and Preparation 3. **2014**, *2*. https://doi.org/10.1016/B978-0-12-407789-8.00003-X.
4. Macovez, R.; Macovez, R. Physical Properties of Organic Fullerene Cocrystals. **2018**, *4* (January), 1–7. https://doi.org/10.3389/fmats.2017.00046.
5. Outline, C. Nanotechnology of Carbon Nanotubes: Sensors, Transistors and Nano-composites. **2018**, 1–24. https://doi.org/10.1016/B978-0-12-811071-3.00001-9.
6. He, H.; Pham-huy, L. A.; Dramou, P.; Xiao, D.; Zuo, P.; Pham-huy, C. Carbon Nanotubes: Applications in Pharmacy and Medicine Carbon Nanotubes: Applications in Pharmacy and Medicine. **2013**, No. September. https://doi.org/10.1155/2013/578290.
7. Iijima, S. Carbon Nanotubes: **2006**, 23–26.
8. Development, B.; Zhou, Y.; Fang, Y.; Ramasamy, R. P. Non-Covalent Functionalization of Carbon Nanotubes for Electrochemical. **2019**. https://doi.org/10.3390/s19020392.
9. Lu, J. P. Elastic Properties of Carbon Nanotubes and Nanoropes. **1997**, 1297–1300.
10. Thomson, N. H.; Kulik, A. J.; Benoit, W.; Zuppiroli, L. Mechanical Properties of Carbon Nanotubes. **1999**, *260*, 255–260.
11. Ruoff, R. S.; Lorents, D. C.; International, S. R. I.; Park, M. Mechanical and Thermal Properties of Carbon Nanotubes. **1995**, *33* (7), 925–930.
12. Xiang, C.; Zhang, Y.; Guo, W.; Liang, X. Biomimetic Carbon Nanotubes for Neurological Disease Therapeutics as Inherent Medication. *Acta Pharm. Sin. B* **2019**. https://doi.org/10.1016/j.apsb.2019.11.003.
13. Chemistry, M. Three-Dimensional SWCNT and MWCNT Hybrid Networks for Extremely High-Loading and High Rate Cathode Materials Three-Dimensional SWCNT and MWCNT Hybrid Networks for Extremely High-Loading and High Rate Cathode Materials. **2019**. https://doi.org/10.1039/x0xx00000x.

14. Sathishkumar, S. D.; Rajmohan, T. Comparison between Single-Walled CNT, Multi-Walled CNT, and Carbon Nanotube-Fiber Pyrograf Comparison between Single-Walled CNT, Multi-Walled CNT, and Carbon Nanotube-Fiber Pyrograf III. https://doi.org/10.1088/1757-899X/305/1/012025.
15. Ahn, K.; Kim, S.; Yu, I. J. Multi-Walled Carbon Nanotube (MWCNT) Dispersion and Aerosolization with Hot Water Atomization without Addition of Any Surfactant. **2011**, 65–69. https://doi.org/10.5491/SHAW.2011.2.1.65.
16. Shi, D.; Mi, G.; Webster, T. J. The Synthesis, Application, and Related Neurotoxicity of Carbon Nanotubes. **2017**, No. February 2018. https://doi.org/10.1016/B978-0-12-804598-5.00011-8.
17. Chem, J. M.; Prasek, J.; Drbohlavova, J.; Chomoucka, J.; Hubalek, J.; Jasek, O.; Kizek, R. Methods for Carbon Nanotubes Synthesis—Review. **2011**, 15872–15884. https://doi.org/10.1039/c1jm12254a.
18. See, C. H.; Harris, A. T. REVIEWS A Review of Carbon Nanotube Synthesis via Fluidized-Bed Chemical Vapor Deposition. **2007**, No. Figure 1, 997–1012. https://doi.org/10.1021/ie060955b.
19. Tanemura, M.; Iwata, K.; Takahashi, K.; Fujimoto, Y.; Okuyama, F.; Sugie, H.; Filip, V.; Tanemura, M.; Iwata, K.; Takahashi, K., et al. Growth of Aligned Carbon Nanotubes by Plasma-Enhanced Chemical Vapor Deposition: Optimization of Growth Parameters Growth of Aligned Carbon Nanotubes by Plasma-Enhanced Chemical Vapor Deposition: Optimization of Growth Parameters. **2005**, *1529* (2001), 1–6. https://doi.org/10.1063/1.1382848.
20. Chem, J. M.; Zhu, T.; Xia, B.; Wen, X.; Lou, D. Arrays of Ultrafine CuS Nanoneedles Supported on a CNT Backbone for Application in Supercapacitors. **2012**, 7851–7855. https://doi.org/10.1039/c2jm30437f.
21. Chen, H.; Wu, R.; Chan, K.; Sun, Y.; Su, P. The Application of CNT/Nafion Composite Material to Low Humidity Sensing Measurement. **2005**, *104*, 80–84. https://doi.org/10.1016/j.snb.2004.04.105.
22. Wang, Y.; Jiang, F.; Chen, J.; Sun, X.; Xian, T.; Yang, H. In Situ Construction of CNT/CuS Hybrids and Their Application in Photodegradation for Removing Organic Dyes. **2020**.
23. Kim, B. S.; Seo, D.; Gwon, H.; Kim, J.; Kang, K. Fabrication of FeF 3 Nanofl Owers on CNT Branches and Their Application to High Power Lithium Rechargeable Batteries. **2010**, *701*, 5260–5264. https://doi.org/10.1002/adma.201002879.
24. Kausar, A.; Rafique, I.; Muhammad, B. Review of Applications of Polymer/Carbon Nanotubes and Epoxy/CNT Composites. **2016**, *2559*. https://doi.org/10.1080/03602559.2016.1163588.
25. Upadhyayula, V. K. K.; Deng, S.; Mitchell, M. C.; Smith, G. B. Science of the Total Environment Application of Carbon Nanotube Technology for Removal of Contaminants in Drinking Water: A Review. *Sci. Total Environ.* **2009**, *408* (1), 1–13. https://doi.org/10.1016/j.scitotenv.2009.09.027.
26. Zhang, Q.; Huang, J.; Qian, W.; Zhang, Y.; Wei, F. The Road for Nanomaterials Industry: A Review of Carbon Nanotube Production, Post-Treatment, and Bulk Applications for Composites and Energy Storage. **2013**, *8*, 1237–1265. https://doi.org/10.1002/smll.201203252.

27. Kuche, K.; Maheshwari, R.; Tambe, V.; Mak, K.; Jogi, H.; Raval, N.; Pichika, M. R.; Tekade, R. K. Carbon Nanotubes (CNTs) Based Advanced Dermal Therapeutics: Current Trends and Future Potential. **2018**, April. https://doi.org/10.1039/C8NR01383G.
28. Singh, P. Synthesis of Carbon Nanotubes and Their Biomedical. **2014**, August.
29. Beg, S.; Rizwan, M.; Sheikh, A. M.; Hasnain, M. S.; Anwer, K.; Kohli, K. Advancement in Carbon Nanotubes: Basics, Biomedical Classification of CNTs. **2011**, 141–163. https://doi.org/10.1111/j.2042-7158.2010.01167.x.
30. Saliev, T. The Advances in Biomedical Applications of Carbon Nanotubes. **2019**.
31. Yang, K.; Maiti, D.; Tong, X.; Mou, X. Z. Carbon Based Nanomaterials for Biomedical Applications: A Recent Study. **2018**. https://doi.org/10.3389/fphar.2018.01401.
32. Raphey, V. R.; Henna, T. K.; Nivitha, K. P.; Mufeedha, P.; Sabu, C.; Pramod, K. Materials Science & Engineering C Advanced Biomedical Applications of Carbon Nanotube. *Mater. Sci. Eng. C* **2019**, *100* (March), 616–630. https://doi.org/10.1016/j.msec.2019.03.043.
33. Li, X.; Fan, Y.; Watari, F. Current Investigations into Carbon Nanotubes for Biomedical Application. **2010**, *022001*. https://doi.org/10.1088/1748-6041/5/2/022001.
34. Sireesha, M.; Babu, V. J.; Ramakrishna, S. Functionalized Carbon Nanotubes in Bio-World : Applications, Limitations and Future Directions. *Mater. Sci. Eng. B* **2017**, *223*, 43–63. https://doi.org/10.1016/j.mseb.2017.06.002.
35. Lee, W. J.; Maiti, N.; Lee, M.; Lim, J.; Han, H. Composite Structures for Energy **2014**, 6818–6830. https://doi.org/10.1039/c4cc00146j.
36. Midilli, A.; Dincer, I.; Ay, M. Green Energy Strategies for Sustainable Development. **2006**, *34*, 3623–3633. https://doi.org/10.1016/j.enpol.2005.08.003.
37. Guo, Z.; Sun, Y.; Pan, S. Integration of Green Energy and Advanced Energy-Efficient Technologies for Municipal Wastewater Treatment Plants. 1–29. https://doi.org/10.3390/ijerph16071282.
38. Bhowmik, C.; Bhowmik, S.; Ray, A.; Murari, K. Optimal Green Energy Planning for Sustainable Development: A Review. *Renew. Sustain. Energy Rev.* **2016**, December, 1–18. https://doi.org/10.1016/j.rser.2016.12.105.
39. Yan, J.; Uddin, M. J.; Dickens, T. J.; Okoli, O. I. Science Direct Carbon Nanotubes (CNTs) Enrich the Solar Cells. *Sol. Energ.* **2013**, *96*, 239–252. https://doi.org/10.1016/j.solener.2013.07.027.
40. Hwang, S.; Batmunkh, M.; Nine, J.; Chung, H. Dye-Sensitized Solar Cell Counter Electrodes Based on Carbon Nanotubes. **2015**, 53–65. https://doi.org/10.1002/cphc.201402570.
41. Hu, X.; Hou, P.; Liu, C.; Cheng, H. Nano Materials Science Carbon Nanotube/Silicon Heterojunctions for Photovoltaic Applications. *Nano Mater. Sci.* **2019**, *1* (3), 156–172. https://doi.org/10.1016/j.nanoms.2019.03.001.
42. Chem, J. M. Paper. **2012**, 25311–25315. https://doi.org/10.1039/c2jm35102a.
43. Cha, S. I.; Koo, B. K.; Seo, S. H.; Lee, D. Y. Pt-Free Transparent Counter Electrodes for Dye-Sensitized Solar Cells Prepared from Carbon Nanotube Micro-Balls . **2010**, *1* (c), 659–662. https://doi.org/10.1039/b918920c.
44. Chen, J.; Tune, D. D.; Ge, K.; Li, H.; Flavel, B. S. Front and Back-Junction Carbon Nanotube-Silicon Solar Cells with an Industrial Architecture. **2020**. https://doi.org/10.1002/adfm.202000484.

45. Al-mamun, M.; Kim, J.; Sung, Y.; Lee, J.; Kim, S. Pt and TCO Free Hybrid Bilayer Silver Nanowire – Graphene Counter Electrode for Dye-Sensitized Solar Cells. *Chem. Phys. Lett.* **2013**, *561–562*, 115–119. https://doi.org/10.1016/j.cplett.2013.01.043.
46. Lee, K. S.; Lee, K.; Wang, H.; Park, N.; Lee, Y.; Park, O. O.; Hyeok, J. Dye-Sensitized Solar Cells with Pt- and TCO-Free Counter Electrodes W. **2010**, 4505–4507. https://doi.org/10.1039/c0cc00432d.
47. Cui, K.; Chiba, T.; Omiya, S.; Thurakitseree, T.; Zhao, P.; Fujii, S. Self-Assembled Microhoneycomb Network of Single-Walled Carbon Nanotubes for Solar Cells. **2013**.
48. Ghartavol, H. M.; Mohammadi, M. R.; Afshar, A.; Li, Y. On the Assessment of Incorporation of CNT–TiO2 Core–Shell Structures into Nanoparticle TiO2 Photoanodes in Dye-Sensitized Solar Cells. **2019**, 1840–1850. https://doi.org/10.1039/c9pp00100j.
49. Gong, M.; Shastry, T. A.; Xie, Y.; Bernardi, M.; Jasion, D.; Luck, K. A.; Marks, T. J.; Grossman, C.; Ren, S.; Hersam, M. C. Polychiral Semiconducting Carbon Nanotube – Fullerene Solar Cells. **2014**.
50. Blackburn, L. Semiconducting Single-Walled Carbon Nanotubes in Solar Energy Harvesting. **2017**. https://doi.org/10.1021/acsenergylett.7b00228.
51. Carbon-based HTL-Free Modular Perovskite Solar Cells with Improved Contact at the Perovskite/Carbon Interfaces.
52. Chen, R.; Feng, Y.; Zhang, C.; Wang, M.; Liu, J.; Ma, C.; Bian, J.; and Shi, A. Y. Materials Chemistry C. **2020**. https://doi.org/10.1039/D0TC02226H
53. Chen, H.; Wei, Z.; He, H.; Zheng, X.; Wong, K. S.; Yang, S. Solvent Engineering Boosts the Efficiency of Paintable Carbon-Based Perovskite Solar Cells to Beyond 14%. **2016**. https://doi.org/10.1002/aenm.201502087.
54. Efficient and Stable Carbon-based Perovskite Solar Cells Enabled by Inorganic Interface of CuSCN and Carbon Nanotubes Xin Wu, Liqiang Xie, Kebin Lin, Jianxun Lu, Kexiang Wang, Wenjing Feng, Bingbing Fan, Penggang Yin and Zhanhua Wei, A. Materials Chemistry A. 2019. https://doi.org/10.1039/C9TA02014D
55. Efficient and Carbon-Based Hole Transport Layer-Free CsPbI2Br Planar Perovskite Solar Cells Using PMMA Modification Xiang Zhang, Yang Zhou, Yuzhu Li, Jiawen Sun, Xubing Lu, Xingsen Gao, Jinwei Gao, Lingling Shui,b Sujuan Wu and Jun-Ming Liua, A. Materials Chemistry C. 2019. https://doi.org/10.1039/C9TC00374F
56. The Role of Carbon-based Materials in Enhancing the Stability of Perovskite Solar Cells, Mahboubeh Hadadian, Jan-Henrik Smått, and Juan-Pablo Correa-Baen, A. Environmental Science. 2020. https://doi.org/10.1039/C9EE04030G.
57. Zhang, W.; Bu, F.; Shen, W.; Qi, X.; Yang, N. Strongly Enhanced Efficiency of Polymer Solar Cells through Unzipped SWNT Hybridization in the Hole. **2020**, 24847–24854. https://doi.org/10.1039/d0ra03461d.
58. Mehmood, U.; Ahmad, W.; Ahmed, S. Sustainable Energy & Fuels Nickel Impregnated Multi-Walled Carbon for Efficient and Platinum-Free Dye-Sensitized Solar. **2019**, *17* (i), 3473–3480. https://doi.org/10.1039/c9se00583h.
59. Mi, L.; Zhang, Y.; Chen, T.; Xu, E.; Jiang, Y. Carbon Electrode Engineering for High Efficiency All-Inorganic Perovskite Solar Cells. **2020**, 12298–12303. https://doi.org/10.1039/d0ra00288g.
60. Maksimova, Y. G. Microorganisms and Carbon Nanotubes: Interaction and Applications (Review). **2019**, *55* (1), 1–12. https://doi.org/10.1134/S0003683819010101.

61. Du, Y.; Ma, F.; Xu, C.; Yu, J.; Li, D.; Feng, Y.; Zhen, L. Nano Energy Nitrogen-Doped Carbon Nanotubes/Reduced Graphene Oxide Nanosheet Hybrids towards Enhanced Cathodic Oxygen Reduction and Power Generation of Microbial Fuel Cells. **2019**, *61* (May), 533–539. https://doi.org/10.1016/j.nanoen.2019.05.001.
62. Yazdi, A. A.; Angelo, L. D.; Windiasti, G.; Lu, X.; Xu, J.; Ahmadian, A.; Angelo, L. D.; Omer, N.; Windiasti, G.; Lu, X., et al. Author's Accepted Manuscript Carbon Nanotube Modification of Microbial Fuel Cell Electrodes Reference: To Appear in: Biosensors and Bioelectronic. *Biosens. Bioelectron.* **2016**. https://doi.org/10.1016/j.bios.2016.05.033.
63. Wei, H.; Wu, X.; Zou, L.; Wen, G.; Liu, D. Amine-Terminated Ionic Liquid Functionalized Carbon Nanotubes for Enhanced Interfacial Electron Transfer of Shewanella Putrefaciens Anode in Microbial Fuel Cells. *J. Power Sources* **2016**, *315*, 192–198. https://doi.org/10.1016/j.jpowsour.2016.03.033.
64. 3D Mesoporous Polysulfone-Carbon Nanotube Anode for Enhanced Bioelectricity Output in Microbial Fuel Cells, Tien-Hoa Nguyen, Yang-Yang Yu, Xin Wang, Jing-Yuan Wang, Hao song, A. ChemComm. 2013, No. 207890. https://doi.org/10.1039/b000000x..
65. Manuscript, A. Materials Chemistry A. **2017**. https://doi.org/10.1039/C6TA09766A.
66. Cheng, H.; Yang, Q.; Liu, C. Hydrogen Storage in Carbon Nanotubes. **2001**, *39*, 1447–1454.
67. Program, H.; Nrel, R. Carbon Nanotube Materials for Hydrogen Storage. **2000**.
68. Kadier, A.; Simayi, Y.; Abdeshahian, P.; Farhana, N.; Chandrasekhar, K.; Sahaid, M. REVIEW A Comprehensive Review of Microbial Electrolysis Cells (MEC) Reactor Designs and Configurations for Sustainable Hydrogen Gas Production. *Alexandria Eng. J.* **2015**. https://doi.org/10.1016/j.aej.2015.10.008.
69. Rivera, I.; Schröder, U.; Patil, S. A. *Microbial Electrolysis for Biohydrogen Production : Technical Aspects and Scale-Up Experiences*; Elsevier B.V., 2019. https://doi.org/10.1016/B978-0-444-64052-9.00036-4.
70. Ngaw, C. K.; Zhao, C.; Wang, B.; Sta, D. Sustainable Energy & Fuels Microbial Fuel Device for Enhanced Hydrogen. **2016**. https://doi.org/10.1039/C6SE00018E.
71. Liu, M.; Zhao, S.; Xiao, X.; Chen, M.; Sun, C.; Yao, Z.; Hu, Z.; Chen, L. Nano Energy Novel 1D Carbon Nanotubes Uniformly Wrapped Nanoscale MgH_2 for Efficient Hydrogen Storage Cycling Performances with Extreme High Gravimetric and Volumetric Capacities. *Nano Energy* **2019**, *61* (April), 540–549. https://doi.org/10.1016/j.nanoen.2019.04.094.
72. Bai, Y.; Shen, B.; Zhang, S.; Zhu, Z.; Sun, S.; Gao, J.; Li, B. Storage of Mechanical Energy Based on Carbon Nanotubes with High Energy Density and Power Density. **2018**, *1800680*, 1–26. https://doi.org/10.1002/adma.201800680.

CHAPTER 8

Carbon Nanotubes: Application for Water Purification and Desalination

AVINASH V. BORGAONKAR[1*] and SHITAL B. POTDAR[2]

[1]*Department of Mechanical Engineering, National Institute of Technology Warangal, Warangal 506004, Telangana, India*

[2]*Department of Chemical Engineering, National Institute of Technology Warangal, Warangal 506004, Telangana, India*

[*]*Corresponding author. E-mail: avi.borgaonkar@gmail.com*

ABSTRACT

In recent years, carbon nanotubes (CNTs) are attracting researchers' attention as they are used globally in wide-ranging applications such as solar and hydrogen storage, fuel cells, lithium batteries, supercapacitors, nanocomposites, gas sensors, pathogens, heavy metals, dyes, pesticides, and water desalination as well as decontamination. This chapter elaborates characteristics and fascinating structures of CNTs, activation of CNT membranes, mechanism of pollutants, and salt removal from water. The CNTs can be effectively used for deionizing light metal salts, to take away organic dyes and enriching heavy metal ions. The CNTs with well-aligned structures can be employed as effective pores in membranes to desalinate and purify the water. The CNTs pores can be modified on the need basis for sensing and refusing the ions. The distinctive nonpolar tip functionalized interior structure of CNTs interacts with the polar water (H_2O) molecules and inhibits the pollutants and salts. The noteworthy properties such as energy efficiency, antifoulant, and self-cleaning nature have proven CNTs as a strong candidate compared to the conventional ones. Such distinguished properties and the reusable nature of CNTs, assisted with membrane technology make them a promising candidate in the area of water purification and desalination. This chapter will also focus on the current challenges in the commercialization of CNT membranes.

8.1 INTRODUCTION

Water is an inevitable natural resource, and nowadays scarcity of the water is increasing. In the twenty-first century, many societies and the world are facing the problem of water scarcity compared to other problems. In the last century, the global consumption of water has increased at greater than twice the growth rate of the population of world. In the near future, increasing population, pollution, and uncertain climate changes will lead to induce a tremendous fall in water supply.[1] Various factors such as input energy required, interest and depreciation, operation and maintenance cost comprise the cost for desalted water. Each one of these contributes nearly one third to the total water cost. In order to reduce the cost for desalinating water, in the area of thermal desalination, efforts are made toward employing low-grade heat as well as waste heat as energy input source for desalination, use of optimum chemical reactions, and scaling up of the plant to increased capacity.[2] In the field of membrane desalination, novel pre-treatment methods are employed such as ultrafiltration (UF), use of energy recovery devices to cut down its cost, and use of high-quality membranes to improve its life.[3] Recently, work is mainly concentrated in the area of hybrid desalination for developing compact, economic, and potable equipment useful for process industries.[4] In the present and future, the scarcity of clean as well as safe drinking water will continue to become as major global problem. In order to assuage these issues, water-purification technologies are needed to be updated. To fulfill the current demand of clean water, different water treatment processes were recommended and adopted.[5] These processes commonly fall into three main categories such as primary (screening, filtering, centrifuge, segregation, settling, coagulation, and flocculation), secondary (aerobic treatment and anaerobic treatment), and tertiary (distilling, crystallization, evaporation, extraction of solvent, oxidation, precipitation, exchange of ions, reverse osmosis [RO], nanofiltration [NF], ultrafiltration [UF], microfiltration [MF], adsorption, electrodialysis, and electrolysis). However, the most of these processes are inefficient to arrest the water pollutants in efficient manner. Some of the processes are energy consuming and operationally expensive, hence these are unaffordable at a commercialized scale. The adsorption processes are simple as well as easier in nature but ineffective to desalinate the saline water.[6]

Membrane technology is an effective-integrated approach for the water treatment and it is reusable. It helps to arrest various types of contaminants and permits the utilization of unconventional water sources (e.g., seawater, wastewater, and brackish water). Also, it requires less floor space and use of chemicals, exhibits high level of automation, and its modular structure is flexible for different scaled systems.[7] The efficiency of the membrane mostly depends upon the membrane material, which tolerates an inalienable trade-off in between the solvent inter-permeation and solute exclusion.[8] In the recent, various membrane separation techniques like RO, UF, NF, MF, distillation, electrodialysis, and dialysis are exists in the market. Out of these, RO, UF, NF, and MF are the pressure-driven membrane processes which have been often used in water treatment systems.[9] Brackish and saline water are desalinated-employing RO process. Approximately, 19% of the world's desalination plants are operating successfully with the installation of the RO membrane.[10]

Most recently, researchers are finding that carbon nanotubes (CNTs) are extremely promising for applications in separation sciences. Currently, the use of nanostructured materials such as CNTs, zeolites, metal and metal oxide nanoparticles, and dendrimers proved to be an impressive breakthrough in the area of water purification.[11] However, the testimonials set forth by CNTs are significant to researchers to produce next-generation membranes comprised with high selectivity, high flux, less fouling (hydrophobic, hydrophilic as well as biofouling).[12] The CNTs possess the capability to control the trade-off between selectivity and flux that associated with the membrane. In order to enhance the membrane efficiency, the CNTs nanoscale structures required to be arranged in well-defined configurations to build the complete integrated systems. These nanotubes are reusable; possess high flux and exhibit chemical selectivity through suitable chemical functionalization.[13] These benefits of CNTs make them a high potential candidate in separation technology. The CNTs are capable to remove arsenic, fluoride, heavy metals, and toxic organic components.[14]

CNT permits water molecules to move in a single chain with a phenomenal permeation percentage, having a magnitude higher than the prediction obtained using Hagen–Poiseuille equation. The water is selected based on the CNT diameter and its chemistry, its specific size, shape, and chemical structure of aquaporin.[15] Inclusion of vertically aligned CNTs and aquaporin inside the membrane matrices, with little quantity, gives flux greater than the existing commercial seawater RO membranes. The high flux may not remarkably improve the energy consumption but it can significantly decrease the footprints and expenditure on wastewater treatment.[16]

8.2 CHARACTERISTICS OF CNTs

8.2.1 CNT FILTERS AS AN ANTIMICROBIAL MATERIALS

Each year, millions of people pass away due to water-borne diseases. In this matter, the elimination of organisms and bacteria presents in the drinking water is very important aspect. CNTs, specifically single-walled carbon nanotubes (SWCNTs), possess strong antimicrobial properties.[17] The direct contact with SWCNTs leads to severe damage to the membrane and afterward results in cell death. SWCNT filter demonstrated high-bacterial retention whereas the MWCNT filter demonstrated a great potential of viral removal at low pressure. The combination of SWCNT–MWCNT hybrid filter helps to achieve effective inactivation of bacteria and at the same time viral retention at low pressure.[18]

The antibacterial ability of a CNT filter was remarkably improved by combining it with silver nanoparticles or silver nanowires. The viral removal capacity of a CNT filter was significantly enhanced with the use of external electric field because of increased viral particle transport. The reported CNT filters have been fabricated with simple filtration of CNT suspensions through a substrate possessing a porous structure.[19] Srivastava and his co-workers[20] prepared a monolithic hollow cylinder having uniform macroscopic nature with radially aligned CNT walls employing a continuous spray pyrolysis method. This helped to achieve a uniform nanoporous structure, which proved to be beneficial for filtration since it had very few blockages. The cylinder also possesses good thermal stability as well as higher mechanical strength. In comparison to conventional membrane filters, the CNT filter exhibited a major advantage of frequently cleaning using simple ultrasonication and autoclaving. After regeneration, these filters retain their full filtering efficiency.

8.2.2 CNTs AS AN ADDITIVES FOR ANTIFOULING MEMBRANES

In case of polymer-based membranes, fouling mainly depends upon the interaction between foulants and membrane surface (i.e., associated with the foulants properties and morphology of the membrane as well as its chemistry). The membrane fouling can be effectively controlled by tuning the membrane surface chemistry.[21] It is well proven that high hydrophilic surfaces provide a better fouling resistance since most of the organic foulants are hydrophobic in nature. The CNTs are also hydrophobic in nature and could be converted to hydrophilic through acid treatment. It was observed that polysulfone and polyethersulfone-blended CNT membranes are highly hydrophilic in nature.[22] The functionalized CNTs employing hydrophilic carboxylic groups exhibited improvement in the fouling resistance. Apart from these hydrophilic isophthaloyl chloride groups,[23] amphilic-polymer functional groups possessing protein-resistant ability[24] demonstrated improvement in the antifouling properties of CNTs. The CNT material properties show significant improvement in the performance of CNT-blended membranes. The experimental findings demonstrated that construction of CNT-membrane using larger diameter CNTs exhibits superior fouling control compared as compared to smaller diameter CNTs. This is because of the superior ability of larger diameter CNTs toward the removal of great extent of organic foulants. Biofouling, the biofilm growth onto the membrane surfaces also can be reduced with the implementation of CNTs. The introduction of CNTs to polymer membranes helps to pass on the biocidal properties to the membrane due to its antibacterial properties, resulting in a better biofouling resistance.[25] The blending of CNTs with conventional polymer membranes provides enhanced tensile strength as compared to pristine membranes. The poly-acrylonitrile and chitosan membranes blended with 2 wt.% MWCNTs results into 97 and 90%, increase in tensile strength, respectively, compared to the plane membranes.[26] Beside, CNTs lead to improve the permeability of water of the polymer membrane. Surface-oxidized MWCNTs can be introduced effectively within the coating layer to develop a high-flux filtration membrane. As the quantity of MWCNTs is increasing, improvement in the water flux has been observed with reduction in the total organic rejection.[27] The improvement in the flux rate leads to production of highly efficient hydrophilic nano-channels for the water passage inside the composite membranes.[28]

8.2.3 ALIGNED CNT MEMBRANES FOR FUTURISTIC SEAWATER (SALINE WATER) DESALINATION

In the near future, seawater desalination will take an attention as an important alternative to mitigate the intensity of global water crisis. The present desalination depends on the polymer RO membranes for restricting and filtering out fine solids and dissolved salts. For this, excessive pressure and more energy is required to carry out the desalination process. In this case, trade-off takes place between water flux and selectivity associated with the design of membrane.

The movement of media throughout the membrane depends on various driving forces. The processes are classified as equilibrium based, non-equilibrium based, pressure driven, and non-pressure driven processes.[9] The summary of few of these techniques associated to their driving forces is schematically represented with Figure 8.1.

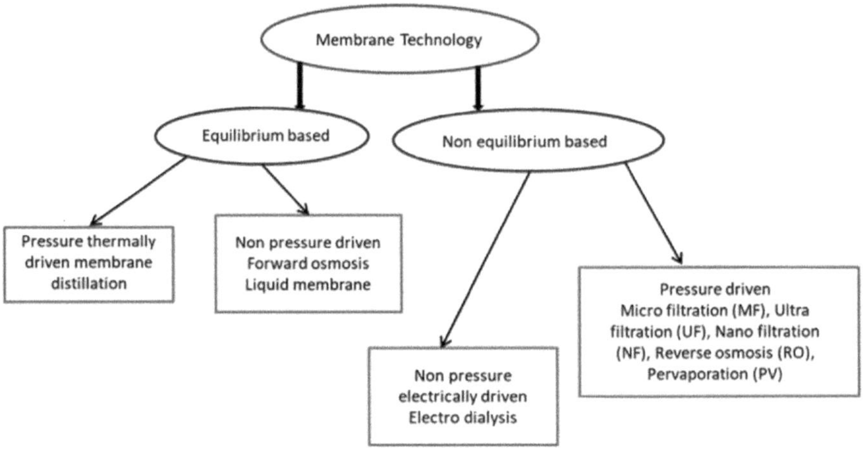

FIGURE 8.1 Schematic representation of few membrane processes.

The published theoretical and experimental investigation proved that the transport of water is faster in CNTs compared to other porous materials. This is due to the smooth as well as hydrophobic inner walls of CNTs. Aligned CNT membrane possesses the ability to execute as high-flux desalination membrane. Holt and his co-workers[29] fabricated aligned nanotube membranes employing catalytic chemical vapour deposition

(CVD) followed with laser etching in order to decapsulate the nanotubes. The water permeability of these membranes found significantly higher compared to the existing commercial polycarbonate membranes; despite they possess small pore sizes.[30] The CNTs tips have been modified employing electrochemical grafting of diazonium salts, and the coefficient of separation of CNT membranes adjusted by changing the applied voltage. With the use of negatively charged groups developed employing plasma treatment, makes the aligned CNT membranes to refuse (as high as) 98% ions. It is observed that the ion rejection was influenced by charged rejection mechanism. However, there are chances of decline in the efficiency of charged rejection mechanism in high-ionic strength environment such as brackish water and seawater. Thus in the near future, manufacturing of fault-free CNT membranes having sub-nanometer pores plays crucial role in the upcoming future desalination technologies.[31]

8.3 FASCINATING STRUCTURE OF CNT

Carbon nanotube is a unique one-dimensional system, which can be visualized as single-rolled sheet of graphite (or graphene). This rolling can be performed at different angles and curvatures to obtain different nanotube properties. Single-walled carbon nanotubes (SWCNTs) made up of cylinders with one/more graphene (lattice) layers. SWNTs and MWNTs are having diameters in the range of 0.8–2 nm and 5–20 nm, respectively, even though MWNTs diameters may exceed 100 nm. The length of CNTs is ranging from 100nm to 0.5 m.[32] CNTs are the strongest as well as stiffest materials in respect of elastic modulus and tensile strength. CNTs exhibit useful properties such as absorption, fluorescence, and Raman spectroscopy properties. In the advanced membrane technologies, CNTs are attracting for water desalination systems, as they offer optimum energy solution for water purification systems.[33] CNT membranes exhibit very smooth water flow through them by reducing a broad spectrum of water pollutants. The inner hollow cavity of CNTs offers a higher probability for desalinating the water. The smooth hydrophobic walls, high-aspect ratios, and inner pore diameter of CNTs provide smooth transport of water molecules.[6]

Inner hollow cavities of CNTs provide high-future prospects to material scientists, engineers, and chemists to expand research with nanoscale test tube. The water molecule passes sometimes along the nanotube axis by means of bursting of clusters of hydrogen-bonded molecules. The water molecules get assembled into one-dimensional diameter-dependent structures.[34] The functionalization of CNTs exposes a novel research direction to acquire the future benefits of CNTs and to improve the separation characteristics without compromising the quality. The attachment of organic moieties results in better adherence of nanotubes in host material which provides better reinforcement of composites. The previously reported literature showed that SWNTs possess antimicrobial activity toward negative and positive bacteria; and the damages forced were assigned either to a physical interaction or oxidative stress that compromises the integrity of cell membrane.[35,36] This could help to use CNTs for inhibiting microbial attachment as well as formation of biofouling on the surfaces. Beside the inhibition and biofouling of the surface, CNTs tip could be functionalized properly to protect hydrophilic and hydrophobic fouling. The tip functionalization of CNTs helps to produce the required physicochemical properties onto surface of the membrane, which leads to fine selective elimination of contaminants based on the physicochemical interaction of the species associated with the functional group present on the CNTs tip.[37]

8.4 SYNTHESIS OF CNT MEMBRANES

Mainly, here are four approaches which are being applied to synthesize the CNT-based membranes:

1. By depositing carbonaceous materials inside the readily existing ordered porous membranes (i.e., anodized alumina). This is also called as template synthesized CNT membranes.[38]
2. Membrane depending on the interstices within the nanotubes in a vertical array of CNTs, known as CNT membrane with the denser array on the outside wall.[20]
3. Ceramic matrix followed by the opening of the CNTs tips using plasma chemistry or encapsulating vertically aligned grown CNTs with the space-filling inert polymer, or with the open-ended CNT membrane.[29]
4. Membranes comprised of nanotubes as the fillers in a polymer matrix, which is also called as mixed-matrix membranes.[21]

8.4.1 CNT MECHANISM OF POLLUTANTS AND SALT REMOVAL FROM WATER

CNTs are photosensitive and capable to generate reactive oxygen species in water. When these are activated with visible light, the amino fullerenes and folderol-generated singlet oxygen possess high selectivity about contaminants carrying rich electron moieties, permitting their degradation in water with little obstruction from background organics.[8] CNTs are substantially explored as alternatives for AC, as they successfully take out both organic as well as metal contaminants. The existing binding sites and CNT/contaminant interaction controls the adsorption of organic contaminants on the CNTs. Binding sites present on the CNTs are highly accessible as compared to AC, which carries inaccessible pores, particularly for bulky molecules.[39] The CNTs π electron rich surface serves as either electron acceptor or donor for various polar aromatics like nitro-aromatics as well as phenols. The graphitic hydrophobic surfaces are the key locations for organic adsorption. The metal ions get adsorbed firstly on the surface functional groups, which could be reversed using pH adjustment, for its reuse.[40] In order to reduce the cost, graphite oxide nanosheets, attracting as a potential candidate that can be produced by exfoliating graphite. This has emerged as substitute nano-adsorbent for organic and metal contaminants.[41] The smooth hydrophobic hollow CNTs inner core allows the regular and spontaneous passage for water molecules having little absorption. The specially aligned CNTs are of specific interest in the fabrication of CNT membranes. The pore diameter demonstrated special effects on the water passages through the membranes comprised of aligned CNTs. A prototype model of CNT membrane has been used for removal of salts and pollutants which shows the movement of water molecules from salinated water through vertically aligned CNT membrane and mixed-matrix CNT membranes (depicted in Fig. 8.2).

8.4.2 ACTIVATED CARBON (AC) FOR WATER PURIFICATION

Activated carbon (AC) is extensively utilized as an adsorbent material in both water treatment units as well as for the point of use (POU) water purification systems. AC is distinguished as a carbonaceous material which possessess a greatly porous structure. This is commonly adapted from the chemical treatment and pyrolysis of various sources such as wood, nutshells, bamboo, coal, and other organic materials.[42] The activation process has been carried out at high temperature using steam or chemical treatment, produces highly porous structure inside the carbonaceous material.[43] The adsorptive nature of AC about organic contaminants was popularly acknowledged and investigated for several years and this study shows that this material is well suited for the removal of contaminants from the contaminated water. Since AC exhibits high performance and is economical, the majority of POU water purification systems is presently consist of AC. A lot of AC POU systems function under the water pressure and flow of water along a developed carbon block, where the density of packed AC particles leads to maximum internal pore sizes, ranges from 0.5 to 10 μm. Other AC POU purification systems function with a gravity phenomenon, where the water flows over larger AC particles to facilitate the adsorption of contaminants.

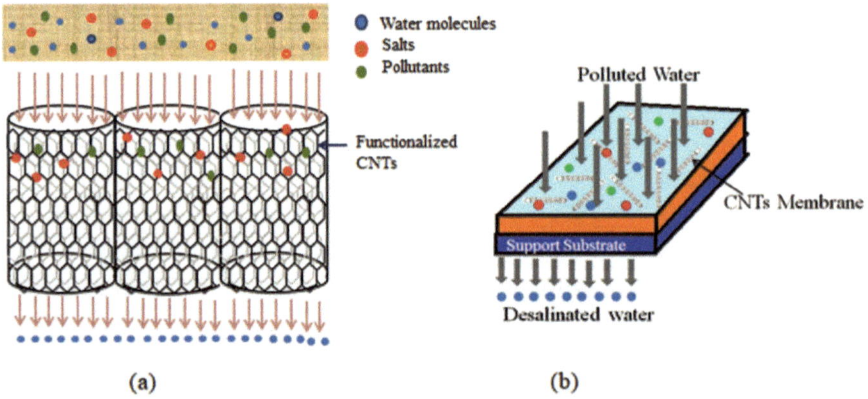

FIGURE 8.2 Prototype of CNT membrane: (a) vertically aligned CNT membrane and (b) mixed-matrix CNT membranes.

8.4.3 FUNCTIONALIZATION OF CNT

The functionalized CNT have been incorporated in the membrane to enhance the dispersity as well as to boost up CNT membrane performance. The functionalized CNT can be grouped into physical and chemical approaches. In the case of chemical functionalization, CNT is chemically bonded with the functional groups. In the case of physical methods, the modifier is wrapped and adsorbed on the CNT/stored in the CNT cavities.[44] CNT is functionalized employing different functional groups like carboxylic, amine, hydroxyl, polymer, and inorganic compounds. Acid treatment has been employed in numerous studies to introduce carboxylic groups on the CNT surface. The carboxylic group possesses hydrophilic properties, which helps to enhance dispersity in the polar organic solvent.[21] The CNT is dispersed with an acid mixture and refluxed at comparatively higher temperature for a said time period. Afterward, the CNT is washed, filtered, and dried. These acid-treated CNT utilized to develop CNT membranes or these can be subsequently chemically treated to associate other functional groups. In the case of CNTs, both tip as well as core can be functionalized as per the requirement as shown in Figure 8.3.

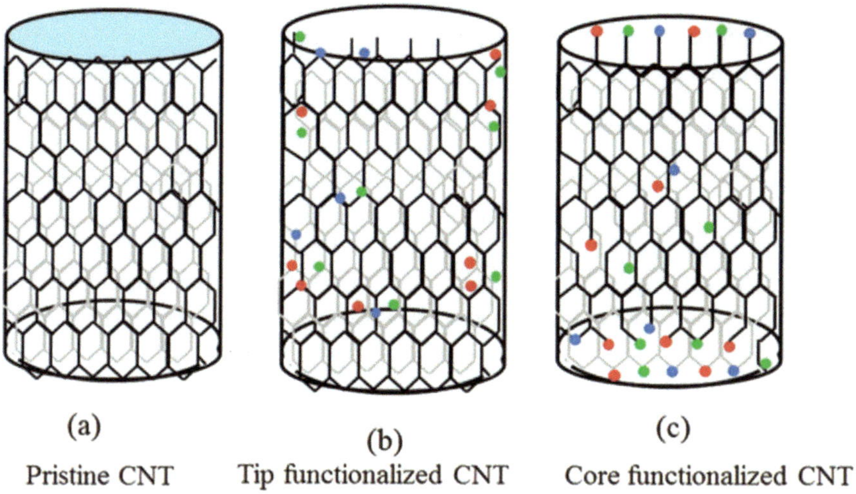

FIGURE 8.3 Functionalization of CNT membrane. (a) pristine CNT, (b) tip-functionalized CNT, and (c) core-functionalized CNT.

8.4.4 MEMBRANE FOULING (CONDITIONS) DESIRABLE POROSITY AND PORE SIZE DISTRIBUTION

Recently, membrane-based processes attaining significant importance among the other separation practices being followed. The quality of a membrane is defined in terms of flux and selectivity. An ideal membrane should generate reasonable flux without compromising the selectivity.[45] The most important drawback associated with the membrane-based processes is the membrane surface fouling caused due to microorganisms.

Due to this, the fouling is termed as biofouling, which leads to decline in the flux as well as its deterioration. Biofouling produces obstruction to the flow of water and also leads to failure of the membrane life.[46] In order to improve the anti-biofouling characteristics of the membrane surface; the nanostructured materials play a very important role. Among the present nanostructured materials, CNTs are the potential candidate. In the recent, to improve the anti-biofouling tendency polysulfone membranes, nanocomposites were fabricated impregnating with both SWCNTs and MWCNTs.[47] It has been observed that with the increase in CNT diameter, decline in the anti-biofouling performance of the membrane surface. It is assumed that the alignment of the CNTs in the membrane host matrix and the homogeneous dispersion plays a significant role in judging the improvement in anti-biofouling characteristics.[48]

Feedwater conditions, type of membrane, membrane material, process operation conditions, and membrane module configuration decide the extent of fouling. The sidewall and tip of the CNTs could be functionalized using desired chemical groups to reduce membrane fouling problems to a great extent. Lot of biofouling problem of the membranes also can be mitigated with the impregnation of CNTs. This could help the membrane to restore its natural and original performance for a longer time period and eventually leads to reduce the cost of membrane-based separation process.[49]

8.5 ENVIRONMENTAL REMEDIATION

Water contaminants like oil, pesticides, fertilizers, and pharmaceuticals have been effectively soaked by a CNT nanostructured sponge (nano-sponge) which contains iron and sulfur. Once its cleaning is done, their magnetic properties help to retrieve them easily. The iron and sulfur lead to increase the sponge size around 2 cm. It also helps to increase the porosity due to beneficial defects, producing buoyancy as well as reusability.[50] Iron, in the ferrocene structural form enables it easier to manage and allow speedy recovery with the use of magnets. Such kinds of nanosponges help to enhance the absorption capability of the toxic organic solvent (dichlorobenzene) from water by almost 3.5times. The nanosponges also can absorb vegetable oil about 150 times of their initial weight. It can also be used to absorb engine oil. Experimentally, it is observed that a magnetic boron-doped MWNT nanosponge absorb the oil from water. The sponge was developed on a substrate using CVD. Boron develops kinks and elbows inside the tubes as they evolve and fosters the formation of the covalent bonds. The results reveal that these nanosponges hold back their elastic property even after 10,000 compressions. The sponges exhibit both super hydrophobic (i.e., forcing them to present at the surface of water) and oleophilic.[51]

The CNTs demonstrate high-adsorption affinities for a widespread range of aliphatic and aromatic contaminants present in the water because of their larger hydrophobic surface areas. They demonstrated identical kind of adsorption capacities as of AC in the presence of natural organic matter. Due to these advantages, they become the promising adsorbents for the elimination of water contaminant as well as for the wastewater treatment plants. Moreover, CNT membranes are recommended as switchable molecular sieves, with permeation and sieving features which help to make them dynamically activated/deactivated with either pore size distribution or external electrostatic fields.[52] Reuse of wastewater is now becoming a usual need, even there is a source of potable water. The CNT membranes should be used in both separate wastewater collection and separation of impurities as well as supply of reusable water for the specific application.

8.6 CHALLENGES IN COMMERCIALIZATION OF CNT MEMBRANES

Even though CNT-based membranes demonstrate tremendous advantageous as an important part of desalination and water purification systems; however, there are various challenges involved with each step. These steps in the membrane fabrication start from growth of CNTs for enhancement of membrane performance and scaling the production.[53] The complex methods used for the CNTs synthesis to produce homogeneous pore size and distribution are quite difficult task. The present CVD method is often incapable for synthesizing controlled size pores with homogeneous distribution. There are difficulties in implementation of bottom up approach for synthesis of fine controlled CNT-chirality with required pore diameters and structural integrity.[54] This method is costly and time taking, proving them unfit for industrial manufacturing. Another limitation is to produce reduced pore size and CNT membranes diameter. The small size pores along with better desalination properties might be possible only for SWCNTs and DWCNTs, which have grown either on epoxy or mixed composite matrices. However, both of these are quite costly, therefore, the larger scale production and sustainable synthesis of such membranes are commercially unfeasible.[55] Twelve to thirteen times, the growth of aligned $CNTs/cm^2$ is a practical and real technological challenge. The development of nanocomposite membrane route has to ensure that CNTs are well aligned and well dispersed. To achieve this proper functionalization of CNTs needs to be done to obtain the fine dispersion in the host matrix.[46] The functionalization of CNTs associated to required functional groups is quite difficult to accomplish, for CNT is scarcely soluble in any of the solvent. In the case of aligned CNT membranes, the major crucial step is the opening of the CNT tips with the high exclusion of small contaminant molecules and salts remain present which is a difficult task.[56] This is because of the shortage of nanotubes with homogeneous sub-nanometer diameters; and the opening is either responsive to charge screening or it decreases the permeability of the membrane. Eventually, the scale-up of the membrane development is a key barrier for both technologies, that is, alignment of CNTs and generation of aquaporin are quite difficult at industrial scale. Although the transportation of ions through CNT channels has been analyzed both computationally and experimentally and noted that ion rejection is insufficient for desalination.[31] Beside the monovalent, salt rejection was not checked, since the plasma-based oxidation or acid treatment may lead to CNT wall thinning and disruption of tube integrity which in turn in failure of membrane channels. Eventually, scale-up in terms of growth of CNTs, alignment of CNTs, nanocomposite formation (tubular or flat sheet), functionalization of CNT tip or sie-wall, these all are extremely complicated and in this a lot of materials and process difficulties are involved.[57] The hydraulic conductivity depends on various parameters such as liquid slip at the solid–liquid boundary, confinement-induced reductions in the liquid viscosity, and sub-continuum changes to the liquid structure. These factors need to be taken into consideration in the case of CNT channels.

8.7 CONCLUSION AND FUTURE SCOPE

The individual membrane technology has their own advantages and limitations, the CNT membranes came into picture as an exemplary water purification technology. It has demonstrated great performances and having great potential to be commercially feasible in the near future. The commercial availability of the CNT membranes must fulfill the certain conditions like desalination capacity, water permeability, solute selection, antifouling, energy savings, robustness, material costs, and compatibility and scalability with industrial requirements. The impurities present with CNTs such as carbonaceous, ash, contaminated metal catalysts, and amorphous carbon need to be removed before its use in membrane development.

The previous study shows that salt rejection capability and water permeability of CNT membranes are estimated using computational modeling instead of experimental investigation. There is scope to perform an experimental study to investigate the CNT membrane efficiency at the operable scale. Further, the complexity for binding of functional groups to CNT membranes can be studied using a cationic membrane and anionic mixture of contaminants. The environmental safety and health issues are significant concerns which need to be determined effectively. Even though there is little possibility to escape CNTs from membranes, any deviations may own leaching of CNTs into the water and eventually in the environment. Hence, safety guidelines need to be prepared a draft of new strategy in this area.

KEYWORDS

- **antifouling**
- **carbon nanotubes**
- **CNT membrane**
- **functionalized CNTs**
- **water desalination**
- **water purification**

REFERENCES

1. Kar, S.; Bindal, R. C.; Tewari, P. K. Carbon Nanotube Membranes for Desalination and Water Purification: Challenges and Opportunities. *Nano Today* 2012, *7* (5), 385–389.
2. Esmaeilion, F. Hybrid Renewable Energy Systems for Desalination. *Appl. Water Sci.* 2020, *10* (3), 1–47.

3. Roy, S.; Ragunath, S. Emerging Membrane Technologies for Water and Energy Sustainability: Future Prospects, Constraints and Challenges. *Energies* 2018, *11* (11), 2997.
4. Atab, M. S.; Smallbone, A. J.; Roskilly, A. P. A Hybrid Reverse Osmosis/Adsorption Desalination Plant for Irrigation and Drinking Water. *Desalination* 2018, *444*, 44–52.
5. Bolisetty, S.; Peydayesh, M.; Mezzenga, R. Sustainable Technologies for Water Purification from Heavy Metals: Review and Analysis. *Chem. Soc. Rev.*, *48* (2), 463–487.
6. Das, R.; Ali, M. E.; Abd Hamid, S. B.; Ramakrishna, S.; Chowdhury, Z. Z. Carbon Nanotube Membranes for Water Purification: A Bright Future in Water Desalination. *Desalination* 2014, *336*, 97–109.
7. Figoli, A.; Criscuoli, A., Eds. *Sustainable Membrane Technology for Water and Wastewater Treatment*; Springer, 2017.
8. Qu, X.; Brame, J.; Li, Q.; Alvarez, P. J. Nanotechnology for a Safe and Sustainable Water Supply: Enabling Integrated Water Treatment and Reuse. *Accounts Chem. Res.* 2013, *46*(3), 834–843.
9. Obotey Ezugbe, E.; Rathilal, S. Membrane Technologies in Wastewater Treatment: A Review. *Membranes* 2020, *10* (5), 89.
10. Shatat, M.; Riffat, S. B. Water Desalination Technologies Utilizing Conventional and Renewable Energy Sources. *Int. J. Low-Carbon Technol.* 2014, *9* (1), 1–19.
11. Kar, S.; Tewari, P. K. Nanotechnology for Domestic Water Purification. In *Nanotechnology in Eco-Efficient Construction*. Woodhead Publishing, 2013; pp 364–427.
12. Pichardo-Romero, D.; Garcia-Arce, Z. P.; Zavala-Ramirez, A.; Castro-Munoz, R. Current Advances in Biofouling Mitigation in Membranes for Water Treatment: An Overview. *Processes* 2020, *8* (2), 182.
13. Saifuddin, N.; Raziah, A. Z.; Junizah, A. R. Carbon Nanotubes: A Review on Structure and Their Interaction with Proteins. *J. Chem.* 2013, 2013.
14. Kar, S.; Bindal, R. C.; Prabhakar, S.; Tewari, P. K.; Dasgupta, K.; Sathiyamoorthy, D. Potential of Carbon Nanotubes in Water Purification: An Approach towards the Development of an Integrated Membrane System. *Int. J. Nucl. Desalination* 2008, 3(2), 143–150.
15. Tajiri, T.; Matsuzaki, R.; Shimamura, Y. Simulation of Water Impregnation through Vertically Aligned CNT Forests Using a Molecular Dynamics Method. *Sci. Rep.* 2016, *6*, 32262.
16. Wang, R.; Chen, D.; Wang, Q.; Ying, Y.; Gao, W.; Xie, L. Recent Advances in Applications of Carbon Nanotubes for Desalination: A Review. *Nanomaterials* 2020, *10* (6), 1203.
17. Kang, S.; Pinault, M.; Pfefferle, L. D.; Elimelech, M. Single-Walled Carbon Nanotubes Exhibit Strong Antimicrobial Activity. *Langmuir* 2007, *23* (17), 8670–8673.
18. Brady-Estevez, A. S.; Schnoor, M. H.; Kang, S.; Elimelech, M. SWNT– MWNT Hybrid Filter Attains High Viral Removal and Bacterial Inactivation. *Langmuir* 2010, *26* (24), 19153–19158.
19. Rahaman, M. S.; Vecitis, C. D.; Elimelech, M. Electrochemical Carbon-Nanotube Filter Performance toward Virus Removal and Inactivation in the Presence of Natural Organic Matter. *Environ. Sci. Technol.* 2012, *46* (3), 1556–1564.
20. Srivastava, A.; Srivastava, O. N.; Talapatra, S.; Vajtai, R.; Ajayan, P. M. Carbon Nanotube Filters. *Nat. Mater.* 2004, *3* (9), 610–614.

21. Choi, J. H.; Jegal, J.; Kim, W. N. Fabrication and Characterization of Multi-Walled Carbon Nanotubes/Polymer Blend Membranes. *J. Membr. Sci.* 2006, *284* (1–2), 406–415.
22. Celik, E.; Park, H.; Choi, H.; Choi, H. Carbon Nanotube Blended Polyethersulfone Membranes for Fouling Control in Water Treatment. *Water Res.* 2011, *45* (1), 274–282.
23. Qiu, S.; Wu, L.; Pan, X.; Zhang, L.; Chen, H.; Gao, C. Preparation and Properties of Functionalized Carbon Nanotube/PSF Blend Ultrafiltration Membranes. *J. Membr. Sci.* 2009, *342* (1–2), 165–172.
24. Liu, Y. L.; Chang, Y.; Chang, Y. H.; Shih, Y. J. Preparation of Amphiphilic Polymer-Functionalized Carbon Nanotubes for Low-Protein-Adsorption Surfaces and Protein-Resistant Membranes. *ACS Appl. Mater. Interf.* 2010, *2* (12), 3642–3647.
25. Tiraferri, A.; Vecitis, C. D.; Elimelech, M. Covalent Binding of Single-Walled Carbon Nanotubes to Polyamide Membranes for Antimicrobial Surface Properties. *ACS Appl. Mater. Interf.* 2011, *3* (8), 2869–2877.
26. Tang, C.; Zhang, Q.; Wang, K.; Fu, Q.; Zhang, C. Water Transport Behavior of Chitosan Porous Membranes Containing Multi-Walled Carbon Nanotubes (MWNTs). *J. Membr. Sci.* 2009, *337* (1–2), 240–247.
27. Majeed, S.; Fierro, D.; Buhr, K.; Wind, J.; Du, B.; Boschetti-de-Fierro, A.; Abetz, V. Multi-Walled Carbon Nanotubes (MWCNTs) Mixed Polyacrylonitrile (PAN) Ultrafiltration Membranes. *J. Membr. Sci.* 2012, *403*, 101–109.
28. Zhao, Y.; Xu, Z.; Shan, M.; Min, C.; Zhou, B.; Li, Y.; Li, B.; Liu, L.; Qian, X. Effect of Graphite Oxide and Multi-Walled Carbon Nanotubes on the Microstructure and Performance of PVDF Membranes. *Separat. Purific. Technol.* 2013, *103*, 78–83.
29. Holt, J. K.; Park, H. G.; Wang, Y.; Stadermann, M.; Artyukhin, A. B.; Grigoropoulos, C. P.; Noy, A.; Bakajin, O. Fast Mass Transport through Sub-2-nanometer Carbon Nanotubes. *Science* 2006, *312* (5776), 1034–1037.
30. Yu, M.; Funke, H. H.; Falconer, J. L.; Noble, R. D. High Density, Vertically-Aligned Carbon Nanotube Membranes. *Nano Lett.* 2009, *9* (1), 225–229.
31. Fornasiero, F.; Park, H. G.; Holt, J. K.; Stadermann, M.; Grigoropoulos, C. P.; Noy, A.; Bakajin, O. Ion Exclusion by Sub-2-nm Carbon Nanotube Pores. *Proc. Natl. Acad. Sci.* 2008, *105* (45), 17250–17255.
32. De Volder, M. F.; Tawfick, S. H.; Baughman, R. H.; Hart, A. J. Carbon Nanotubes: Present and Future Commercial Applications. *Science* 2013, *339* (6119), 535–539.
33. Aqel, A.; Abou El-Nour, K. M.; Ammar, R. A.; Al-Warthan, A. Carbon Nanotubes, Science and Technology Part (I) Structure, Synthesis and Characterisation. *Arab. J. Chem.* 2012, *5* (1), 1–23.
34. Hirsch, A.; Vostrowsky, O. Functionalization of Carbon Nanotubes. In *Functional Molecular Nanostructures*; Springer: Berlin, Heidelberg, 2005; pp 193–237.
35. Tewari, P. Nanocomposite Membranes in Water Treatment. *Nanocomp. Membr. Technol.* 2015.
36. Dasgupta, K.; Kar, S.; Venugopalan, R.; Bindal, R. C.; Prabhakar, S.; Tewari, P. K.; Bhattacharya, S.; Gupta, S. K.; Sathiyamoorthy, D. Self-Standing Geometry of Aligned Carbon Nanotubes with High Surface Area. *Mater. Lett.* 2008, *62* (12–13), 1989–1992.
37. Liu, X.; Zhang, S.; Pan, B. Potential of Carbon Nanotubes in Water Treatment. *Recent Progr. Carbon Nanotube Res.* 2012, *201110* (5772), 51332.

38. Miller, S. A.; Young, V. Y.; Martin, C. R. Electroosmotic Flow in Template-Prepared Carbon Nanotube Membranes. *J. Am. Chem. Soc.* 2001, *123* (49), 12335–12342.
39. Pan, B.; Xing, B. Adsorption Mechanisms of Organic Chemicals on Carbon Nanotubes. *Environ. Sci. Technol.* 2008, *42* (24), 9005–9013.
40. Rao, G. P.; Lu, C.; Su, F. Sorption of Divalent Metal Ions from Aqueous Solution by Carbon Nanotubes: A Review. *Separat. Purific. Technol.* 2007, *58* (1), 224–231.
41. Gao, W.; Majumder, M.; Alemany, L. B.; Narayanan, T. N.; Ibarra, M. A.; Pradhan, B. K.; Ajayan, P. M. Engineered Graphite Oxide Materials for Application in Water Purification. *ACS Appl. Mater. Interf.* 2011, *3* (6), 1821–1826.
42. Ando, N.; Matsui, Y.; Kurotobi, R.; Nakano, Y.; Matsushita, T.; Ohno, K. Comparison of Natural Organic Matter Adsorption Capacities of Super-Powdered Activated Carbon and Powdered Activated Carbon. *Water Res.* 2010, *44*(14), 4127–4136.
43. Yahya, M. A.; Al-Qodah, Z.; Ngah, C. Z. Agricultural Bio-Waste Materials as Potential Sustainable Precursors Used for Activated Carbon Production: A Review. *Renew. Sustain. Energy Rev.* 2015, *46*, 218–235.
44. Ma, P. C.; Siddiqui, N. a.; Marom, G.; Kim, J.-K. Dispersion and Functionalization of Carbon Nanotubes for Polymer-Based Nanocomposites: A Review. *Compos. Part A Appl. Sci. Manuf.* 2010, *41* (10), 1345–1367.
45. Nguyen, T.; Roddick, F. A.; Fan, L. Biofouling of Water Treatment Membranes: A Review of the Underlying Causes, Monitoring Techniques and Control Measures. *Membranes* 2012, *2* (4), 804–840.
46. Kar, S.; Subramanian, M.; Pal, A.; Ghosh, A. K.; Bindal, R. C.; Prabhakar, S.; Nuwad, J.; Pillai, C. G. S.; Chattopadhyay, S.; Tewari, P. K. Preparation, Characterisation and Performance Evaluation of Anti-Bio-Fouling Property of Carbon Nanotube-Polysulfone Nanocomposite Membranes. *AIP Conf. Proc.* 2013, *1538* (1), 181–185.
47. Pandele, A. M.; Serbanescu, O. S.; Voicu, S. I. Polysulfone Composite Membranes with Carbonaceous Structure. Synthesis and Applications. *Coatings* 2020, *10* (7), 609.
48. Lee, J.; Jeong, S.; Liu, Z. Progress and Challenges of Carbon Nanotube Membrane in Water Treatment. *Crit. Rev. Environ. Sci. Technol.* 2016, *46* (11–12), 999–1046.
49. Kucera, J. Biofouling of Polyamide Membranes: Fouling Mechanisms, Current Mitigation and Cleaning Strategies, and Future Prospects. *Membranes* 2019, *9* (9), 111.
50. Amin, M. T.; Alazba, A. A.; Manzoor, U. A Review of Removal of Pollutants from Water/Wastewater Using Different Types of Nanomaterials. *Adv. Mater. Sci. Eng.* 2014, *2014*.
51. Camilli, L.; Pisani, C.; Gautron, E.; Scarselli, M.; Castrucci, P.; D'Orazio, F.; Passacantando, M.; Moscone, D.; De Crescenzi, M. A Three-Dimensional Carbon Nanotube Network for Water Treatment. *Nanotechnology* 2014, *25* (6), 065701.
52. Fasano, M.; Chiavazzo, E.; Asinari, P. Water Transport Control in Carbon Nanotube Arrays. *Nanoscale Res. Lett.* 2014, *9* (1), 559.
53. Ihsanullah, 2019. Carbon Nanotube Membranes for Water Purification: Developments, Challenges, and Prospects for the Future. *Separat. Purific. Technol.* 2019, *209*, 307–337.
54. Prasek, J.; Drbohlavova, J.; Chomoucka, J.; Hubalek, J.; Jasek, O.; Adam, V.; Kizek, R. Methods for Carbon Nanotubes Synthesis. *J. Mater. Chem.* 2011, *21* (40), 15872–15884.

55. Khalid, A.; Al-Juhani, A. A.; Al-Hamouz, O. C.; Laoui, T.; Khan, Z.; Atieh, M. A. Preparation and Properties of Nanocomposite Polysulfone/Multi-Walled Carbon Nanotubes Membranes for Desalination. *Desalination* 2015, *367*, 134–144.
56. Li, S.; Liao, G.; Liu, Z.; Pan, Y.; Wu, Q.; Weng, Y.; Zhang, X.; Yang, Z.; Tsui, O. K. Enhanced Water Flux in Vertically Aligned Carbon Nanotube Arrays and Polyethersulfone Composite Membranes. *J. Mater. Chem. A* 2014, *2* (31), 12171–12176.
57. Noy, A.; Park, H. G.; Fornasiero, F.; Holt, J. K.; Grigoropoulos, C. P.; Bakajin, O. Nanofluidics in Carbon Nanotubes. *Nano Today* 2007, *2* (6), 22–29.

CHAPTER 9

Sensing of Environmentally Hazardous Chemicals by Carbon Nanotubes

SIJO FRANCIS[1], REMYA VIJAYAN[2], EBEY P. KOSHY[1], and BEENA MATHEW[2]

[1]*St. Joseph's College, Moolamattom, Kerala, India*

[2]*School of Chemical Sciences, Mahatma Gandhi University, Kottayam, Kerala, India*

[*]*Corresponding author. E-mail: srsijofrancis@gmail.com*

ABSTRACT

Environmental contamination by heavy metals, pesticide residues, dye staffs, and other hazardous chemicals is a serious threat to public health. Carbon nanotubes are allotropic forms of carbon which are used for the pollution remediation mission. Carbon nanotubes (CNTs), CNT-nanocomposite, CNT–bimetallic hybrid nanocomposites, etc. can be used to detect and estimate these environment polluting chemicals. CNT-based electrochemical sensors are well known for their sensing applications. Voltammetric detections of heavy metals are done to sense the presence of poisonous chemicals in air, soil, and water. Electrochemical biosensors with various interfaces like nanoparticles, CNTs, and metal oxides are found in the literature. Polymer-based CNT composites are efficient candidates for environmental remediation. This chapter briefly narrates the role of carbon nanotubes in sensing environmentally hazardous chemicals.

9.1 INTRODUCTION

Chemicals that cause physical or biological harm are called hazardous chemicals. They include acids, phenolics, polyaromatic hydrocarbons, pesticides, dyes, heavy metals, volatile organic chemicals, and chlorinated organic matter, etc. Polynuclear aromatics like phenanthroline are used to detoxify algae *Pseudokirchneriella subcapitata*.[1]

Dyes are generally coloring materials and they are used routinely as analytical reagents. Methyl orange, methyl red, methylene blue, phenolphthalein, etc. are the common examples. They are normally nonbiodegradable. Industrial outlet produces tremendous amount of colored matter as waste as well as reaction by-products. Mining processes produce plenty of organic waste which is harmful to the environment. Drugs and personal care products like ibuprofen, triclosan, etc. have become polluting chemicals because of over usage. SWCNTs are a better option to remove these drugs from the environment.[2]

Waste from industries, laboratories, factories, lathe, etc. contributes to major portion of contaminant in the environment. Pesticides, herbicides, fungicides, insecticides, fertilizers, etc. are used for different purposes in agriculture industry. Pesticides are used in agriculture to destroy pests. For the production of crops in good yield with less effort, fertilizers are used. In order to controlthe diseases as well as to inhibit the growth and multiplication of weeds, and crop-harming microorganisms many chemicals are used. Many of them are nonbiodegradable and contain nitro functional groups. Biological waste produces many polluting gases like methane which are dangerous to both animals and human beings. They consume the dissolved oxygen in water which makes the aquatic life horrible. Microorganisms that are useful in agriculture may be killed by the over usage of chemicals in fields. Weeds may grow incessantly as a result of implementation of fertilizers in abnormal rates leading to eutrophication due to over nutrition. Algal bloom destroys the harmony of aquatic ecosystem.

Environmental contamination by heavy metals is a serious threat to public health. Heavy metal pollutants are non-biodegradable and hence they last for a long life time in the environment. Most of the heavy metal ions are carcinogenic due to the production of free radicals. Lead from lead batteries, paints, and PVC pipes can be toxic to kidneys, and causes Alzheimer's disease and senile dementia, etc. Cadmium from paints, pigments, and photoconductors causes renal toxicity and pulmonary disorders. Mercury from natural and manmade sources affects skin, kidneys, nerves, and digestive tract. Arsenic produced mainly from fertilizers and pesticides adversely affect the central nervous system. Chromium used for tanning in leather industry may adversely affect the reproductive systems.

Zinc from cosmetics and pigments negatively affect the respiratory system. Heavy metal copper produces many allergies. The toxicity mechanism of heavy metal ions is through their unwanted interaction with regular proteins in unnecessary sites, inhibition of action of enzyme, and impaired antioxidant metabolism. Free radical generation may cause DNA damage and lipid oxidation. Destruction of antioxidant defense mechanism may also happen.[3]

These environmental problems are to be effectively addressed in a green and eco-friendly manner. Zeolites, clay materials, biochar, etc. are used for the removal of these hazardous chemicals. Low cost and efficiency in the removal of hazardous contaminants are the need of the hour. In this chapter, we highlight the sensing of environmentally hazardous chemicals using carbon nanotubes (CNTs).

9.2 CNT BASICS

One of the allotropic forms of carbon is called CNT. CNTs are formed by rolling graphitic sheets with large length-to-diameter ratio. CNT has two models – single-walled carbon nanotubes (SWCNTs) and multiwalled carbon nanotubes (MWCNTs). SWCNTs have a single rolled graphene sheet with diameters of 0.4–2 nm and length up to 20 cm. MWCNTs have several SWCNTs rolled together and have diameter of 100 nm. The distance between the layers in MWCNTs is approximately 0.33 nm. CNTs possess distinctive mechanical, thermal, and electrical properties. They have large surface area, light weight, and high porosity. They show good interactions to attract contaminants. The mechanism of removal of contaminants may be through adsorption at the surface of the CNTs.[4] The syntheses of CNTs include arc discharge, laser ablation, chemical vapor deposition, etc.

(a) SWCNT (b) MWCNT

9.3 SENSING OF CHEMICALS

Sensor technology is employed to collect some hidden information in the nature using a characteristic property of materials. Carbon nanotube-based biosensors are used for the detection of biomolecules and material structures and are important in health care and clinical medicine. CNT biosensors have high sensitivity, fast response time, lower redox potential, high stability, and longer life time. The hollow structure of CNTs is well enough to adsorb enzymes. Functionalized CNTs can act as more efficient biomolecules or bioreceptors for proteins, oligonucleotides, and micro-organisms. Functionalization of CNTs can be done by nanomaterials, polymers, DNA, proteins, etc. Enzyme-fixed CNT electrodes are used to study redox reactions in different medium.

9.3.1 BIOMOLECULES

CNTs possess inherent luminescence properties and can be used as optical biosensor in the sensing of biomolecules using IR light. CNT optical sensor is again used for the detection of glucose levels in human body. Simultaneous detection of ascorbic acid, dopamine, uric acid, and folic acid can be done using gold nanoclusters from human urine real samples by the method of electrochemical oxidation cyclic voltammetry.[5] Carbon nanotubes decorated by iron atoms are sensitive and selective for CO, CO_2, and H_2O molecules. By using theoretical methods, the CO_2 gas sensing on armchair CNTs is found better than that on zigzag CNTs by the mechanism of chemisorptions.[6] Electrochemical biosensors are classified into three types, potentiometric-based, amperometric-based, and impedimetric-based. Electrochemical DNA biosensor based on silver nanoparticles/poly(trans-3-(3-pyridyl) acrylic acid)/multiwalled carbon nanotubes is used for sensing Adriamycin, an electroactive indicator and a drug used for different types of cancers. The composite film is composed of metal nanoparticles-CNTs–conductive polymeric film which is a sensitive DNA sensor.[7] High electrical conductivity is the peculiarity of the synthesized nanocomposite and this has high biocompatibility.[8] DNA sensor, a biosensor which allows the sensing of oligonucleotide DNA by a CNTs–conductive polymeric film using differential pulse voltammetry (DPV) and the electroactive indicator adriamycin.[7] Piezoresistive effect of CNTs can be utilized by the piezoresistive CNTs-based sensors.[9]

9.3.2 METHANOL FUEL CELLS

Methanol fuel cells are devices used for the direct conversion of chemical energy to electricity by utilizing the renewable fuel methanol. The electrochemical energy conversion systems in which CNTs are incorporated enhance electrocatalytic activity significantly. MWCNT and SWCNT are the supporting carbon materials for Pt catalysts and act as efficient anode materials in methanol fuel cells. Among the two, SWNTs significantly alter the morphology and the electronic structure of the nanoparticles compared with MWNTs and their cyclic voltammetry measurement gives the highest current density and offers the better stability to the catalyst for methanol electrooxidation.[10] We can see that the distribution of nanoparticles on CNTs strongly alters the structure according to the pretreatment and surface properties of the CNTs. Functionalization of CNTs by conducting polymers like polyindole, modify the hybrids of nanoparticles preserving their intrinsic properties. Surface studies by electron spectroscopy for chemical analysis (ESCA) shows polyindole on MWCNT favor the homogeneous dispersion of uniform-sized platinum nanoparticles.[11] Instead of commercial Pt-based cathode catalysts, nitrogen-doped MWCNT can be used as cathode for direct methanol fuel cells and this metal-free cathode catalyst displays higher peak power density than the available carbon-supported Pt catalyst.[12] Palladium nanoflowers deposited on MWCNT decorated with polypyrrole are also good electrocatalysts with better stability and anodic performance for liquid fuel cells.[13] Pt–Polypyrrole–MWCNT-nanocomposite acts as catalyst for methanol oxidation examined using cyclic voltammetry.[13] Enzymes like nicotinamide adenine dinucleotide, glucose oxidase, aflatoxin oxidase, cholesterol oxidase, urease, lactic acid oxidase, horseradish peroxidase, etc. are used to detect different analytes. CNTs-based biosensors for the quantitative and qualitative detection of glucose are in worldwide usage.

9.3.3 PESTICIDES

Pesticides used in agriculture may cause vulnerable impacts on the environment and they are considered as priority pollutants. In drinking water, the allowed limit of single active component is 0.1 μg/L and the cumulative effect of all ingredients must be less than 0.5 μg/L. Based on the nature of pest that are controlled by pesticides, they are classified as insecticides, herbicides, and fungicides. Dichlorodiphenyltrichloroethane and benzene hexachloride are banned in many countries because of serious health impacts. CNTs are used to reduce negative involvement of pesticides in the environment. Chromophore in the pesticides can be adsorbed or the activity of specific functional groups may be destroyed by the carbon derivatives. Organochlorine, organophosphorus, carbamate, etc. are the major functional groups in pesticides.[14]

Among the pesticides, organophosphorus pesticides are most common. Organochlorines make them highly efficient but cause bioaccumulation as well. α-Hexachloro cyclohexane, chlorpyrifos, methyl paraoxon, methyl parathion, and ethyl paraoxon are some examples of pesticides. Heterogenous photocatalysis arise with less hazardous by-products and also because of the high efficiency of photocatalysts for the environmental purification of waste waters. Degradation of organopesticidesin the presence of light follow first-order kinetics with Langmuir adsorption protocol.[15] Organophoshate pesticides like paraoxon can be detoxified using imidazole group functionalized magnetic carbon nanotube catalyst. Separation and recycling of the catalyst can be easily performed owing to the magnetic behavior of the catalyst. Covalent functionalized imidazole has the capacity to cleave P–O bond in the organophosphate pesticide.[16] Hydrolysis of organophosphorus pesticides like parathion or paraoxon derivatives toward p-nitrophenol can be established by utilizing chemical species like amino acids, serine, histamine, and glutamic acid that are bound to TiO_2. Following the electrochemical techniques, we were able to separate these toxic compounds from real-time samples within 10 min.[17]

Silver-based CNT compounds stimulate the activity of visible light and thus degradation of pesticides, preferably organophosphate pesticides, are availed by modern carbon-based functional materials.[14] Membrane reactors are used to remove liquid effluents by the catalytic degradation of 2,4-dichlorophenoxyacetic acid, a pesticide used in agriculture.[18] The pesticide diuron can be removed by CNT at pH 7.[19] Chlorine-based pesticide chlordane can be removed with the help of SWCNT. Pesticide-contaminated soil and water in urban areas causing waste water and sewage sludge are filled with these pollutants. A range of pesticide residues can be removed by CNT-based technologies. Functionalization improves the activity of CNTs. Pesticide recovery depends on the external diameter and length of the CNTs.[20] Diquat dibromide removal by CNT and the mechanism of uptake of the pesticide on the surface of carbonaceous materials is by hydrophobic interactions, π–π interactions.[21]

9.3.4 BIOTOXINS

Harmful algal blooms are the effects of eutrophication and global warming. Water-assisted chemical vapor deposition method is good for the preparation of multiwalled CNTs and is used as a biosensor for Microcystin-LR. Oxygen carrying functional groups when added to MWCNT give good result for the immobilization of different biomolecules.[22] C-N-TiO_2 system that is active in visible light is also efficient to remove hepatotoxin microcystin-LR by a catalytic strategy.[23]

9.3.5 ORGANIC DYES

Dyes are colored chemicals and are non-biodegradable in nature due to presence of aromatic chains and high molecular weight. Single-walled and multiwalled carbon nanotubes help the efficient removal of various organic dyes with the assistance of different technologies. The incorporation of functional groups with CNTs would enhance the adsorption capability of dyes. Congo red dye can be removed within 60 min by pseudo-second-order adsorption kinetics.[24] MWCNT functionalized by cysteamine composite with carboxyl group and thiol groups is used for the removal of Amido black 10B dye.[25] MWCNT acts as catalyst for the decomposition of Acid Red 14 in the presence of light. The contaminant removal by nanotechnology uses mainly the mechanism called advanced oxidation process. Adsorption on CNTs and removal of pollutants using nanostructured membranes based on CNTs are quite common.[26] The manganese-based catalysts are highly selective in the reduction of NO with NH_3. Green synthesized multilayered Fe_2O_3@MnOx@CNTs have MnOx@CNTs core-shell structure.[27]

Methyl orange (MO) is an organic dye that will cause water contamination as it is incessantly used in laboratory purposes. Recent report shows that removal of this dye in water medium is possible by the cumulative effect of ultrasound and multiwalled carbon nanotubes. The performance of the adsorbent MWCNT is enhanced by the application of ultrasonic radiation. Several parameters like pH, concentration of the adsorbent, reaction span, etc. have great influence on the dye removal process. The porous structure with high surface area enabled the removal of methyl orange with the help of ultrasound-assisted strategy within few seconds from water phase.[28]

Advance oxidation processes are used in photochemistry to break the dye molecules to small nonhazardous moieties. Various photocatalytic processes will remove harmful contaminants in wastewater and air. Absorptions of either ultraviolet or visible light photons by the semiconductor that possess appropriate energy gaps are used. MWCNTs coated with titanium oxide particles can degrade the dye Rhodamine B with 96.44% efficiency.[29]

Hazardous Azo dyes from paper and textile industry is an environmental pollutant. The azo dye Acid Red 14 can be decolorized from aqueous medium using MWCNT. The extent of degradation largely depends on H_2O_2 concentration, the catalyst and pH of the reaction. The λmax of the Acid Red at 514 nm in the UV-Vis spectra is decreased in intensity indicating the degradation of the dye. The pseudo-first-order kinetics is accomplished in the degradation reaction.[30]

Iron catalysts supported on MWCNTs were used for Fenton-like oxidation which enables the removal of organic contaminants like Acid Orange II.[26]

Ag–ZnO–CNT nanocatalysts have superior catalytic activity for the degradation of dye Acid Orange 7 than the nanocomposite Ag–ZnO and the pure semiconductor nanoparticles ZnO. The added catalytic effect is attributed to the increased separation of electron-hole pairs in the semiconductor–CNT interface. The increased number of reactive free radicals formed in the CNT–semiconductor and increased surface area also contribute to the increased photoactivity of the CNT-based material.[31] For environmental remediation, TiO_2 catalyst is used in visible light. The bimetallic hybrid nanocomposites composed of Silver–TiO_2–MCNT. CNTs–TiO_2-based heterostructures formed by adding carboxylic groups to CNTs have absorption at 400 nm and have increased photocatalytic activity.[32] Technologies stationed mainly on adsorption, catalysis (photo and nonphoto), and membranes, etc. are productively adapted in CNTs-based waste-water treatments. ZnO/CNT-nanocomposite is used for the selective detection of hydrazine in the presence of phenol from wastewater.[33]

The mechanism of dye adsorption on CNTs is the bonded and nonbonded interactions between the functional groups on the dye molecules and CNTs. They are covalent bonding, hydrogen bonding, hydrophobic interactions, $\pi - \pi$ interactions, and electrostatic interactions. Zeta potential measurements showed that a positive charge of about 10–30 mV is situated on CNTs. Thus, anionic dyes like methyl orange are more susceptible to adsorption than cationic dyes due to electrostatic attraction. There are reports of weak adsorption of cationic dyes (Methylene blue and Acid Red 183) on CNTs and are explained by $\pi - \pi$ interactions between the chromophores of two moieties. The reduction in adsorption capacity is due to mismatch of chromophore functional group and nonplanar special geometry of methylene blue and Acid Red 183, respectively. If the anionic dye has a planar geometry, the adsorption phenomenon will be enhanced multifold.

In the case of CNT nanocomposites formed from Fe_3O_4, the binding of methylene blue on pure CNT is found to be larger than the composite. Pure CNTs generally undergo agglomeration process and also they are hydrophobic in nature. On nanocomposite formation, both the effects are increased leading to lower adsorption. Iron carbide (Fe_3C)-CNT-nanocomposites also showed less methylene blue adsorption capacity than bare CNTs. While comparing nanocomposites CNT–Fe_3O_4 and CNT–Fe_3C, the magnetic CNT component is attributed toward the lower adsorption capacity.[34] Large-scale usage of CNT-based technologies is not commercially accepted because of the high-cost parameters.[35] Catalytic ozonation is used for the decomposition of ozone to hydroxyl radicals by the Fe-based materials. Fe-based catalysts are generally three types.[36] Bimetallic hybrid nanocomposites like Ag–TiO_2-CNT have been used for the removal of oxides of nitrogen and many easily volatile organic pollutants. CNT–TiO_2 combination has photocatalytic activity under visible light.[32]

9.3.6 HEAVY METALS

The heavy metal detection is a challenge for various industries, environment, and for medical fields. Heavy metals have high atomic weight and cause an increased risk to the environment. Toxic heavy metals with detrimental effects include zinc, copper, manganese, and nickel. The next class of dangerous heavy metals is cadmium, mercury, and lead. Some toxic semimetals are also known. They are arsenic and selenium. According to the US environmental protection agency, the permissible limits of cadmium, lead, and copper are 0.027, 0.0452, and 0.024 μM, respectively.

The exposure or uptake of heavy metals at abnormal concentrations may lead to adverse health effects and various diseases. This led to the development of modern and on the spot devices for their analytical detection with high accuracy. Various spectroscopic techniques like fluorescence spectrometry, mass spectrometry, atomic absorption spectrometry, and coupled plasma spectrometry, etc. are used as means for the detection of heavy metals.

Nowadays, electrochemical sensing is employed over these spectroscopic techniques because they are simple, rapid, and sensitive analytical methods.

Nondecomposable heavy metals in food, soil, and water samples can be detected through graphene and CNT-based electrochemical sensors. Premier challenge associated with electroanalysis is the electrode modification to get best sensing performance. In electrochemical sensing, carbon-based electrodes, such as glassy carbon electrode, and screen-printed carbon electrode are widely utilized for heavy metal detection. Simple and green synthesized 3D Graphene-MWCNTs-based electrochemical sensor can successfully be utilized for the detection of heavy metal ions like Zn, Cd, Pb, and Cu. MWCNTs and graphene together provide conductivity and partial hydrophilicity. Fabrication of convenient sensor electrode needs a prerequisite of high stability of the electrode film to get a cost-effective sensor.[37]

Toxic metal ions in the environment can be sensed by gold nanoparticles-based biosensing tools. Inorganic contaminates can be colorimetrically detected using unmodified and un-functionalized gold nanoparticles. The pink color of gold nanoparticles changed to purple-blue on agglomeration. The binding of metal ions on the surface causes a slight change in color. Peak broadening and shift in Surface Plasmon Resonance (SPR) peak are observed in general. SPR is a size-dependent characteristic of noble metal nanoparticles. A colorimetric sensor will also depend mainly on aggregation and dielectric constant of the system under consideration. Functionalization of gold nanoparticles is conducted for the selective binding of the targeted metal ions on their surface. The attachment of the target may be explained by mechanisms like physical adsorption, covalent coupling, or molecular recognition. As^{3+}, Hg^{2+}, Cr^{3+}, Cu^{2+},Pb^{2+}, and Cd^{2+}, etc. can be detected by GNP-based colorimetric sensors.[38]

Mad hatter syndrome and Minamata disease are caused by the heavy metal Hg due to its toxicity. The origin of pollution is both from natural and anthropogenic sources. Hg poisoning is dangerous to kids, pregnant and breast feeding women. Carbon nanotubes guaranteed the specific and sensitive detection of Hg in solution. Solvated aqueous mercury ions can be converted to methylmercury and it has the capacity to be bioaccumulated. Widely adopted fast and sensitive techniques used for finding traces of Hg measurement are Direct Mercury Analyzer and Atomic Fluorescence Spectroscopy, etc. Hybrid nanocomposites are used as electrochemical nanosensors that are developed from metal nanoparticles and CNTs. The mechanisms associated with this nanosensing of aqueous Hg^{2+} ions may be primarily adsorption and binding of it on CNTs.[39]

CNTS are efficient adsorbents for chromium removal from real water samples. Monolayer adsorption of Cr (III) and Cr (VI) are reported and the kinetic models followed are Langmuir isotherm and pseudo-second-order kinetics, respectively. pH is the main factor that controls the adsorption process and optimum pH ranges are 5–8 and 1–4 for Cr (III) and Cr(VI), sequentially.[40] Bare CNT and CNTs functionalized with –COOH are good enough to remove chromium (III) from aquatic media.[41]

9.4 CATALYSIS

CNT-based catalysts have homogeneous uniform sites on their surface. The activity of CNTs-based carbocatalysts can be enhanced by inculcating heteroatoms to it. CNTs are conducting materials that facilitate electron transport from organic material to persulfates. In N-doping, the N atom can accept electrons from C atoms of organic substrates toward the oxidants through the surface of carbon catalyst.[42]

Water contamination by various aromatic dye stuffs and phenolic substrates is a serious threat that affects globally. Immobilization of the redox enzyme, peroxidase on nanocarbon materials like CNT enhances the capacity of peroxidase-catalyzed reactions. The extraordinary perspectives of CNTs like its stability, mechanical strength, biocompatibility, and chemical inertness enabled its usage in food and pharmaceutical samples. The immobilization of biomolecules (enzymes) like laccase and lignin peroxidase on the surface of CNT makes it useful in fallow water treatments. Various phenolic compounds can be fruitfully removed on this type of immobilized carbon nanomaterials. When CNT is decorated with metal oxides, such as Fe_2O_3, Al_2O_3, TiO_2, MnO_2, etc. and on which ginger peroxidase are immobilized, it results in an efficient removal of water contaminants.[43] MWCNTs-immobilized enzyme has 93% catalytic activity than the support chitosan and dispersion of nanotubes.[44]

9.5 MEMBRANES

Membrane filters based on CNTS are highly useful in filtering various chemical and biological contaminants in water samples.[45] Nanocomposite membranes containing CNTs are useful for the filtration of chemicals like triclosan, acetaminophen, and ibuprofen from pharmaceuticals and personal care products. Here, rather than mechanical filtration, adsorptive membrane filtration dependent on size is employed. The interaction between the CNTs and pharmaceuticals and personal care products are mainly concerned here and a pseudo-second-order kinetic equation is pursued here.[46] Organic contaminants from drinking water resources can be excreted by adsorption and oxidation with the help of carbon nanotubes followed by additional oxidation by H_2O_2. Model organic compound phenol, antibiotic tetracycline, organic dye methyl orange and the taste and smell builder geosmin are studied by this in situ oxidation reaction in conjugation with CNT electrochemical filter.[47] A fourfold carbon nanotube membrane stack has good efficiency and low energy consumption at neutral pH.[48] CNT membranes immobilized by $CuFe_2O_4$ particles are used to remove As(III) under various conditions from wastewater.[49] Separation based on membrane technology is used for desalination and waste-water treatment with additional membrane properties. Outstanding properties of CNT composite membranes are due to the encouragement of adsorption by CNTs owing to their high specific area, chemical inertness, and high mechanistic characteristics.[50]

9.6 CONCLUSION

Environmental cleanup using carbon nanotubes is a modern approach for the removal of organic, inorganic, and biological contaminants. Because of the excellent adsorption properties of CNTs, water contamination due to the unwanted chemical constituents can be eliminated.

KEYWORDS

- **carbon nanotubes**
- **heavy metals**
- **nanocomposites**
- **sensing**
- **biosensors**
- **electroanalytic techniques**

REFERENCES

1. Glomstad, B., et al. Carbon Nanotube Properties Influence Adsorption of Phenanthrene and Subsequent Bioavailability and Toxicity to Pseudokirchneriella Subcapitata. *Environ. Sci. Technol.* **2016,** *50* (5), 2660–2668.
2. Sun, H., et al. Simultaneous Removal Of Nitrogen and Pharmaceutical and Personal Care Products from the Effluent of Waste Water Treatment Plants Using Aerated Solid-Phase Denitrification System. *Bioresour. Technology* **2019,** *287*, 121389.
3. Gumpu, M. B., et al. A Review on Detection of Heavy Metal Ions in Water–an Electrochemical Approach. *Sens. Actuat. B Chem.* **2015,** *213*, 515–533.
4. Ghasempour, R.; Narei, H. CNT Basics and Characteristics. In *Carbon Nanotube-Reinforced Polymers*; Elsevier, 2018 pp 1–24.
5. Abdelwahab, A. A.; Shim, T-B. Simultaneous Determination of Ascorbic Acid, Dopamine, Uric Acid and Folic Acid Based on Activated Graphene/MWCNT Nanocomposite Loaded Au Nanoclusters. *Sens. Actuat.: Chem.* **2015,** *221*, 659-665.
6. Tit, N., et al. Detection of CO2 using CNT-based sensors: Role of Fe catalyst on sensitivity and selectivity. *Mater. Chem. Phys.* **2017,** *186*, 353–364.
7. Zhang, Y., Zhang, K.; Ma, H. Electrochemical DNA Biosensor Based on Silver Nanoparticles/Poly (3-(3-pyridyl) Acrylic Acid)/Carbon Nanotubes Modified Electrode. *Analy. Biochem.* **2009,** *387* (1), 13–19.
8. Abdelwahab, A. A.; Shim, Y-B. Simultaneous Determination of Ascorbic Acid, Dopamine, Uric Acid and folic Acid Based on Activated Graphene/MWCNT Nanocomposite Loaded Au Nanoclusters. *Sens. Actuat. B Chem.* **2015,** *221*, 659–665.
9. Yang, N., et al. Carbon Nanotube Based Biosensors. *Sens. Actuat. B Chem.* **2015,** *207*, 690–715.
10. Chen, Y., et al. Electro-oxidation of Methanol at the Different Carbon Materials Supported Pt Nano-Particles. *Int. . Hydrogen Energ.* **2010,** *35* (19), 10109–10117.
11. Wang, R-X., et al. Pt Nanocatalysts on a Polyindole-Functionalized Carbon Nanotube Composite with High Performance for Methanol Electrooxidation. *J. Power Sources* **2015,** *287*, 341–348.
12. Kruusenberg, I., et al. Highly Active Nitrogen-Doped Nanocarbon Electrocatalysts for Alkaline Direct Methanol Fuel Cell. *J. Power Sources* **2015,** *281*, 94–102.
13. Fard, L. A.; Ojani, R.; Raoof, J. B. Electrodeposition of Three-Dimensional Pd Nanoflowers on a PPy@ MWCNTs with Superior Electrocatalytic Activity for Methanol Electrooxidation. *Int. J. Hydrogen Energ.* **2016,** *41* (40), 17987–17994.
14. Debnath, D.; Gupta, A. K.; Ghosal, P. S. Recent Advances in the Development of Tailored Functional Materials for the Treatment of Pesticides in Aqueous Media: A Review. *J. Indust. Eng. Chem.* **2019,** *70*, 51-69.
15. Rani, M.; Shanker, U. Effective Adsorption and Enhanced Degradation of Various Pesticides from Aqueous Solution by Prussian Blue Nanorods. *J. Environ. Chem. Eng.* **2018,** *6* (1), 1512–1521.
16. Blaskievicz, S. F., et al. Magnetic Nanocatalysts Derived from Carbon Nanotubes Functionalized with Imidazole: Towards Pesticide Degradation. *Appl. Cataly. B Environ.* **2020,** *264*, 118496.

17. Qiu, L., et al. Electrochemical Detection of Organophosphorus Pesticides Based on Amino Acids Conjugated Nanoenzyme Modified Electrodes. *Sens. Actuat. B Chem.* **2019,** *286*, 386–393.
18. Dumée, L. F., et al. Hybrid Thin Film Nano-Composite Membrane Reactors for Simultaneous Separation and Degradation of Pesticides. *J. Membr. Sci.* **2017,** *528*, 217–224.
19. Hladik, M. L.; Calhoun, D. L. Analysis of the Herbicide Diuron, Three Diuron Degradates, and Six Neonicotinoid Insecticides in Water–Method Details and Application to Two Georgia Streams. *Scientific Investigations Report 2012* **2012,** 5206.
20. Sarkar, B., et al. Sustainable Sludge Management by Removing Emerging Contaminants from Urban Wastewater Using Carbon Nanotubes. *Industrial and Municipal Sludge*. Butterworth-Heinemann, 2019; pp 553–571.
21. Dichiara, A. B.; Harlander, S. F.; Rogers, R. E. Fixed Bed Adsorption of Diquat Dibromide from Aqueous Solution Using Carbon Nanotubes. *RSC Adv.* **2015,** *5* (76), 61508–61512.
22. Han, C., et al. A Multiwalled-Carbon-Nanotube-Based Biosensor for Monitoring Microcystin-LR in Sources of Drinking Water Supplies. *Adv. Funct. Mater.* **2013,** *23* (14), 1807–1816.
23. Liu, G., et al. Enhanced Visible Light Photocatalytic Activity of CN-codoped TiO_2 Films for the Degradation of Microcystin-LR. *J. Mol. Cataly. A Chem.* **2013,** *372*, 58–65.
24. Zare, Karim, et al. Enhanced Removal of Toxic Congo Red Dye Using Multi Walled Carbon Nanotubes: Kinetic, Equilibrium Studies and Its Comparison with Other Adsorbents. *J. Mol. Liquids* **2015,** *212*, 266–271.
25. Sadegh, H., et al. Synthesis of MWCNT-COOH-Cysteamine Composite and Its Application for Dye Removal. *J. Mol. Liquids* **2016,** *215*, 221–228.
26. Divyapriya, G.; Nambi, I. M.; Senthilnathan, J. Nanocatalysts in Fenton Based Advanced Oxidation Process for Water and Wastewater Treatment. *J. Bionanosci.* **2016,** *10* (5), 356–368.
27. Cai, S., et al. Design of Multi-Shell Fe_2O_3@ MnO x@ CNTs for the Selective Catalytic Reduction of NO with NH_3: Improvement of Catalytic Activity and SO_2 Tolerance. *Nanoscale* **2016,** *8* (6), 3588–3598.
28. Ghaedi, M., et al. Removal of Methyl Orange by Multiwall Carbon Nanotube Accelerated by Ultrasound Devise: Optimized Experimental Design. *Adv. Powder Technol.* **2015,** *26* (4), 1087–1093.
29. Cotto-Maldonado, M. C., et al. Photocatalytic Degradation of Rhodamine-B under UV-Visible Light Irradiation Using Different Nanostructured Catalysts. *Chem. Sci. Int. J.* **2013,** 178–202.
30. Moradi, R.;Mahanpoor, K. Photocatalytic Degradation of Azo Dye Acid Red 14 from Aqueous Solutions Using MWCNTs Nanocatalyst. *Arch. Hygiene Sci.* **2018,** *7* (2), 71–80.
31. Moradi, M.; Haghighi, M.; Allahyari, S. Precipitation Dispersion of Ag–ZnO Nanocatalyst Over Functionalized Multiwall Carbon Nanotube Used in Degradation of Acid Orange from Wastewater. *Process Safe. Environ. Protect.* **2017,** *107*, 414–427.
32. Truppi, A., et al. Visible-Light-Active TiO2-Based Hybrid Nanocatalysts for Environmental Applications. *Catalysts* **2017,** *7* (4), 100.

33. Karimi-Maleh, H., et al. A High Sensitive Biosensor Based on FePt/CNTs Nanocomposite/N-(4-hydroxyphenyl)-3, 5-dinitrobenzamide Modified Carbon Paste Electrode for Simultaneous Determination of Glutathione and Piroxicam. *Biosens. Bioelectron.* **2014,** *60*, 1–7.
34. Tan, K. B., et al. Adsorption of Dyes by Nanomaterials: Recent Developments and Adsorption Mechanisms. *Separat. Purific. Technol.* **2015,** *150*, 229–242.
35. Sarkar, B., et al. Designer Carbon Nanotubes for Contaminant Removal in Water and Wastewater: A Critical Review. *Sci. Total Environ.* **2018,** *612*, 561–581.
36. Wang, J.; Bai, Z. Fe-Based Catalysts for Heterogeneous Catalytic Ozonation of Emerging Contaminants in Water and Wastewater. *Chem. Eng. J.* **2017,** *312*, 79–98.
37. Pandey, S. K., et al. Nanocarbon-Based Electrochemical Detection of Heavy Metals. *Electroanalysis* **2016,** *28* (10), 2472–2488.
38. Priyadarshini, E.; Pradhan, N. Gold Nanoparticles as Efficient Sensors in Colorimetric Detection of Toxic Metal Ions: A Review. *Sens. Actuat. B Chem.* **2017,** *238*, 888–902.
39. Pokhrel, L. R., et al. Novel Carbon Nanotube (CNT)-Based Ultrasensitive Sensors for Trace Mercury (II) Detection in Water: A Review. *Sci. Total Environ.* **2017,** *574*, 1379–1388.
40. Anastopoulos, I., et al. A Review for Chromium Removal by Carbon Nanotubes. *Chem. Ecol.* **2017,** *33* (6), 572–588.
41. Atieh, M. A., et al. Removal of Chromium (III) from Water by Using Modified and Nonmodified Carbon Nanotubes. *J. Nanomater.* **2010,** *2010*.
42. Lee, H., et al. Activation of Persulfates by Carbon Nanotubes: Oxidation of Organic Compounds by Nonradical Mechanism. *Chem. Eng. J.* **2015,** *266*, 28–33.
43. Jun, L. Y., et al. An Overview of Immobilized Enzyme Technologies for Dye and Phenolic Removal from Wastewater. *J. Environ. Chem. Eng.* **2019,** *7* (2), 102961.
44. Jamie, A., et al. Immobilization and Enhanced Catalytic Activity of Lipase on Modified MWCNT for Oily Wastewater Treatment. *Environ. Progress Sustain. Energ.* **2016,** *35* (5), 1441–1449.
45. Jame, S. A.; Zhou, Z. Electrochemical Carbon Nanotube Filters for Water and Wastewater Treatment. *Nanotechnol. Rev.* **2016,** *5* (1), 41–50.
46. Wang, Y., et al. Carbon Nanotube Composite Membranes for Microfiltration of Pharmaceuticals and Personal Care Products: Capabilities and Potential Mechanisms. *J. Membr. Sci.* **2015,** *479*, 165–174.
47. Liu, Yanbiao, et al. Electrochemical Wastewater Treatment with Carbon Nanotube Filters Coupled with in Situ Generated H_2O_2. *Environ. Sci. Water Res. Technol.* **2015,** *1* (6), 769–778.
48. Gao, Guandao, et al. Carbon Nanotube Membrane Stack for Flow-Through Sequential Regenerative Electro-Fenton. *Environ. Sci. Technol.* **2015,** *49* (4), 2375–2383.
49. Luan, H., et al. Adsorptive Filtration of As (III) from Drinking Water by $CuFe_2O_4$ Particles Embedded in Carbon Nanotube Membranes. *J. Chem. Technol. Biotechnol.* **2019,** *94* (9), 2816–2825.
50. Ma, L., et al. Fabrication and Water Treatment Application of Carbon Nanotubes (CNTs)-Based Composite Membranes: A Review. *Membranes* **2017,** *7* (1), 16.

CHAPTER 10

Carbon Nanotubes-Based Electrochemical Sensors for Environmentally Hazardous Chemicals

PIYUSH KUMAR SONKAR* and NARVADESHWAR

Department of Chemistry, MMV, Banaras Hindu University, Varanasi, Uttar Pradesh, India

**Corresponding author. E-mail: piyush.sonkar37@gmail.com, piyushchem.mmv@bhu.ac.in*

ABSTRACT

The environmental pollution of air, water, and natural resources has been increasing rapidly due to extensive industrialization and urbanization. The increasing level of pollution in the environment is the major global concern. It increases due to excessive use of pesticides, toxic wastes, synthetic fertilizers, biological warfare agents, etc. Monitoring of these pollutants is essential for sustainability of the environment. Electrochemical method is one of the important tools to determine the level of these toxic chemicals. Carbon nanotubes (CNTs)-based modified electrodes may play an essential role to monitor these environmental pollutants. By taking the advantages of extraordinary CNTs' physicochemical and electrocatalytic properties, a large number of electrochemical sensing platforms have been studied for sensing environmental pollutants. The detailed description of CNTs-based electrochemical sensors of different pollutants, such as heavy metals, pesticides, chemical fertilizers, food contaminants, microbial pollutants, etc. have been discussed in this chapter. Thus, this chapter provides a comprehensive overview of the recent development of CNTs-based modified electrodes for electrochemical sensing of environmentally hazardous chemicals and their future prospects.

10.1 INTRODUCTION

As the family of carbon nanomaterials, carbon nanotubes (CNTs) have attracted the attention of many researchers worldwide since their discovery in 1991.[1,2] Owing to the unique electronic, mechanical, optical properties, and chemical versatility, CNTs gained a lot of advantages in electroanalysis and electrocatalysis.[2–5] The sensing platforms based on the CNTs are designed to overcome the several limitations where traditional detection systems fail (such as limitations in size, sensitivity, selectivity, etc.).[4–6] Typically, the traditional detection methods which include atomic absorption spectrometry (AAS), atomic emission spectrometry (AES), inductively coupled plasma mass spectrometry (ICP-MS) are employed for the detection of different types of pollutants and heavy metals.[1,7,8] Although all of them possess high sensitivity. Yet, it is a time-consuming and expensive process to detect the samples due to the involvement of complex sample-pretreatments. Furthermore, it is also not appropriate for on-site detection.[7] However, compared with the above three techniques, an electrochemical methods based on the use of CNTs can offer several advantages such as low cost, simple, easy to handle, and possible to on-site detection.[5,9,10] There are several hazardous chemicals present in environmental samples, food, pharmaceutical, pathological, agriculture, chemical, and biological samples.[7,8] Thus, in this chapter, we have briefly covered the CNTs-based sensing platforms that can be used for sensitive, selective, and rapid monitoring of environmental pollutants like water pollutants, soil pollutants, air pollutants, heavy metals, pesticides, microbial toxins, etc.[7,11–14] In addition, various applications of CNTs-based electrochemical sensors and future prospects are also discussed.

10.1.1 ENVIRONMENTALLY HAZARDOUS CHEMICALS

Due to the rapid urbanization and changing lifestyle of people, incremental environmental pollution is becoming a serious concern.[7,15] The major anthropogenic activities that contribute toward the deterioration of environments due to industrial effluents, improper disposal of toxic wastes, enormous use of pesticides, and synthetic fertilizers.[7] Typically, environmental pollutants include inorganic pollutants (heavy metal ions, nitrite, etc.), gaseous analytes (CO_2, SO_2, CO, NO_x, NH_3, methanol vapor, etc.), organic pollutants (phenolic compound, aromatic amine, nitroaromatic explosives, pesticides, etc.), endocrine-disrupting chemicals (estrogen, bisphenol, and nonylphenol) and biological substances, such as pathogens and antibiotics.[16–18] In addition, toxicity levels of soil, water, and air monitoring are also necessary. The toxicity of these pollutants has severe effects on human health due to the ingestion of contaminated food and drinking water, as well as the inhalation of polluted atmosphere.[19] Therefore, it is necessary for scientists to develop reliable, highly sensitive, user-friendly, cost-effective, and a portable sensor for constant on-site monitoring and detection of different pollutants in different mediums.

10.1.2 CARBON NANOTUBES

CNTs are one of the unique nanosized allotropes of carbon.[6,20] It withdraws the attention of researchers worldwide due to their specific geometry, surface morphology, optical and electrical properties.[1,2,6] CNTs have high electrical conductivity, mechanical and thermal stability. Due to unique properties of CNTs, it has a wide range of applications in the field of an electrochemical sensor, biosensor, drug delivery, nanomedicine, semiconducting devices, energy storage, nanoelectronics, etc.[21,22] The structure of the CNTs belongs to the fullerene family, having sp^2 hybridized carbon arranged in hexagonal honeycomb lattice rolled in tubular arrangement.[1,2,6] On the basis of concentric nanotubes, CNTs are classified into two types: single-walled carbon nanotubes **(SWCNTs)** and multiwalled carbon nanotubes **(MWCNTs)**. SWCNT is a rolled structure of a single layer of graphite, similar to a cylinder with a nanometer-sized diameter (0.4–3 nm) and micrometer-sized (1–100 μm) length.[7,23] On the other hand, MWCNTs contains multiple concentric cylindrical tubes sharing the same central axis. The geometric structure of the different CNTs is summarized in Figure 10.1. The specific structure of CNTs provides a unique physical, electrical, and chemical characteristics.[20] Due to the presence of C-C bonds, CNTs have very high mechanical stability. Hence, CNTs are the strongest and stiffest fibers.[20] In addition, they have very high thermal stability in vacuum and air. They show high electrical conductivity with a very less heating effect.[24,25]

Based on the folding orientation of graphene sheath, CNTs are categorized into three classes: **Chiral** (n,m), **Armchair** (n,n), and **Zigzag** (n,0).[27,28] Here, m and n represent the integer along the chiral vector. The electrical properties of CNTs can be metallic or semiconducting based on the chirality and tube diameter.[20] CNTs have nearly one-dimensional arrangements, so the dielectric behavior of CNTs is highly anisotropic. The band gap of the MWCNTs is zero while SWCNTs are in the range of 0.4 to about 2 eV,[29] resulting in the different electrical properties of MWCNTs and SWCNTs. These unique properties of CNTs make it a potential candidate for the selection of electrode material to develop an electrochemical sensor for the detection of environmentally hazardous chemicals.

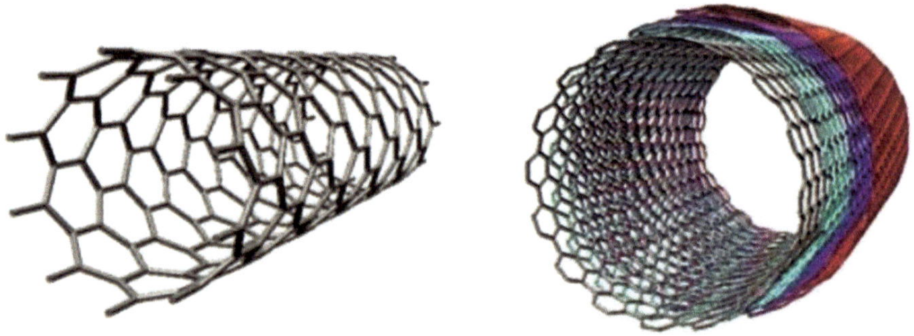

FIGURE 10.1 Structure of different types of CNTs: SWCNT (a) and MWCNTs (b).
Source: Reproduced with permission from Ref. [26]. © 2018 Elsevier.

10.2 CNTs MODIFIED +ELECTRODES FOR SENSING OF ENVIRONMENTALLY HAZARDOUS CHEMICALS

CNTs are significant materials for the development of modified electrodes because of their unique properties. The high electrical conductivity and surface area provide a unique platform for improved catalytic activities. In addition, CNTs synergistically improve the catalytic activities of metal nanoparticle, metal complexes, metal oxides, catalytic centers, and enzymes, etc.[5,9,30] Based on the nature of catalysis and its constitution, CNTs modified electrodes are of two types: sensor and biosensor. The detailed description of the CNTs modified electrochemical sensor and biosensor are discussed below.

10.2.1 CNTs MODIFIED ELECTROCHEMICAL SENSOR

An electrochemical sensor is based on real-time monitoring by an introduction of chemically selective layers with electrochemical transducers.[31,32]

Development of the electrochemical sensor based on CNTs is the promising area of the research worldwide. CNTs have been introduced in electrochemical sensing to improve the sensitivity and reduce the over-potential.[1,32] CNTs modified electrodes have been explored extensively due to low cost, high electron transport, improved kinetics, and suitable biocompatibility.[33] Hence, CNTs-based electrochemical sensors generally have fast electron transfer kinetics, high sensitivities, low detection limits compared with other modified electrodes.[5,34] The electrochemical performance of CNTs modified electrode may vary on the method of preparation, surface functionalization, method of electrode immobilization, and electron mediators.[34] Several CNTs modified electrodes are documented in the literature as electrochemical sensors.[34,35] For example, nonenzymatic CNTs-based modified electrode used for glucose determination by Sonkar et al. [9], CNTs used for the simultaneous detection of phenolic pollutants by Govindhan et al.,[34] highly dispersed MWCNTs modified manganese salen used for the simultaneous electrochemical sensing of vitamin B_2 and B_6 by Sonkar et al.,[5] electrochemical sensors based on CNTs used for the determination of dopamine and uric acid by Yang et al.,[35] electrochemical sensing of rifampicin using meso-tetrakis (4-hydroxyphenyl) porphyrinato cobalt(II) anchored CNTs by Sonkar et al.,[36] etc.[37–39] These examples are suggesting that CNTs have a wide range of applications as electrochemical sensors.

10.2.2 CNTs MODIFIED ELECTROCHEMICAL BIOSENSOR

Electrochemical biosensors are the modified version of the electrochemical sensors. It may be specific for a biomolecule.[40,41] They can be prepared using biological entities, such as proteins, enzymes, nucleic acid, etc. with suitable transducers.[42,43] Electrochemical biosensors have very high selectivity and sensitivity for particular analyte. CNTs modified electrochemical biosensors have a variety of applications in the determination of a variety of chemicals, biochemicals, and pathogens.[44–47] For example, An electrochemical biosensor based on MWCNTs was used for the detection of salivary uric acid by Shi et al.,[48] ethanol biosensor based on polyfluorene-g-poly (ethylene glycol) and MWCNTs used by Bekmezci et al.,[41] glucose biosensor based on CNTs/chitosan matrix used by Liu et al.,[40] graphene oxide quantum dots/carboxylated CNTs for pollutants by Zhu et al.,[42] etc.[44–47] It is clear from the above discussion that CNTs modified electrodes are useful for the development of electrochemical sensors and biosensors. Here, in this chapter, we will discuss in detail about different CNTs-based electrochemical sensors/biosensors for environmentally hazardous chemicals.

10.3 CNTs MODIFIED ELECTRODES FOR SENSING OF ENVIRONMENTALLY HAZARDOUS CHEMICALS

Environmental pollution is considered a global threat due to rapid industrialization and urbanization.[49] The pollution level of air, water, and other natural resources has been increasing day by day form the past few decades.[49] The level of these toxic chemicals is increasing due to some of the anthropogenic activities, such as mismanagement of waste disposal, bioaccumulation, biological warfare substances, extensive use of synthetic fertilizers, and pesticides.[42,49] Hence, monitoring of these toxic chemicals is essential to reduce environmental deterioration, animal and plant diseases. CNTs-based modified electrodes play an essential role in sensing of environmentally hazardous chemicals. Sensing of different toxic chemicals using CNTs modified electrodes is summarized in Figure 10.2.

The detailed descriptions of the CNTs modified electrodes for different environmentally hazardous chemicals are discussed in the following sections.

10.3.1 ELECTROCHEMICAL SENSING OF WATER POLLUTANTS

The level of toxic chemicals in water bodies is increasing rapidly. The toxic chemicals, such as heavy metals, pesticides, chemical fertilizers, nitrites, phenolic compounds, dyes, industrial waste, etc. heavily pollute the water bodies.[7,50] There is no strict boundary for pollutants to contaminate any particular natural resources. For example, pesticides and chemical fertilizers harm water bodies as well as soil and air too. Here, we are discussing some specific pollutants as water pollutants. However, they may also be contaminants for air, soil, and other natural resources. CNTs modified electrodes are one of significant tools to monitor the level of hazardous chemicals in water bodies. For example, Govindhan et al. reported the use of CNTs for simultaneous detection of phenolic pollutants.[34] They detected catechol, p-cresol, and p-nitrophenol with limit of detection (LOD) of 2.3, 3.7, and 7.7 nM, and sensitivities of 135.08, 94.90, and 110.38 mA/

mM, respectively.[34] Anbumannan et al. documented α-MnO_2-modified MWCNTs (MWCNTs–MnO_2) electrochemical sensor for the detection of p-nitrophenol in a water sample with LOD (0.644 μM) and sensitivity (0.186 μA/μM/cm^2).[51] TEM images of the MWCNTs and MWCNTs/MnO_2 are shown in Figure 10.3, it indicates the MnO_2 immobilization on the surface of MWCNTs.[51] Further, Zhu et al. studied the electrochemical sensing of 2,4-dichlorophenol using poly(eosin-Y)/hydroxylated MWCNTs.[52] They reported very low LOD (1.5 nM) for 2,4-dichlorophenol with high sensitivity and reproducibility.[52] Sun et al. reported the removal of methyl orange dye from CNTs modified electrode in water samples.[53] Liu et al. reported CNTs powder microelectrodes for nitrite sensing with LOD (0.1 μM) and sensitivity (6.8 A/cm^2/M).[54] They determined nitrite using steady-state polarization curve (Figure 10.4).[54] Recently, Atta et al. reported simultaneous determination of dihydroxybenzene isomers using crown ether modified polymer/CNTs-based sensor.[19] They have tested three isomers of dihydroxybenzene: resorcinol (LOD: 0.427 nM), hydroquinone (LOD: 0.156 nM), and catechol (LOD: 0.118 nM) as water pollutants.[19]

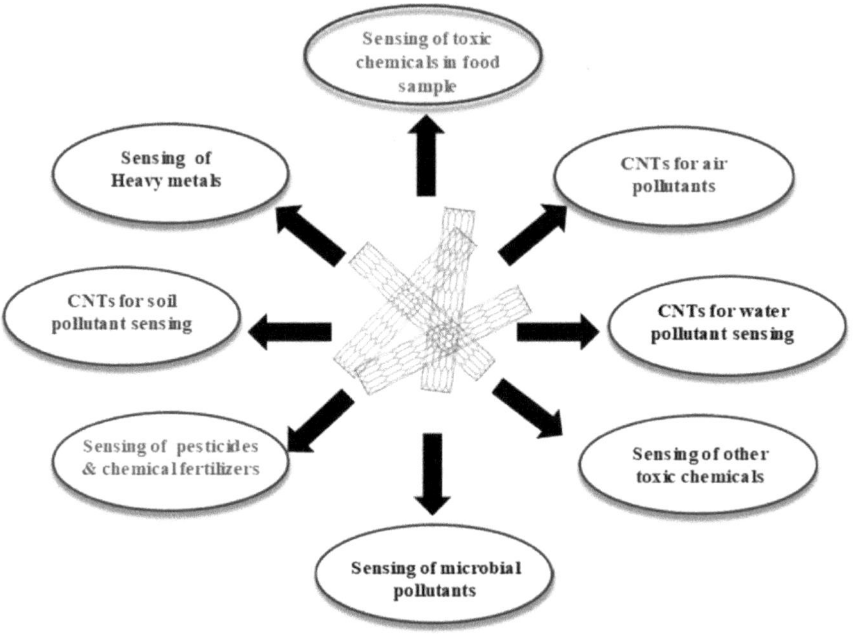

FIGURE 10.2 Electrochemical sensing of different environmental pollutants using CNTs modified electrodes.

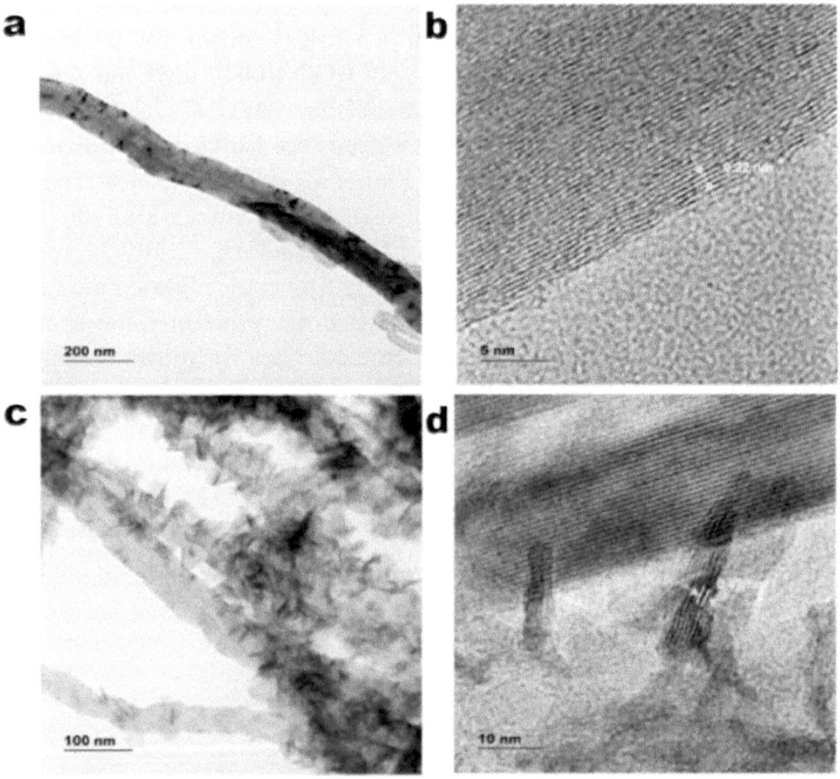

FIGURE 10.3 Shows the transmission electron microscope (TEM) images of MWCNTs (a, b) and MWCNTs–MnO_2 composite material.

Source: Reproduced with permission from Ref. [51]. © 2019 Elsevier.

Further, Turdean et al. reported Iron(III) protoporphyrin SWCNTs modified electrodes for hydrogen peroxide and nitrite detection.[55] In addition, there are a number of CNTs-based modified electrodes are recognized for sensing of hazardous chemicals in water bodies, such as nitrite,[56] pharmaceutical pollutants,[57,58] sulfite,[59] heavy metals,[60] pesticides,[61] surfactants,[62] etc.[7,13,14]

It is clear from the above discussion that CNTs-based modified electrode has numerous applications in the sensing of a variety of environmentally hazardous chemicals affecting the water bodies. Further, the detailed description of the sensing of heavy metals, pesticides, and synthetic fertilizers are discussed in the next sections.

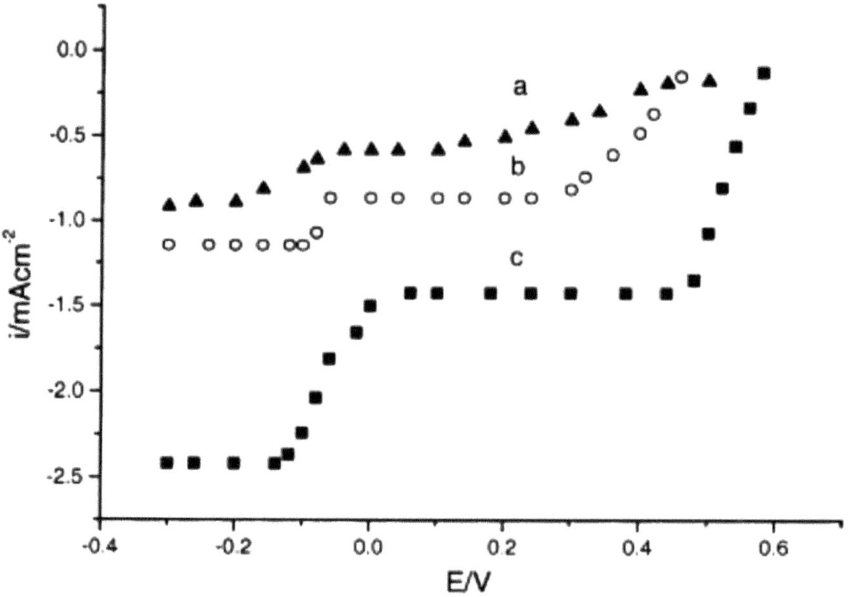

FIGURE 10.4 Represents the steady-state polarization measurements for nitrite reduction at CNTs powder-modified electrode (in 0.05 M H_2SO_4 + 5 mM $NaNO_2$): (a) no pretreatment, (b) with anodic treatment, (c) with anodic treatment and $Os(bpy)_3^{2+}$ pre-adsorption.

Source: Reproduced with permission from Ref. [54]. © 2002 Elsevier.

10.3.1.1 ELECTROCHEMICAL DETERMINATION OF HEAVY METALS

Heavy metals, such as lead (Pb), cadmium (Cd), mercury (Hg), chromium (Cr), copper (Cu) and arsenic (As), etc. are considered to be the major environmental pollutants due to their high toxicity in the environment and adverse effect on human health.[63–65] Pollutants occur mainly as a result of mining, melting operations, industrial processes, use of metal compounds in domestic and agricultural activities. Even the presence of a small amount of these heavy metals in the environment can lead to an adverse effect on human health.[11,12] Considering this, it is necessary to regulate the concentrations of these metal ions in water, soil, food, and many more. Several heavy metal electrochemical sensors are known using CNTs-based electrodes.[11,12,66,67] For example, Jeromiyas et al. reported bismuth nanoparticles decorated graphenated CNTs-modified electrodes for mercury detection with low LOD (0.2 nM).[11,12] Fu et al. reported the

MWCNTs modified electrodes for sensing of zinc, cadmium, and lead using a screen-printed electrode.[11] The level of Zn^{2+}, Cd^{2+}, and Pb^{2+} in the real sample (lake water) were recorded to be 17.26, 6.31, and 8.13 μg/L, respectively.[11] In another work, Moyo et al. documented an amperometric biosensor based on horseradish peroxidase (HRP) immobilized MWCNTs modified glassy carbon (GC) electrode for the determination of lead (LOD: 2.5 μg/L) and copper (4.2 μg/L) ions in aqueous solution.[66] Fabrication of the CNTs modified electrode with HRP is shown in Figure 10.5.[66]

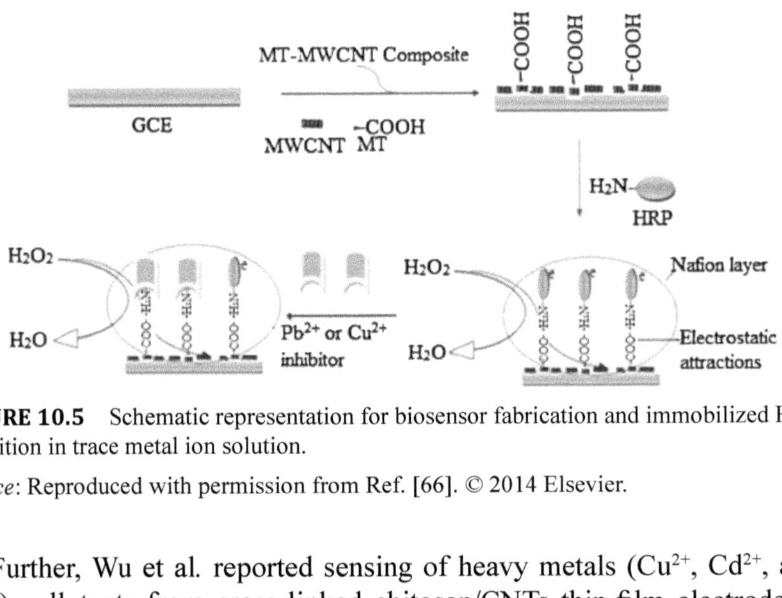

FIGURE 10.5 Schematic representation for biosensor fabrication and immobilized HRP inhibition in trace metal ion solution.

Source: Reproduced with permission from Ref. [66]. © 2014 Elsevier.

Further, Wu et al. reported sensing of heavy metals (Cu^{2+}, Cd^{2+}, and Pb^{2+}) pollutants from cross-linked chitosan/CNTs thin-film electrodes.[67] The LOD value for Cu^{2+}, Cd^{2+} and Pb^{2+} are 0.1, 0.8 and 0.6 ppm, respectively using chitosan/CNTs modified electrode.[67] Wang et al. reported thiacalixarene covalently functionalized MWCNTs modified electrode for the determination of ultratrace Pb^{2+} ions.[68] They have reported very low LOD 0.04 nM with high selectivity for Pb^{2+} ions (Figure 10.6).[68]

In addition, there are several other CNTs-based heavy metal electrochemical sensors are known for arsenic,[60,69–71] cadmium,[72–74] mercury,[75–77] lead,[78,79] copper,[80] etc.[63–65] So it is clear from these examples that CNTs have a variety of applications in sensing of heavy metals. The above discussion suggesting that CNTs modified electrodes have a significant role in sensing of heavy metals with very low detection limits, high sensitivities, and selectivities.

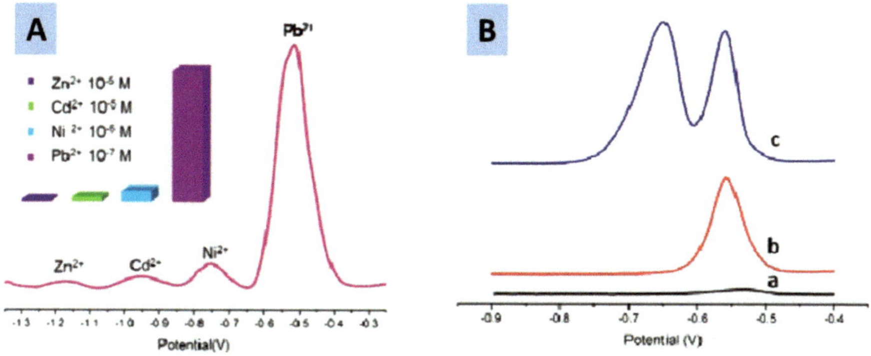

FIGURE 10.6 Differential pulse anodic stripping voltammetry (DPASV) measured signals after 15 min accumulation of: (A) 10^{-7} M lead ion-containing 100-fold excess amounts of zinc, nickel, and cadmium ions; (B) 10^{-7} M lead ion containing 10^{-7} M tin ion at different glassy carbon (GC) electrode: (a) Bare GC, (b) MWCNTs-GC, and (c) Thiacalixarene–MWCNTs–GC. DPV scanned from −0.9 to −0.3 V with an amplitude of 0.05 V.

Source: Reproduced with permission from Ref. [68]. © 2012 American Chemical Society.

10.3.2 ELECTROCHEMICAL SENSING OF AIR POLLUTANTS

There are several gaseous molecules that affect the quality of air and make it unfit for plants, animals, and the environment.[81,82] In this section, we will discuss the sensing of some specific gases hazardous to the environment. Electrochemical sensing of gaseous molecules is always a tough task for the researchers due to its low concentration and fast diffusion.[81,82] However, few electrochemical sensors have been documented in the literature for sensing of gaseous hazardous chemicals using CNTs.[81,82] For example, Meng et al. reported an electronic chip based on CNTs microelectrode array for indoor air pollutants sensing.[81] They have detected indoor air pollutants formaldehyde, toluene, and ammonia gas using CNTs-based microelectrode chips. The highly porous 3D network of CNTs provides a remarkable capacitive response for the detection of formaldehyde, toluene, and ammonia gas on the surface of gold electrode. SEM images of the CNTs modified electronic chip are shown in Figure 10.7. Also, sensing of formaldehyde, toluene, and ammonia gas are summarized in Figure 10.8.[81] Further, Pedrosa et al. reported copper nanoparticles and CNTs-based electrochemical sensor for the determination of tricresyl-phosphate in aqueous and air samples.[82] The measured LOD and linear range for cresol is 0.6 and 10–160 μM, respectively. This method finds an alternative way to monitor the tricresylphosphate as cresol in the air samples.[82] Xie et al. documented MWCNTs gas sensors modified with amino group for the determination of formaldehyde.[83] In addition, some other electrochemical sensors are also known for gaseous pollutants, such as CO_2,[84] CO,[85] hydrogen,[86] SO_2,[87] hexane,[88] ammonia,[88] etc.[89–91] Hence, CNTs modified electrodes are efficient sensing platforms for gaseous molecules. Due to a wide range of application in the sensing, CNTs is a promising material for sensing of gaseous pollutants with a low limit of detection and high sensitivity.

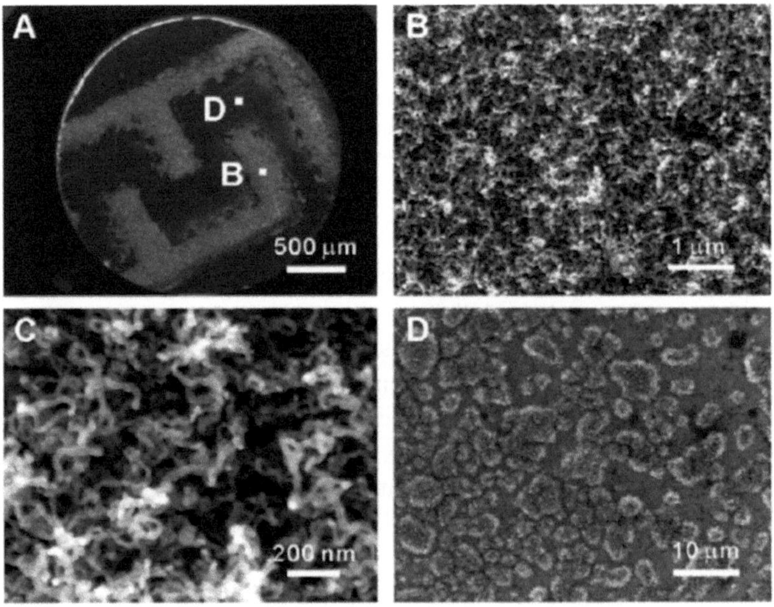

FIGURE 10.7 (A) Scanning electron microscope (SEM) image of an electronic chip having self-oriented MWCNTs. (B) Self-oriented MWCNTs developed gold electrode, representing that gold platform is suitable for CNT growth. (C) Enlarged view of panel B. (D) SEM image representing scattered island-like development of MWCNTs on a silica platform (low-resolution image).

Source: Reproduced with permission from Ref. [81]. © 2011 Elsevier.

Further, the detailed applications of the CNTs modified electrodes for the determination of soil pollutants including pesticides and synthetic fertilizers were discussed in the next section.

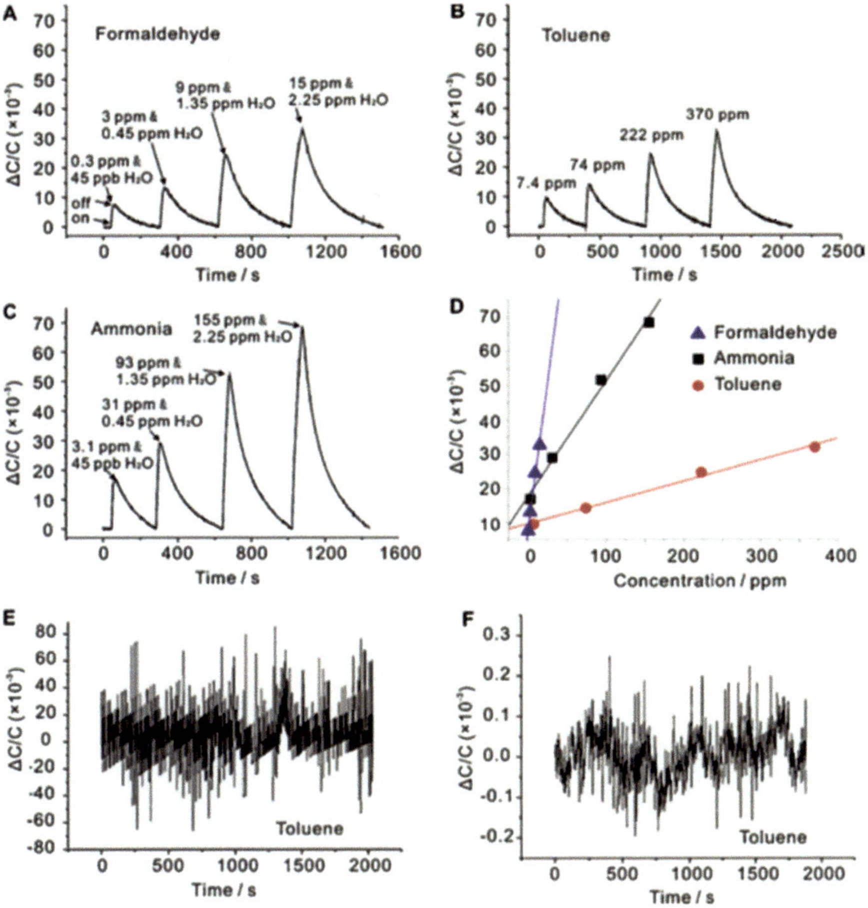

FIGURE 10.8 Represents the relative capacitance changes ($\Delta C/C$) of capacitive electronic chip with self-oriented MWCNTs microelectrode array in the presence of indoor air pollutants at varying concentrations of: (A) formaldehyde (HCHO), (B) toluene (C_7H_8), (C) ammonia (NH_3), (D) graph of $\Delta C/C$ *vs* concentration obtained from formaldehyde, toluene and ammonia, (E) $\Delta C/C$ of electronic chip in the absence of MWCNTs measured for toluene, (F) $\Delta C/C$ of electronic chip with catalyst in the absence of MWCNTs measured for toluene.

Source: Reproduced with permission from Ref. [81]. © 2011 Elsevier.

10.3.3 ELECTROCHEMICAL SENSING OF SOIL POLLUTANTS

The level of hazardous chemicals in soil is increasing very rapidly due to industrial effluents, poor waste management, explosives, pesticides, synthetic fertilizers, chlorinated hydrocarbons (e.g., perchloroethylene and trichloroethylene), etc.[92,93] From the past few decades, explosives such as trinitrotoluene, nitroaromatic compounds, and their derivatives were identified as hazardous chemicals for the environment as well as for human health.[92] There are different electrochemical CNTs-based sensors are known for these chemicals. For example, Hrapovic et al. developed an electrochemical sensor for the determination of explosive nitroaromatic compounds using metal nanoparticles (Pt, Au, and Cu)-CNTs composites.[92] Hnaien et al. developed a microbial biosensor based on SWCNTs modified microelectrodes for trichloroethylene (LOD: 20 μg/L) level in soils and groundwaters. It provides a successful recovery (100%–103%) in the real samples.[94] The details of the specific soil pollutants, such as pesticides, herbicides, and synthetic fertilizers are discussed below.

10.3.3.1 SENSING OF PESTICIDES, HERBICIDES, AND CHEMICAL FERTILIZERS

Agricultural pollutants, such as pesticides, herbicides, and chemical fertilizers affect the environment badly.[16] The common chemicals of this category are organophosphate, organochlorine, malathion, urea, propham, chlorpyrifos, monocrotophos, endosulfan, diuran, etc.[16–18] Researchers reported different electrochemical sensors for the sensing of these hazardous chemicals. For example; Wang et al. reported acetylcholinesterase biosensor of organophosphate detection using CNTs modified electrodes.[95] A compound trichlorfon was selected as a base compound for indirect sensing of organophosphate pesticides (LOD: 0.01 μg/L).[95] Chauhan et al. reported electrochemical sensing of chlorpyrifos, malathion, monocrotophos, and endosulfan pesticides with a very low detection limits (0.1 nmol/L for all four molecules). Saber-Tehrani et al. reported oxadiazole derivative MWCNTs-based electrode for hydrazine determination (LOD: 0.16 μM) with high reproducibility and stability.[96] Rahemi et al. reported a bentazone herbicide sensor using CNTs β-Cyclodextrin.[97] They used cyclic voltammetry techniques for the determination of bentazone with low LOD (1.6 μmol/L) and wide calibration range (10–80 μmol/L).[97]

Leniart et al. reported a herbicide propham sensor based on glassy carbon-modified MWCNTs electrode using square-wave voltammetry technique (LOD: 0.76 μM) and square-wave adsorptive stripping voltammetry technique (LOD: 0.37 μM).[17] The electrochemical sensing of propham is summarized in Figure 10.9. They have successfully applied it for testing of river water samples.[17] There are many other electrochemical sensors based on CNTs are also known for detecting agricultural pollutants.[16,18,98–100] However, a detailed description of each one is beyond the scope of this chapter.

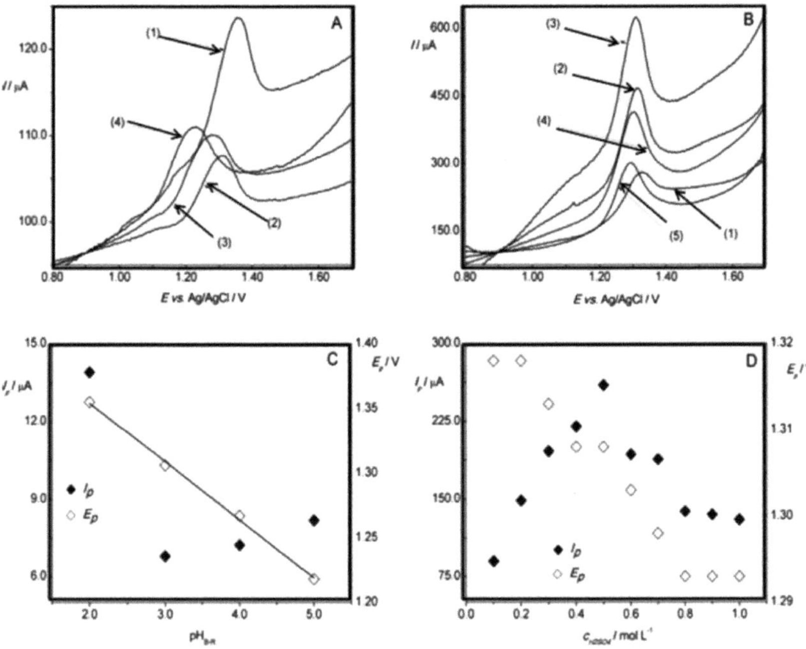

FIGURE 10.9 Square-wave voltammograms of propham (9.13×10^{-5} M) at GC-MWCNTs in (A) Britton–Robinson (B–R) buffers (1) pH= 2.0, (2) pH= 3.0, (3) pH= 4.0, (4) pH= 5.0) and in (B) H_2SO_4 solutions (1) 0.1 M, (2) 0.3 M, (3) 0.5 M, (4) 0.7 M, (5) 1.0 M), and corresponding dependency of square-wave voltammograms peak currents (Ip) (◊, left y-axis) and peak potentials (Ep) (♦, right y-axis) on pH of B–R buffers (C) and concentration of H_2SO_4 solutions (D).

Source: Reproduced with permission from Ref. [17]. © 2016 Elsevier.

10.3.4 ELECTROCHEMICAL SENSING OF MICROBIAL POLLUTANTS

The level of microbial pollutants (including microbes and its toxins) is greatly affecting the environment as well as the health of plants and animals.[101,102] Several diseases are caused by microbes which greatly affect the plants and animals and producing adverse effects on the environment.[94,103–105] Coronavirus disease (Covid-19) is one of the most highlighted examples of microbial pollutants; it affected almost every dimension of the environment.[106,107] So, it is necessary to monitor the level of microbial pollutants and toxins generated by them. There are few CNTs-based electrochemical biosensors are known for the sensing for microbial pollutants. For example, Queiros et al. developed plastic antibodies immobilized CNTs for In situ determination of bacterial toxins.[102] They have selected microcystins as a target molecule for the sensing of toxins released by cyanobacteria in water. This CNTs-based biosensor shows a very low detection limit of toxin (below 1 nM) with high reproducibility and selectivity.[102] Bhardwaj et al. reported a paper-based electrochemical sensor with antibody-SWCNTs modified electrode for sensing of food-borne pathogens.[101] The antibody-SWCNTs modified electrode is capable of sensing *Staphylococcus aureus*. Fabrication of the antibody–SWCNTs modified electrode is summarized in Figure 10.10.[101]

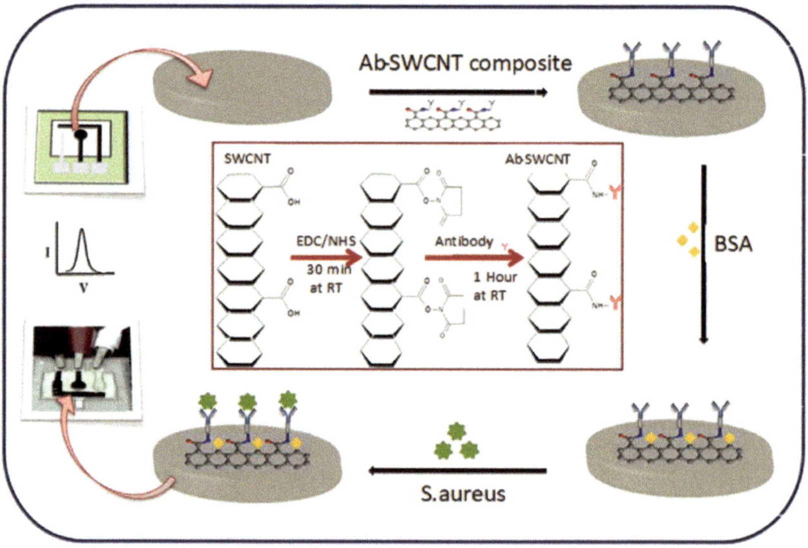

FIGURE 10.10 Shows the fabrication of antibody–CNTs bioconjugate: coupling of antibodies to CNTs using *N*-(3-dimethylaminopropyl)-*N'*-ethylcarbodiimidehydrochloride (EDC)/*N*-hydroxysuccinimide (NHS) coupling reagent and sensing procedure.

Source: Reproduced with permission from Ref. [101]. © 2017 Elsevier.

Further, Guner et al. developed immunosensor for sensing of *Escherichia coli* O157: H7 using CNTs–gold nanoparticles modified electrodes.[104]

Therefore, it is clear that CNTs modified electrodes have significant application in the sensing of microbial pollutants.[94,103–105] Some other hazardous chemical sensors based on CNTs modified electrodes are discussed in the following sections.

10.3.6 ELECTROCHEMICAL SENSING OF OTHER HAZARDOUS CHEMICALS

We have already discussed the major environmental hazardous chemicals and their electrochemical sensing using different CNTs modified electrodes. Here, in this section, we will discuss some specific pollutants such as radiochemicals, food contaminants, pharmaceutical pollutants, and tobacco, etc.[108–110] There are some specific electrochemical sensors based on CNTs are known such as chitosan–acrylic acid/MWCNTs-based nanocomposite for theranostic ^{47}Sc separation from the neutron-irradiated titanium target,[109] chitosan/carboxylated CNTs nanocomposite for the isolation of uranium(VI) from aqueous solution,[108] silver nanoparticles/MWCNTs modified electrode for sensing of cyanide in food samples,[111] corn zein CNTs-based electrochemical sensor for gliadin (it causes the autoimmune disorder, fatigue, weight loss, and chronic diarrhea),[110] electrochemical sensing of nicotine using MWCNTs modified electrode,[112] CNTs-based sensor for adsorptive stripping voltammetry,[113] etc.[114–118] These examples are suggesting that CNTs modified electrodes have significant applications in the field of electrochemical sensing.

10.4 APPLICATION OF CNTs MODIFIED ELECTROCHEMICAL SENSORS

We have discussed the different CNTs modified electrodes for almost every type of environmental pollutants. These CNTs modified electrodes are useful for the development of electrochemical sensing devices for monitoring of pollutant levels. It may be applicable for the determination of hazardous chemicals in water bodies, microbial toxins, air, and soil samples. In addition, it may also be useful for the determination of pharmaceutical pollutants, drug-level monitoring, radiochemical determination, food contaminants, etc. There are some electrochemical devices developed for the determination of hazardous chemicals. For example, self-powered environmental sensor based on CNTs,[119] CNTs-based gas sensor,[120] DNA-immobilized CNTs-based sensors using metal oxide semiconductor circuitry,[30] etc.[26,49,121] In addition, there are several other applications of the CNTs modified electrodes are known beyond the electrochemical sensing of hazardous chemicals. It is useful in the nanoelectronics, semiconductors, fuel cells, biofuel cells, solar cells, supercapacitors, material chemistry, drug delivery, pharmaceuticals, electrocatalysis, etc.[1,5,122–125] However, the detailed discussion of these applications is beyond the scope of this chapter. For the detailed studies of CNTs and other applications, readers are directed to further readings.[126–130]

10.5 SUMMARY

The increasing level of hazardous chemicals in the environment is a serious global issue for sustainable development. Hence, monitoring of these hazardous chemicals is essential. The electrochemical method is one of the significant methods to monitor these hazardous chemicals. CNTs-based modified electrodes play an essential role in sensing of these environmentally hazardous chemicals. In this chapter, a brief review of literature based on CNTs modified electrodes for electrochemical sensing of hazardous chemicals was discussed. The outstanding properties of CNTs, such as high electrical conductivity, thermal and chemical stability make them a potential candidate for the development of an effective electrochemical sensor. CNTs modified electrodes have a wide range of application in electrochemical sensing of water contaminants, air pollutants, soil pollutants, microbial toxins, food contaminants, agricultural pollutants, radiochemicals, explosives, etc. CNTs-based electrochemical sensors provide low detection limits, high sensitivities, and reproducibilities. It is useful for the development of reliable electrochemical sensing devices for environmental pollutants. Conclusively, CNTs-based modified electrodes are the potential candidate for the development of effective sensing devices with high precision for environmentally hazardous chemicals.

10.6 ACKNOWLEDGMENTS

The generous funding of university grant commission, New Delhi, India (UGC: M-14-58), and IoE Seed grant, Banaras Hindu University, Varanasi,

India (IoE BHU Scheme No. 6031) was gratefully acknowledged. We are thankful to Dr. P. K. Rastogi, GLA University, Mathura for their help and support.

KEYWORDS

- **carbon nanotubes**
- **hazardous chemicals**
- **electrochemical**
- **sensor**
- **environmental pollutants**
- **modified electrode**

REFERENCES

1. Bianco, A.; Kostarelos, K.; Prato, M. Applications of Carbon Nanotubes in Drug Delivery. *Curr. Opin. Chem. Biol.* **2005,** *9*, 674–679.
2. Iijima, S.; Ichihashi, T. Single-shell Carbon Nanotubes of 1-nm Diameter. *Nature* **1993,** *363*, 603–605.
3. Iijima, S.; Ajayan, P.; Ichihashi, T. Growth Model for Carbon Nanotubes. *Phys. Rev. Lett.* **1992,** *69*, 3100–3103.
4. Sonkar, P. K.; Ganesan, V.; Yadav, D. K.; Gupta, R. Dual Electrocatalytic Behavior of Oxovanadium(IV) Salen Immobilized Carbon Materials Towards Cysteine Oxidation and Cystine Reduction: Graphene Versus Single Walled Carbon Nanotubes. *ChemistrySelect* **2016,** *1*, 6726–6734.
5. Sonkar, P. K.; Ganesan, V.; Sen Gupta, S. K.; Yadav, D. K.; Gupta, R.; Yadav, M. Highly Dispersed Multiwalled Carbon Nanotubes Coupled Manganese Salen Nanostructure for Simultaneous Electrochemical Sensing of Vitamin B_2 and B_6. *J. Electroanal. Chem.* **2017,** *807*, 235–243.
6. Sonkar, P. K.; Prakash, K.; Yadav, M.; Ganesan, V.; Sankar, M.; Gupta, R.; Yadav, D. K. Co(II)-Porphyrin-Decorated Carbon Nanotubes as Catalysts for Oxygen Reduction Reactions: An Approach for Fuel Cell Improvement. *J. Mater. Chem. A* **2017,** *5*, 6263–6276.
7. Y. Wang, S. Hu, Applications of Carbon Nanotubes and Graphene for Electrochemical Sensing of Environmental Pollutants. *J. Nanosci. Nanotechnol.* **2016,** *16*, 7852–7872.
8. Trojanowicz, M. Analytical Applications of Carbon Nanotubes: A Review. *TrAC Trends Anal. Chem.* **2006,** *25*, 480–489.

9. Sonkar, P. K.; Ganesan, V.; John, S. A.; Yadav, D. K.; Gupta, R. Non-Enzymatic Electrochemical Sensing Platform Based on Metal Complex Immobilized Carbon Nanotubes for Glucose Determination. *RSC Adv.* **2016,** *6*, 107094–107103.
10. Sonkar, P. K.; Ganesan, V.; Singh, R. K.; Yadav, D. K.; Gupta, R.; Yadav, M. Single Walled Carbon Nanotubes Decorated Vanadyl Phthalocyanine Composite for Electrochemical Oxygen Reduction in Fuel Cells. *Ind. J. Chem. A* **2017,** *56A*, 821–825.
11. Fu, L.; Li, X.; Yu, J.; Ye, J. Facile and Simultaneous Stripping Determination of Zinc, Cadmium and Lead on Disposable Multiwalled Carbon Nanotubes Modified Screen-Printed Electrode. *Electroanalysis* **2013,** *25*, 567–572.
12. Jeromiyas, N.; Elaiyappillai, E.; Kumar, A. S.; Huang, S.-T.; Mani, V. Bismuth Nanoparticles Decorated Graphenated Carbon Nanotubes Modified Screen-Printed Electrode for Mercury Detection. *J. Taiwan Inst. Chem. E.* **2019,** *95*, 466–474.
13. Pavinatto, A.; Mercante, L. A.; Leandro, C. S.; Mattoso, L. H.; Correa, D. S. Layer-by-layer Assembled Films of Chitosan and Multi-Walled Carbon Nanotubes for the Electrochemical Detection of 17α-ethinylestradiol. *J. Electroanal. Chem.* **2015,** *755*, 215–220.
14. Zheng, Q.; Yang, P.; Xu, H.; Liu, J.; Jin, L. A Simple and Sensitive Method for the Determination of 4-n-Octylphenol Based on Multi-Walled Carbon Nanotubes Modified Glassy Carbon Electrode. *J. Environ. Sci.* **2012,** *24*, 1717–1722.
15. Mauter, M. S.; Elimelech, M. Environmental Applications of Carbon-Based Nanomaterials. *Environ. Sci. Technol.* **2008,** *42*, 5843–5859.
16. Alghuthaymi, M.; Mostafa, M.; Abd-Elsalam, K. A. Carbon Nanotubes: An Efficient Sorbent for Herbicide Sensing and Remediation. In *Carbon Nanomaterials for Agri-Food and Environmental Applications*; Elsevier, **2020**; pp 429–457.
17. Leniart, A.; Brycht, M.; Burnat, B.; Skrzypek, S. Voltammetric Determination of the Herbicide Propham on Glassy Carbon Electrode Modified with Multi-Walled Carbon Nanotubes. *Sens. Actuat. B* **2016,** *231*, 54–63.
18. Sharma, P.; Bhalla, V.; Tuteja, S.; Kukkar, M.; Suri, C R. Rapid Extraction and Quantitative Detection of the Herbicide Diuron in Surface Water by a Hapten-Functionalized Carbon Nanotubes Based Electrochemical Analyzer. *Analyst* **2012,** *137*, 2495–2502.
19. Atta, N. F.; Galal, A.; El-Gohary, A. R. M. New Insight for Simultaneous Determination of Hazardous Di-Hydroxybenzene Isomers at Crown Ether Modified Polymer/Carbon Nanotubes Composite Sensor. *J. Hazard. Mater.* **2020,** *388*, 122038.
20. Wang, Y.; Yeow, J. T. A Review of Carbon Nanotubes-Based Gas Sensors. *J. Sens.* **2009,** *2009*,.
21. Upadhyayula, V. K. K.; Deng, S.; Mitchell, M. C.; Smith, G. B. Application of Carbon Nanotube Technology for Removal of Contaminants in Drinking Water: A Review. *Sci. Total Environ.* **2009,** *408*, 1–13.
22. Ajayan, P. M.; Zhou, O. Z. *Applications of Carbon Nanotubes, Carbon Nanotubes*; Springer, 2001; pp 391–425.
23. Dresselhaus, M. S.; Dresselhaus, G.; Eklund, P. C. *Science of Fullerenes and Carbon Nanotubes: Their Properties and Applications*; Elsevier, 1996.
24. Qian, D.; Wagner, G. J.; Liu, W. K.; Yu, M. F.; Ruoff, R. S. Mechanics of Carbon Nanotubes. *Appl. Mech. Rev.* **2002,** *55*, 495–533.
25. Ruoff, R. S.; Lorents, D. C. Mechanical and Thermal Properties of Carbon Nanotubes. *Carbon* **1995,** *33*, 925–930.

26. Gupta, S.; Murthy, C.; Prabha, C. R. Recent Advances in Carbon Nanotube Based Electrochemical Biosensors. *Int. J. Biol. Macromol.* **2018,** *108*, 687–703.
27. Prasek, J.; Drbohlavova, J.; Chomoucka, J.; Hubalek, J.; Jasek, O.; Adam, V.; Kizek, R. Methods for Carbon Nanotubes Synthesis. *J. Mater. Chem.* **2011,** *21*, 15872–15884.
28. Li, C.; Chou, T. W. A Structural Mechanics Approach for the Analysis of Carbon Nanotubes. *Int. J. Solids Struct.* **2003,** *40*, 2487–2499.
29. Shah, K. A.; Tali, B. A. Synthesis of Carbon Nanotubes by Catalytic Chemical Vapour Deposition: A Review on Carbon Sources, Catalysts and Substrates. *Mater. Sci. Semicond. Process*. **2016,** *41*, 67–82.
30. Chen, C. L.; Yang, C.-F.; Agarwal, V.; Kim, T.; Sonkusale, S.; Busnaina, A.; Chen, M.; Dokmeci, M. R. DNA-Decorated Carbon-Nanotube-Based Chemical Sensors on Complementary Metal Oxide Semiconductor Circuitry. *Nanotechnology* **2010,** *21*, 095504.
31. Bandodkar, A. J.; Wang, J. Non-Invasive Wearable Electrochemical Sensors: A Review. *Trends Biotechnol.* **2014,** *32*, 363–371.
32. Agüí, L.; Yáñez-Sedeño, P.; Pingarrón, J. M. Role of Carbon Nanotubes in Electroanalytical Chemistry: A Review. *Anal. Chim. Acta* **2008,** *622*, 11–47.
33. Jacobs, C. B.; Peairs, M. J.; Venton, B. J. Carbon Nanotube Based Electrochemical Sensors for Biomolecules. *Anal. Chim. Acta* **2010,** *662*, 105–127.
34. Govindhan, M.; Lafleur, T.; Adhikari, B. R.; Chen, A. Electrochemical Sensor Based on Carbon Nanotubes for the Simultaneous Detection of Phenolic Pollutants. *Electroanalysis* **2015,** *27*, 902–909.
35. Yang, Y.; Li, M.; Zhu, Z. A Novel Electrochemical Sensor Based on Carbon Nanotubes Array for Selective Detection of Dopamine or Uric Acid. *Talanta* **2019,** *201*, 295–300.
36. Sonkar, P. K.; Yadav, M.; Prakash, K.; Ganesan, V.; Sankar, M.; Yadav, D. K.; Gupta, R. Electrochemical Sensing of Rifampicin in Pharmaceutical Samples Using Meso-tetrakis (4-hydroxyphenyl) Porphyrinato Cobalt(II) Anchored Carbon Nanotubes. *J. Appl. Electrochem.* **2018,** *48*, 937–946.
37. Baghayeri, M.; Beitollahi, H.; Akbari, A.; Farhadi, S. Highly Sensitive Nanostructured Electrochemical Sensor Based on Carbon Nanotubes-Pt Nanoparticles Paste Electrode for Simultaneous Determination of Levodopa and Tyramine. *Russ. J. Electrochem.* **2018,** *54*, 292–301.
38. Ahammad, A. S.; Akter, T.; Al Mamun, A.; Islam, T.; Hasan, M. M.; Mamun, M.; Faraezi, S.; Monira, F.; Saha, J. K. Cost-Effective Electrochemical Sensor Based on Carbon Nanotube Modified-Pencil Electrode for the Simultaneous Determination of Hydroquinone and Catechol. *J. Electrochem. Soc.* **2018,** *165*, B390.
39. Wu, K.; Sun, Y.; Hu, S. Development of an Amperometric Indole-3-Acetic Acid Sensor Based on Carbon Nanotubes Film Coated Glassy Carbon Electrode. *Sens. Actuators B* **2003,** *96*, 658–662.
40. Liu, Y.; Wang, M.; Zhao, F.; Xu, Z.; Dong, S. The Direct Electron Transfer of Glucose Oxidase and Glucose Biosensor Based on Carbon Nanotubes/Chitosan Matrix. *Biosens. Bioelectron.* **2005,** *21*, 984–988.
41. Bekmezci, S. A.; Soylemez, S.; Yilmaz, G.; Udum, Y. A.; Yagci, Y.; Toppare, L. A New Ethanol Biosensor Based on Polyfluorene-g-poly (Ethylene Glycol) and Multiwalled Carbon Nanotubes. *Eur. Polym. J.* **2020,** *122*, 109300.

42. Zhu, X.; Wu, G.; Lu, N.; Yuan, X.; Li, B. A Miniaturized Electrochemical Toxicity Biosensor Based on Graphene Oxide Quantum Dots/Carboxylated Carbon Nanotubes for Assessment of Priority Pollutants. *J. Hazard. Mater.* **2017,** *324*, 272–280.
43. Wang, J. Carbon Nanotube Based Electrochemical Biosensors: A Review. *Electroanalysis* **2005,** *17*, 7–14.
44. Perez, B.; Pumera, M.; Del Valle, M.; Merkoci, A.; Alegret, S. Glucose Biosensor Based on Carbon Nanotube Epoxy Composites. *J. Nanosci. Nanotechnol.* **2005,** *5*, 1694–1698.
45. Xu, L.; Zhu, Y.; Yang, X.; Li, C. Amperometric Biosensor Based on Carbon Nanotubes Coated with Polyaniline/Dendrimer-Encapsulated Pt Nanoparticles for Glucose Detection. *Mater. Sci. Eng. C* **2009,** *29*, 1306–1310.
46. Lopez, B. P.; Merkoci, A. Improvement of the Electrochemical Detection of Catechol by the Use of a Carbon Nanotube Based Biosensor. *Analyst* **2009,** *134*, 60–64.
47. Shahrokhian, S.; Salimian, R.; Kalhor, H. R. A Simple Label-Free Electrochemical DNA Biosensor Based on Carbon Nanotube-DNA Interaction. *RSC Adv.* **2016,** *6*, 15592–15598.
48. Shi, W.; Li, J.; Wu, J.; Wei, Q.; Chen, C.; Bao, N.; Yu, C.; Gu, H. An Electrochemical Biosensor Based on Multi-Wall Carbon Nanotube–Modified Screen-Printed Electrode Immobilized by Uricase for the Detection of Salivary Uric Acid. *Anal. Bioanal. Chem.* **2020,**, 1–9.
49. Ramnani, P.; Saucedo, N. M.; Mulchandani, A. Carbon Nanomaterial-Based Electrochemical Biosensors for Label-Free Sensing of Environmental Pollutants. *Chemosphere* **2016,** *143*, 85–98.
50. Rasheed, T.; Nabeel, F.; Adeel, M.; Rizwan, K.; Bilal, M.; Iqbal, H. M. Carbon Nanotubes-Based Cues: A Pathway to Future Sensing and Detection of Hazardous Pollutants. *J. Mol. Liq.* **2019,** *292*, 111425.
51. Anbumannan, V.; Dinesh, M.; Kumar, R. R.; Suresh, K. Hierarchical α-MnO_2 Wrapped MWCNTs Sensor for Low Level Detection of p-Nitrophenol in Water. *Ceram. Int.* **2019,** *45*, 23097–23103.
52. Zhu, X.; Zhang, K.; Wang, C.; Guan, J.; Yuan, X.; Li, B. Quantitative Determination and Toxicity Evaluation of 2, 4-Dichlorophenol Using Poly (eosin-Y)/Hydroxylated Multi-Walled Carbon Nanotubes Modified Electrode. *Sci. Rep.* **2016,** *6*, 38657.
53. Sun, Y.; Wang, G.; Dong, Q.; Qian, B.; Meng, Y.; Qiu, J. Electrolysis Removal of Methyl Orange Dye from Water by Electrospun Activated Carbon Fibers Modified with Carbon Nanotubes. *Chem. Eng. J.* **2014,** *253*, 73–77.
54. Liu, P.; Hu, J. Carbon Nanotube Powder Microelectrodes for Nitrite Detection. *Sens. Actuat. B.* **2002,** 84, 194–199.
55. Turdean, G. L.; Popescu, I. C.; Curulli, A.; Palleschi, G. Iron(III) Protoporphyrin IX-Single-Wall Carbon Nanotubes Modified Electrodes for Hydrogen Peroxide and Nitrite Detection. *Electrochim. Acta* **2006,** *51*, 6435–6441.
56. Lu, S.; Hummel, M.; Kang, S.; Gu, Z. Selective Voltammetric Determination of Nitrite Using Cobalt Phthalocyanine Modified on Multiwalled Carbon Nanotubes. *J. Electrochem. Soc.* **2020,** *167*, 046515.
57. Teker, T.; Aslanoglu, M. Sensitive and Selective Determination of Paracetamol Using a Composite of Carbon Nanotubes and Nanoparticles of Samarium Oxide and Zirconium Oxide. *Microchem. J.* **2020,**, 105234.

58. Kantize, K.; Booysen, I. N.; Mambanda, A. Electrochemical Sensing of Acetaminophen Using Nanocomposites Comprised of Cobalt Phthalocyanines and Multiwalled Carbon Nanotubes. *J. Electroanal. Chem.* **2019,** *850,* 113391.
59. Sudha, V.; Kumar, S. M. S.; Thangamuthu, R. Simultaneous Electrochemical Sensing of Sulphite and Nitrite on Acid-Functionalized Multi-Walled Carbon Nanotubes Modified Electrodes. *J. Alloys Compd.* **2018,** *749,* 990–999.
60. Xiao, L.; Wildgoose, G. G.; Compton, R. G. Sensitive Electrochemical Detection of Arsenic(III) Using Gold Nanoparticle Modified Carbon Nanotubes via Anodic Stripping Voltammetry. *Anal. Chim. Acta* **2008,** *620,* 44–49.
61. Manisankar, P.; Sundari, P. A.; Sasikumar, R.; Palaniappan, S. Electroanalysis of Some Common Pesticides Using Conducting Polymer/Multiwalled Carbon Nanotubes Modified Glassy Carbon Electrode. *Talanta* **2008,** *76,* 1022–1028.
62. Najafi, M.; Maleki, L.; Rafati, A. A. Novel Surfactant Selective Electrochemical Sensors Based on Single Walled Carbon Nanotubes. *J. Mol. Liq.* **2011,** *159,* 226–229.
63. Wang, T.; Yue, W. Carbon Nanotubes Heavy Metal Detection with Stripping Voltammetry: A Review Paper. *Electroanalysis* **2017,** *29,* 2178–2189.
64. Gao, C.; Guo, Z.; Liu, J.-H.; Huang, X.-J. The New Age of Carbon Nanotubes: An Updated Review of Functionalized Carbon Nanotubes in Electrochemical Sensors. *Nanoscale* **2012,** *4,* 1948–1963.
65. Wanekaya, A. K. Applications of Nanoscale Carbon-Based Materials in Heavy Metal Sensing and Detection. *Analyst* **2011,** *136,* 4383–4391.
66. Moyo, M.; Okonkwo, J. O.; Agyei, N. M. An Amperometric Biosensor Based on Horseradish Peroxidase Immobilized Onto Maize Tassel-Multi-Walled Carbon Nanotubes Modified Glassy Carbon Electrode for Determination of Heavy Metal Ions in Aqueous Solution. *Enzyme Microb. Technol.* **2014,** *56,* 28–34.
67. Wu, K. H.; Lo, H. M.; Wang, J. C.; Yu, S.-Y.; Yan, B. D. Electrochemical Detection of Heavy Metal Pollutant Using Crosslinked Chitosan/Carbon Nanotubes Thin Film Electrodes. *Mater. Express* **2017,** *7,* 15–24.
68. Wang, L.; Wang, X.; Shi, G.; Peng, C.; Ding, Y. Thiacalixarene Covalently Functionalized Multiwalled Carbon Nanotubes as Chemically Modified Electrode Material for Detection of Ultratrace Pb^{2+} Ions. *Anal. Chem.* **2012,** *84,* 10560–10567.
69. Xu, H.; Zeng, L.; Xing, S.; Xian, Y.; Jin, L. Microwave-Irradiated Synthesized Platinum Nanoparticles/Carbon Nanotubes for Oxidative Determination of Trace Arsenic(III). *Electrochem. Commun.* **2008,** *10,* 551–554.
70. Liu, Y.; Wei, W. Jointly Modified Single-Walled Carbon Nanotubes on Low Resistance Monolayer Modified Electrode for Arsenic(III) Detection. *J. Electroanal. Chem.* **2008,** *624,* 299–304.
71. Yusof, N. A.; Daud, N.; Tee, T. W.; Abdullah, A. H. Electrocatalytic Characteristic of Carbon Nanotubes/Glutamine/Nafion Modified Platinum Electrode in Development of Sensor for Determination of As(III). *Int. J. Electrochem. Sci.* **2011,** *6,* 2385–2397.
72. Afkhami, A.; Khoshsafar, H.; Bagheri, H.; Madrakian, T. Construction of a Carbon Ionic Liquid Paste Electrode Based on Multi-Walled Carbon Nanotubes-Synthesized Schiff Base Composite for Trace Electrochemical Detection of Cadmium. *Mater. Sci. Eng. C* **2014,** *35,* 8–14.
73. Ramalingam, M.; Ponnusamy, V.; Sangilimuthu, S. N. A Nanocomposite Consisting of Porous Graphitic Carbon Nitride Nanosheets and Oxidized Multiwalled Carbon

Nanotubes for Simultaneous Stripping Voltammetric Determination of Cadmium(II), Mercury(II), Lead(II) and Zinc(II). *Microchim. Acta* **2019,** *186*, 1–10.
74. Chen, K.; Li, J.; Zhang, L.; Xing, R.; Jiao, T.; Gao, F.; Peng, Q. Facile Synthesis of Self-Assembled Carbon Nanotubes/Dye Composite Films for Sensitive Electrochemical Determination of Cd(II) Ions. *Nanotechnology* **2018,** *29*, 445603.
75. Khani, H.; Rofouei, M. K.; Arab, P.; Gupta, V. K.; Vafaei, Z. Multi-Walled Carbon Nanotubes-Ionic Liquid-Carbon Paste Electrode as a Super Selectivity Sensor: Application to Potentiometric Monitoring of Mercury Ion(II). *J. Hazard. Mater.* **2010,** *183*, 402–409.
76. Nguyen, D. T.; Dai Tran, L.; Le Nguyen, H.; Nguyen, B. H.; Van Hieu, N. Modified Interdigitated Arrays By Novel Poly (1, 8-Diaminonaphthalene)/Carbon Nanotubes Composite for Selective Detection of Mercury(II). *Talanta* **2011,** *85*, 2445–2450.
77. Wei, J.; Yang, D.; Chen, H.; Gao, Y.; Li, H. Stripping Voltammetric Determination of Mercury(II) Based on SWCNT-PhSH Modified Gold Electrode. *Sens. Actuat. B* **2014,** *190*, 968–974.
78. Zhu, Y.; Zeng, G.; Zhang, Y.; Tang, L.; Chen, J.; Cheng, M.; Zhang, L.; He, L.; Guo, Y.; He, X. L. Highly Sensitive Electrochemical Sensor Using a MWCNTs/GNPs-Modified Electrode for Lead(II) Detection Based on Pb^{2+}-Induced G-rich DNA Conformation. *Analyst* **2014,** *139*, 5014–5020.
79. Anandhakumar, S.; Mathiyarasu, J. Detection of Lead(II) Using an Glassy Carbon Electrode Modified with Nafion, Carbon Nanotubes and Benzo-18-crown-6. *Microchim. Acta* **2013,** *180*, 1065–1071.
80. Niu, S.; Han, B.; Cao, W.; Zhang, S. Sensitive DNA Biosensor Improved by Luteolin Copper(II) as Indicator Based on Silver Nanoparticles and Carbon Nanotubes Modified Electrode. *Anal. Chim. Acta* **2009,** *651*, 42–47.
81. Meng, F. L.; Zhang, L.; Jia, Y.; Liu, J. Y.; Sun, Y. F.; Luo, T.; Li, M. Q.; Liu, J. H.; Huang, X. J. Electronic Chip Based on Self-Oriented Carbon Nanotube Microelectrode Array to Enhance the Sensitivity of Indoor Air Pollutants Capacitive Detection. *Sens. Actuat. B* **2011,** *153*, 103–109.
82. V.A. Pedrosa, R. Epur, J. Benton, R.A. Overfelt, A.L. Simonian, Copper Nanoparticles and Carbon Nanotubes-Based Electrochemical Sensing System for Fast Identification of Tricresyl-Phosphate in Aqueous Samples and Air. *Sens. Actuat. B* **2011,** *140*, 92–97.
83. Xie, H.; Sheng, C.; Chen, X.; Wang, X.; Li, Z.; Zhou, J. Multi-Wall Carbon Nanotube Gas Sensors Modified with Amino-Group to Detect Low Concentration of Formaldehyde. *Sens. Actuat. B* **2012,** *168*, 34–38.
84. Ong, K. G.; Grimes, C. A. A Carbon Nanotube-Based Sensor for CO_2 Monitoring. *Sensors* **2001,** *1*, 193–205.
85. Santhosh, P.; Manesh, K.; Gopalan, A.; Lee, K. P. Novel Amperometric Carbon Monoxide Sensor Based on Multi-Wall Carbon Nanotubes Grafted with Polydiphenylamine-Fabrication and Performance. *Sens. Actuat. B* **2007,** *125*, 92–99.
86. Kumar, M. K.; Reddy, A. L. M.; Ramaprabhu, S. Exfoliated Single-Walled Carbon Nanotube-Based Hydrogen Sensor. *Sens. Actuators B* **2008,** *130*, 653–660.
87. Hoang, N. D.; Van Cat, V.; Nam, M. H.; Phan, V. N.; Le, A. T.; Van Quy, N. Enhanced SO_2 Sensing Characteristics of Multi-Wall Carbon Nanotubes Based Mass-Type Sensor Using Two-Step Purification Process. *Sens. Actuators A* **2019,** *295*, 696–702.

88. Kim, S.; Lee, K. H.; Lee, J. Y.; Kim, K. K.; Choa, Y. H.; Lim, J. H. Single-Walled Carbon Nanotube-Based Chemi-Capacitive Sensor for Hexane and Ammonia. *Electron. Mater. Lett.* **2019,** *15*, 712–719.
89. Contes-de Jesus, E.; Li, J.; Cabrera, C. R. Latest Advances in Modified/Functionalized Carbon Nanotube-Based Gas Sensors. *Syntheses App. Carbon Nanotubes Comp.* **2013,**, 337–366.
90. Zhang, W.; Liu, Y.; Wang, M. L. DNA-Functionalized Single-Walled Carbon Nanotube-Based Sensor Array for Gas Monitoring. *Smart Struct. Syst.* **2013,** *12*, 73–95.
91. Liu, C. K.; Wu, J. M.; Shih, H. C. Application of Plasma Modified Multi-Wall Carbon Nanotubes to Ethanol Vapor Detection. *Sens. Actuators B* **2010,** *150*, 641–648.
92. Hrapovic, S.; Majid, E.; Liu, Y.; Male, K.; Luong, J. H. Metallic Nanoparticle–Carbon Nanotube Composites for Electrochemical Determination of Explosive Nitroaromatic Compounds. *Anal. Chem.* **2006,** *78*, 5504–5512.
93. Wang, J. Electrochemical Sensing of Explosives. *Electroanalysis* **2007,** *19*, 415–423.
94. Hnaien, M.; Bourigua, S.; Bessueille, F.; Bausells, J.; Errachid, A.; Lagarde, F.; Jaffrezic-Renault, N. Impedimetric Microbial Biosensor Based on Single Wall Carbon Nanotube Modified Microelectrodes for Trichloroethylene Detection. *Electrochim. Acta* **2011,** *56*, 10353–10358.
95. Wang, X.; Zhai, C.; Sun, X. Acetylcholinesterase Biosensor Design Based on the Gold Nanoparticles and Multi-Wall Carbon Nanotubes Modified Electrode for Amperometric Detection of Organophosphate Pesticides. *Int. Agric. Eng. J.* **2011,** *20*, 47–51.
96. Chauhan, N.; Pundir, C. S. An Amperometric Acetylcholinesterase Sensor Based on Fe_3O_4 Nanoparticle/Multi-Walled Carbon Nanotube-Modified ITO-Coated Glass Plate for the Detection of Pesticides. *Electrochim. Acta* **2012,** *67*, 79–86.
97. Rahemi, V.; Garrido, J.; Borges, F.; Brett, C.; Garrido, E. Electrochemical Determination of the Herbicide Bentazone Using a Carbon Nanotube β-Cyclodextrin Modified Electrode. *Electroanalysis* **2013,** *25*, 2360–2366.
98. Dong, M.; Ma, Y.; Zhao, E.; Qian, C.; Han, L.; Jiang, S. Using Multiwalled Carbon Nanotubes as Solid Phase Extraction Adsorbents for Determination of Chloroacetanilide Herbicides in Water. *Microchim. Acta* **2009,** *165*, 123–128.
99. Ribeiro, W. F.; Selva, T. M. G.; Lopes, I. C.; Coelho, E. C. S.; Lemos, S. G.; de Abreu, F. C.; do Nascimento, V. B.; de Araújo, M.C.U. Electroanalytical Determination of Carbendazim by Square Wave Adsorptive Stripping Voltammetry with a Multiwalled Carbon Nanotubes Modified Electrode. *Anal. Methods* **2011,** *3*, 1202–1206.
100. Demir, E.; İnam, R. Square Wave Voltammetric Determination of Fomesafen Herbicide Using Modified Nanostructure Carbon Paste Electrode as a Sensor and Application to Food Samples. *Food Anal. Methods* **2017,** *10*, 74–82.
101. Bhardwaj, J.; Devarakonda, S.; Kumar, S.; Jang, J. Development of a Paper-Based Electrochemical Immunosensor Using an Antibody-Single Walled Carbon Nanotubes Bio-Conjugate Modified Electrode for Label-Free Detection of Foodborne Pathogens. *Sens. Actuat. B* **2017,** *253*, 115–123.
102. Queirós, R. B.; Guedes, A.; Marques, P.; Noronha, J.; Sales, M. G. F. Recycling Old Screen-Printed Electrodes with Newly Designed Plastic Antibodies on the Wall of Carbon Nanotubes as Sensory Element for In Situ Detection of Bacterial Toxins in Water. *Sens. Actuators B* **2013,** *189*, 21–29.

103. Prieto-Simón, B.; Bandaru, N.; Saint, C.; Voelcker, N. Tailored Carbon Nanotube Immunosensors for the Detection of Microbial Contamination. *Biosens. Bioelectron.* **2015,** *67*, 642–648.
104. Güner, A.; Cevik, E.; Şenel, M.; Alpsoy, L. An Electrochemical Immunosensor for Sensitive Detection of Escherichia coli O157: H7 by Using Chitosan, MWCNT, Polypyrrole with Gold Nanoparticles Hybrid Sensing Platform. *Food Chem.* **2017,** *229*, 358–365.
105. Inbaraj, B. S.; Chen, S. Nanomaterial-Based Sensors for Detection of Foodborne Bacterial Pathogens and Toxins as Well as Pork Adulteration in Meat Products. *J. Food Drug Anal.* **2016,** *24*, 15–28.
106. Zambrano-Monserrate, M. A.; Ruano, M. A.; Sanchez-Alcalde, L. Indirect Effects of Covid-19 on the Environment. *Sci. Total Environ.* **2020,**, 138813.
107. Chakraborty, I.; Maity, P. COVID-19 Outbreak: Migration, Effects on Society, Global Environment and Prevention. *Sci. Total Environ.* **2020,**, 138882.
108. Tang, X.; Zhou, L.; Le, Z.; Wang, Y.; Liu, Z.; Huang, G.; Adesina, A. A. Preparation of Porous Chitosan/Carboxylated Carbon Nanotube Composite Aerogels for the Efficient Removal of Uranium(VI) from Aqueous Solution. *Int. J. Biol. Macromol.* **2020,**.
109. Gizawy, M. A.; Shamsel-Din, H. A.; Abdelmonem, I. M.; Ibrahim, M. I. A.; Mohamed, L. A.; Metwally, E. Synthesis of Chitosan-Acrylic Acid/Multiwalled Carbon Nanotubes Composite for Theranostic ^{47}Sc Separation from Neutron Irradiated Titanium Target. *Int. J. Biol. Macromol.* **2020,** *163*, 79–86.
110. Rouf, T. B.; Díaz-Amaya,S.; Stanciu, L.; Kokini, J. Application of Corn Zein as an Anchoring Molecule in a Carbon Nanotube Enhanced Electrochemical Sensor for the Detection of Gliadin. *Food Control* **2020,** *117*, 107350.
111. Zhang, H.; Sun, D.; Cao, T. Electrochemical Sensor Based on Silver Nanoparticles/ Multi Walled Carbon Nanotubes Modified Glassy Carbon Electrode to Detect Cyanide in Food Products. *Int. J. Electrochem. Sci.* **2020,** *15*, 3434–3444.
112. Xiong, H.; Zhao, Y.; Liu, P.; Zhang, X.; Wang, S. Electrochemical Properties and the Determination of Nicotine at a Multi-Walled Carbon Nanotubes Modified Glassy Carbon Electrode. *Microchim. Acta* **2010,** *168*, 31–36.
113. Lo, T.W.B.; Aldous, L.; Compton, R. G. The Use of Nano-Carbon as an Alternative to Multi-Walled Carbon Nanotubes in Modified Electrodes for Adsorptive Stripping Voltammetry. *Sens. Actuators B* **2012,** *162*, 361–368.
114. Majumdar, S.; Thakur, D.; Chowdhury, D. DNA Carbon-Nanodots Based Electrochemical Biosensor for Detection of Mutagenic Nitrosamines. *ACS Appl. Bio Mater.* **2020,** *3*, 1796–1803.
115. Xi, W.; Zhai, J.; Zhang, Y.; Tian, L.; Zhang, Z. Integrating Brucine with Carbon Nanotubes Toward Electrochemical Sensing of Hydroxylamine. *Microchim. Acta* **2020,** *187*, 343.
116. Filik, H.; Avan, A. A. Neutral Red Interlinked Gold Nanoparticles/Multiwalled Carbon Nanotubes Modified Electrochemical Sensor for Simultaneous Speciation and Detection of Chromium(VI) and Vanadium(V) in Water Samples. *Microchem. J.* **2020,**, 105242.
117. Tajik, S.; Beitollahi, H.; Dourandish, Z.; Zhang, K.; Le, Q. V.; Nguyen, T. P.; Kim, S. Y.; Shokouhimehr, M. Recent Advances in the Electrochemical Sensing

of Venlafaxine: An Antidepressant Drug and Environmental Contaminant. *Sensors* **2020,** 20, 3675.
118. Torrinha, A.; Oliveira, T. M.; Ribeiro, F. W.; Correia, A. N.; Lima-Neto, P.; Morais, S. Application of Nanostructured Carbon-Based Electrochemical (Bio)Sensors for Screening of Emerging Pharmaceutical Pollutants in Waters and Aquatic Species: A Review. *Nanomaterials* **2020,** *10*, 1268.
119. Lee, M.; Bae, J.; Lee, J.; Lee, C. S.; Hong, S.; Wang, Z. L. Self-Powered Environmental Sensor System Driven by Nanogenerators. *Energy Environ. Sci.* **2011,** *4*, 3359–3363.
120. Evans, G. P.; Buckley, D. J.; Adedigba, A. L.; Sankar, G.; Skipper, N. T.; Parkin, I. P. Controlling the Cross-Sensitivity of Carbon Nanotube-Based Gas Sensors to Water Using Zeolites. *ACS Appl. Mater. Interf.* **2016,** *8*, 28096–28104.
121. Dinh, T.; Phan, H. P.; Nguyen, T. K.; Qamar, A.; Foisal, A. R. M.; Viet, T. N.; Tran, C.-D.; Zhu, Y.; Nguyen, N.-T.; Dao, D. V. Environment-Friendly Carbon Nanotube Based Flexible Electronics for Noninvasive and Wearable Healthcare. *J. Mater. Chem. C* **2016,** *4*, 10061–10068.
122. Vairavapandian, D.; Vichchulada, P.; Lay, M. D. Preparation and Modification of Carbon Nanotubes: Review of Recent Advances and Applications in Catalysis and Sensing. *Anal. Chim. Acta* **2008,** *626*, 119–129.
123. Zhang, B. T.; Zheng, X.; Li, H. F.; Lin, J. M. Application of Carbon-Based Nanomaterials in Sample Preparation: A Review. *Anal. Chim. Acta* **2013,** *784*, 1–17.
124. Yadav, M.; Sonkar, P. K.; Prakash, K.; Ganesan, V.; Sankar, M.; Yadav, D. K.; Gupta, R. Insight into Efficient Bifunctional Catalysis: Oxygen Reduction and Oxygen Evolution Reactions Using MWCNTs Based Composites with 5,10,15,20-tetrakis (3′,5′-dimethoxyphenyl)porphyrinato Cobalt(II) and 5,10,15,20-tetrakis(3′,5′-dihydroxyphenyl) porphyrinato Cobalt(II). *Int. J. Hydrogen Energ.* **2020,** *45*, 9710–9722.
125. Schnorr, J. M.; Swager, T. M.; Emerging Applications of Carbon Nanotubes. *Chem. Mater.* **2011,** *23*, 646–657.

FURTHER READINGS

126. Terrones, M. Science and Technology of the Twenty-First Century: Synthesis, Properties, and Applications of Carbon Nanotubes. *Annu. Rev. Mater. Res.* **2003,** *33,* 419–501.
127. Ando, Y.; Iijima, S. Preparation of Carbon Nanotubes by Arc-Discharge Evaporation. *Jpn. J. Appl. Phys.* **1993,** *2* (32), L107–L107.
128. Yakobson, B. I.; Avouris, P. Mechanical Properties of Carbon Nanotubes. *Carbon Nanotubes;* Springer, 2001; pp 287–327.
129. Koyama, S.; Haniu, H.; Osaka, K.; Koyama, H.; Kuroiwa, N.; Endo, M.; Kim, Y. A.; Hayashi, T. Medical Application of Carbon Nanotube Filled Nanocomposites: The Microcatheter. *Small* **2006,** *2,* 1406–1411.
130. Frackowiak, E.; Beguin, F. Electrochemical Storage of Energy in Carbon Nanotubes and Nanostructured Carbons. *Carbon* **2002,** *40,* 1775–1787.

CHAPTER 11

Green Synthesis of Carbon Nanotubes

NIMITHA K. C.[1], JIJI ABRAHAM[1*], ARUNIMA REGHUNADHAN[2,3], SONEY C. GEORGE[4], and SABU THOMAS[3]

[1]*Vimala College (Autonomous) Thrissur, Kerala, India*

[2]*Department of Chemistry, TKM College of Engineering, Karicode, Kollam*

[3]*International and Inter University Centre for Nanoscience and Nanotechnology, Mahatma Gandhi University, P.D. Hills, Kottayam, Kerala, India*

[4]*Centre for Nanoscience and Nanotechnology, Amal Jyothi College of Engineering, Kottayam, Kerala, India*

[*]*Corresponding author. E-mail: jijiabraham02@gmail.com*

ABSTRACT

Carbon nanotube, the one-atom thick graphene sheet rolled up into cylindrical form has been considered as an innovative material with fascinating properties. General methods adopted for the synthesis of carbon nanotube include the carbon arc-discharge, chemical vapor deposition often abbreviated as CVD, and laser ablation method, etc. By considering the environmental issues, for all the synthetic chemistry researchers, there is an urge to shift toward the greener routes and so is with the synthesis of CNTs. Natural resources and some waste materials are found to be suitable as good source of carbon. This chapter spotlights on the green synthetic aspects of CNT from various sustainable starting points like plant resources, waste materials, and various green catalytic methods. At the end of the chapter, a brief discussion is made on the various characterization techniques adopted for the characterization of CNT obtained through green synthesis.

11.1 INTRODUCTION

11.1.1 CARBON NANOTUBE: THE FASCINATING MATERIAL

The carbon nanotubes (CNTs) fit into comparatively a new-fangled class of nanomaterials that have been publicly known for almost 30 years, but their history is a little bit longer. CNTs were at the outset scrutinized and reported in 1952 by Radushkevich and Lukyanovich and later in 1976, the single-(or double) walled carbon nanotubes were observed by Oberlin et al. In more recent history, the discovery of CNTs is attributed to Iijima as the first scientist who described the multiwalled carbon nanotubes (MWNTs) preparation process in 1991. Carbon nanotubes are promising material in the nanotechnology. It is seen as unitary of the allotrope of carbon in between fullerene and graphene. It is a one dimensional nanomaterial. CNTs have been discovered by Iijima in 1991. It is a one-atom thick graphene sheet (hexagonal sp^2 carbon layer) rolled up into cylindrical form. Each carbon atom is bonded to three other carbon atom by covalent bonds. In one category of carbon nanotubes, they have cylindrical shape with closed end. In another type CNTs have cylindrical shape with closed ends. Three different types of nanotubes are known: single-walled carbon nanotube (SWCNT): structure composed of only one cylinder, double-walled carbon nanotube (DWCNT): structure composed of two cylinders, and MWCNT: structure composed of more than two cylinders. A single-walled nanotube (SWNT) is assumed to be formed by rolling a sheet of graphene into a cylinder along a (m,n) lattice vector in the graphene plane (Figure 11.1). The (m,n) indices determine the diameter and chirality, which are key parameters of a nanotube.[1] Based on the method of rolling of graphene sheet, there are three different crystallographic configurations for SWCNT, namely, zigzag, armchair, and chiral. Two opposite C-C bonds of each hexagon are parallel to tube axis in the zigzag configurations. But in the arm chair configuration, two opposite C-C bonds of each hexagon are perpendicular to tube axis. The opposite C-C bonds lie at an angle to the tube axis in all other configurations and these are known as chiral type. As mentioned earlier, SWCNTs were first observed by Iijima and Ichihashi in 1993. Carbon nanotubes have high aspect ratio greater than 1000. These CNTs have diameter in the nanomater range and length in the micrometer range. The number of walls of MWCNT can vary from 2 to 100 and usually grow perpendicular to the substrate.

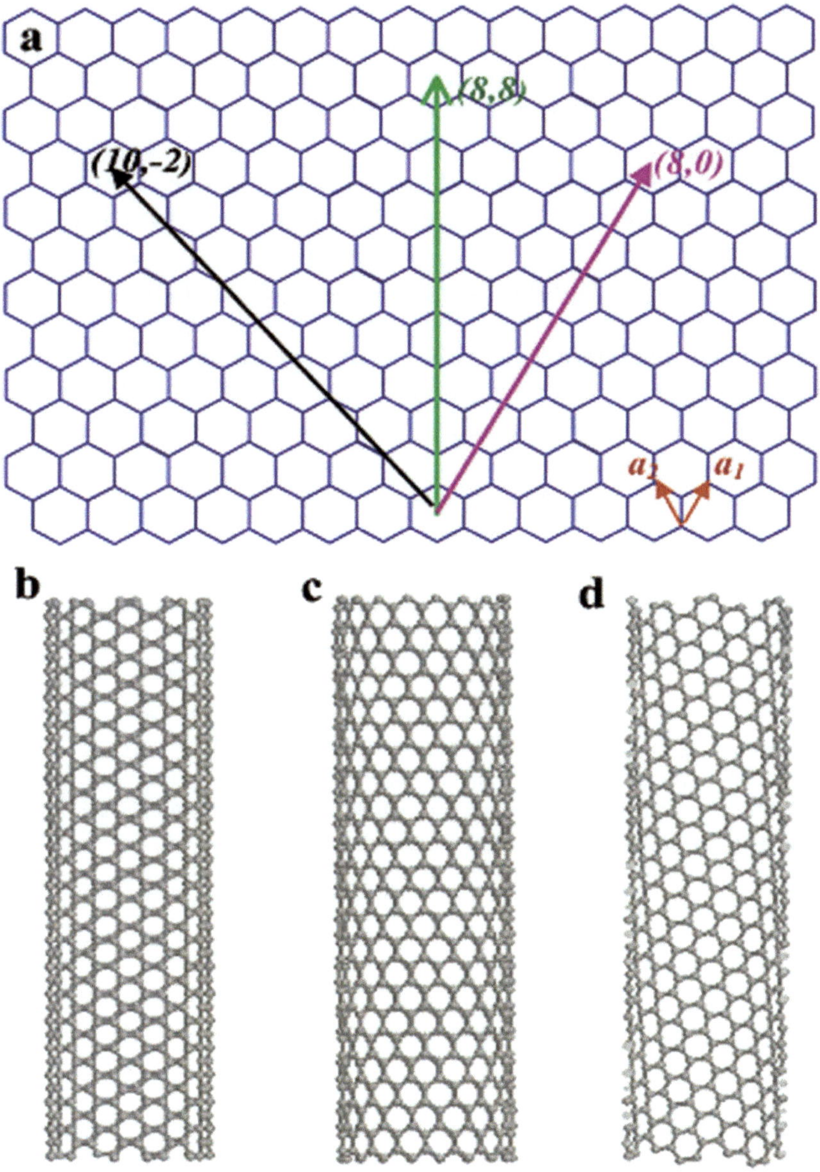

FIGURE 11.1 (a) Schematic representation of honeycomb structure of a graphene sheet. Single-walled carbon nanotubes can be formed by folding the sheet along lattice vectors. The two basis vectors **a**1 and **a**2 are shown. Folding of the (8,8), (8,0), and (10,−2) vectors leads to armchair (b), zigzag (c), and chiral (d) tubes

Source: Reprinted with permission from Ref. 4. © 2002 American Chemical Society.

Extensive research has been carried out in CNTs because of their outstanding properties and commercial applications. Remarkable properties of CNTs are derived from the curved sp^2 graphene layers by impressing extra quantum confinement and topological constraints in the circumferential track of the cylinders.[2] Carbon nanotubes have got excellent electrical, mechanical, electronic, chemical, and optical properties. Carbon–carbon covalent bonds present in CNTs make them excellent thermal conducting material. Youngs modulus of SWCNT is higher than that of the commercial carbon fibers which indicate the high mechanical strength (the C–C covalent bond in CNTs is one of the strongest bonds found in nature) of SWCNT. Carbon nanotubes have remarkable electrical properties—semiconducting, metallic or semimetallic depending on the method of production which controls the degree of graphitization, the diameter of the tube, and the helicity, that is, wrapping angles. Electrical properties of CNTs are structure-dependent, all armchair SWCNTs and one-third of all zigzag nanotubes are metallic, the remaining are semiconducting (Table 11.1).[3]

TABLE 11.1 Some of the Extraordinary Properties of Carbon Nanotubes Ref. [6].

Mechanical properties	Young's modulus of SWCNTs 1 TPa
	Young's modulus of MWCNTs 1–1.2TPa
	Tensile strength of SWCNT ropes 60 GPa
	Tensile strength of MWCNT 0.15TPa
Thermal properties at room temperature	Thermal conductivity of SWCNT 1750–5800 W/mk
	Thermal conductivity of MWCNT 43000 W/mK
Electronic properties	SWCNT band gap When n–m is divisible by 3 (0 eV,metallic) When n–m is not divisible by 3 (0.4–2 eV, semiconducting
	MWCNTs band gap 0 eV (non-semiconducting
Electrical properties	Typical resistivity of SWCNT and MWCNTs = 10^{-6} Ωm
	Typical maximum current density = 10^7–10^9 Acm^{-2}
	Typical quantized conductance (measured) = 12.9 $k\Omega^{-1}$

11.1.2 GENERAL METHODS IMPLEMENTED IN NANOTUBE SYNTHESIS

11.1.2.1 CARBON ARC-DISCHARGE METHOD

Iijima synthesized the carbon nanotube by arc-discharge method in 1991. Arc-discharge is the electrical breakdown of a gas to generate plasma.

Several synthesis parameters like setup modification, power supplies, arc current, catalyst, pressure, grain size, electrode geometry, and temperature affect nanotube production. There are different methods of growth mechanism for CNT in this method: growth in vapor phase, liquid phase, solid phase, and crystal phase.[4] Figure 11.2 represents the schematic of an arc discharge setup.

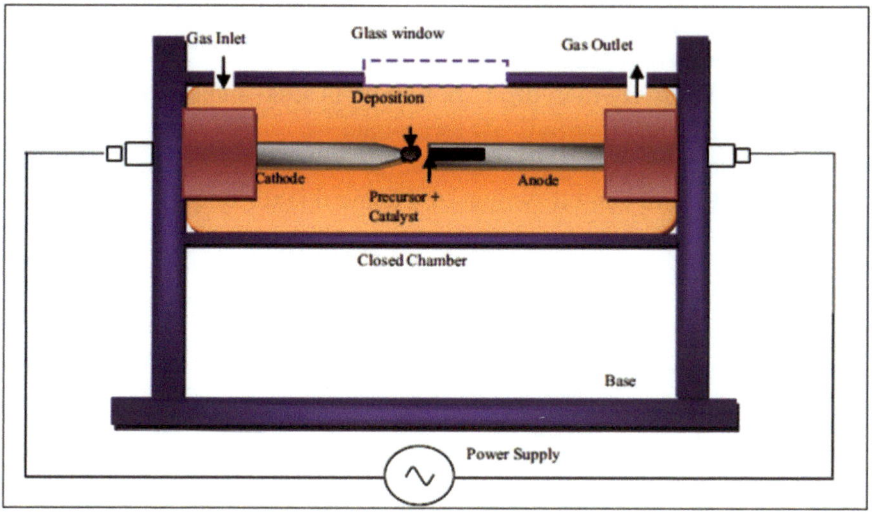

FIGURE 11.2 Schematic of an arc discharge setup.

Source: Reprinted with permission from Ref. 7. © 2014 Elsevier.

11.1.2.2 CHEMICAL VAPOR DEPOSITION

CVD technique is a widely accepted technique because it is a simple and cost-effective technique. Here the catalyst is supported and is deposited on a substrate and the nucleation of catalyst is done through chemical etching or thermal annealing. A transition metal, such as Fe, Co, Ni, or their alloys, is commonly used as catalyst, while Al_2O_3, MgO, and SiO_2 are the best oxide supports. The preprepared supporting material is placed in a tubular furnace for growth process. Then tubular furnace is heated to 600°C–1200°C. A mixture of hydrocarbon gas (ethylene, methane, and acetylene, etc.) and processing gas (nitrogen, hydrogen, and argon) are purged into reaction chamber for a given period of time (usually 15–60 min). CNTs are grown on the surface of the catalyst and CNT-deposited catalyst is collected from the side walls of the reactor after cooling the system to room temperature. There are different types of CVD techniques which include water-assisted CVD, oxygen-assisted CVD, hot-filament CVD, microwave plasma CVD, or radiofrequency CVD. This is the most effective method for the synthesis of doped CNT.[5] Figure 11.3 shows the schematic diagram of a CVD setup. Important advantages of CVD technique include:

- It is a simple, economically feasible method for the large-scale synthesis of CNT.
- A wide variety of hydrocarbons both in solid and liquid form can be used as precursor in CVD technique.
- Aligned or entangled, straight, or coiled nanotubes or a desired architecture can be synthesized using this technique.

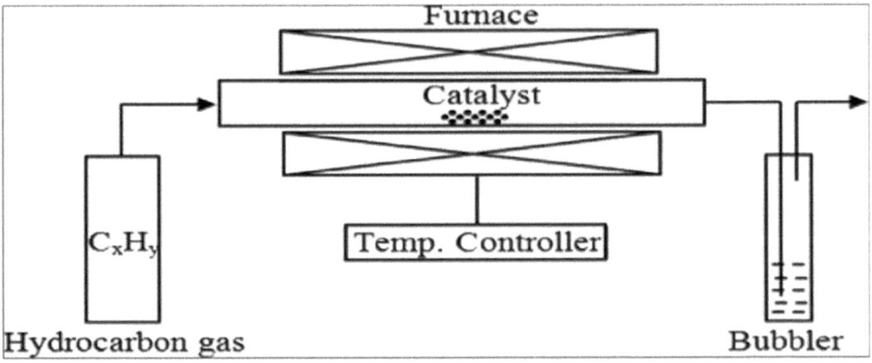

FIGURE 11.3 Schematic drawing of a CVD setup in its simplest form.

Source: Reprinted with permission from Ref. 6. © 2015 Elsevier.

11.1.2.3 PLASMA SYNTHESIS OF CNT

Plasma contains three different kinds of particles: free electrons, positively charged ions, and nonionized atom. There are two different types of plasma: low-temperature plasma and high-temperature plasma. Low-temperature plasma technique has several advantages, such as highly distributed active species, reduced energy requirements enhanced catalyst activation, shortened operation time, and decreased environmental pollution.[6] Microwave plasma-enhanced chemical vapor deposition (MPE-CVD) was used for the direct growth of aligned carbon nanotubes with open ends on the silicon wafer in single step. Through this method, it is possible to synthesize CNT at a temperature below 700°C and this method is suitable for the fabrication of many electronic devices.[7]

11.1.2.4 LASER ABLATION METHOD

Pulsed laser deposition (PLD) is an effective strategy for the preparation of multilayer structure and thin films. Laser ablation experiments were done in an electrical tube furnace which is heated to a temperature of 1200 °C. It offers extremely high temperature. Graphite is used as precursor and a metallic catalyst is used to accelerate the reaction. Upon laser irradiation, graphite and metallic salts are carbon–metal vapor. Vaporized particles are heated again by laser beam and the temperature reaches to tens of kilokelvin and plasma plume is formed. Plasma plume is further expanded. During the expansion of plasma it cools down, carbon and catalyst metal atoms condense into larger structures. CNTs synthesized by PLV technique have excellent purity, structural integrity and are long. Usually, 1064 or 532 nm Nd:YAG lasers are applied.[8] Figure 11.4 represents the reactor for carbon nanotube synthesis by the laser ablation method.

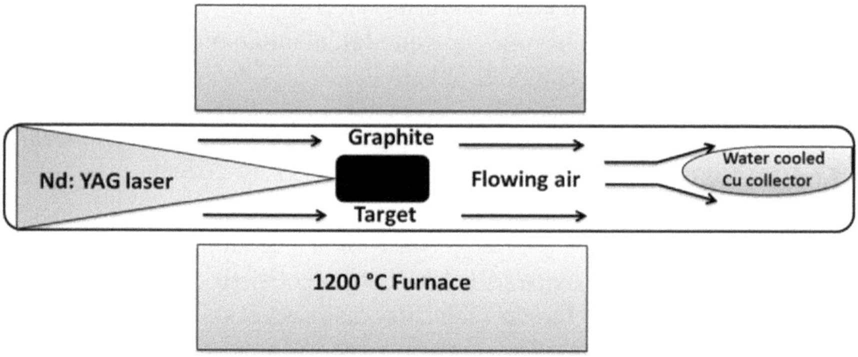

FIGURE 11.4 The laser ablation method reactor for carbon nanotube synthesis.

11.1.2.5 PYROLYSIS

Pyrolysis is indeed one of the thermochemical processes in which, in the absence of oxygen, matter is fragmented at a high temperature into smaller components. It is a two-stage process. CNT precursors are generated by pyrolysis in the very first step, and these precursors are later deposited on catalysts. Waste materials like sawdust, organic solid wastes, plastic wastes, etc. can be used for the synthesis of CNT.[9] There are several pyrolysis techniques which include fixed-bed pyrolysis, spray pyrolysis, fluidized-bed pyrolysis, etc.

11.1.2.6 MOLTEN SALT ELECTROLYTIC METHOD

The molten salt production method for carbon nanotubes was discovered by Hsu et al.[10] The method is comparatively straightforward, fast, and harmless for the production of large-scale synthesis of CNT. In molten salt electrolytic method, graphite is used as cathode, fused ionic salt as electrolyte.[11] Carbon nanotubes are formed by the interaction of the metal with cathode. The carbon cathode erodes during the electrolysis and a mixture of CNT and carbon nanoparticles of different structures is obtained. Electrolytic methods have following advantages: simple apparatus, low energy consumption for electrolysis process, cheap raw materials, ability to control the structure and morphologies of the carbon structures. But this method has some disadvantages, such as (i) cracking and damage of graphite cathode during electrolysis, (ii) accumulation of electrolysis products: chlorine gas (anode), alkaline metal (cathode), and carbon nanomaterials (cathode) in bath.[12]

11.1.3 NECESSITY FOR GREEN ALTERNATIVES

Green nanotechnology is categorized as a technology that seeks to enhance clean technologies to lessen the potential risks to human health and the environment. Green chemistry uses a set of techniques which minimize the use of hazardous materials and use a clean and safe synthesis strategy. Scientists and politicians have given much importance to 21st century, so this century is called "Century of Environment."[13] Greener environmental friendly techniques have been getting increased interest because of the severe environmental problems associated with chemical methods of synthesis. Even though several techniques have been used for the chemical synthesis of CNT, they are expensive and involve the use of hazardous chemicals. Green synthesis of nanomaterials involves the use of biological route using naturally available reagents like vitamins, sugars, plant polyphenols, and agricultural residues, biodegradable polymers, and microorganisms, and microwave heating. Advantages of green synthesis strategy include

- Eliminates the use of expensive chemicals.
- Can produce less hazardous products.
- No need of high temperature, pressure, energy, and toxic chemicals.
- Safer solvents and auxiliary substances.
- Produces less toxic waste.
- Economical and Eco-friendly approach.

11.2 ENVIRONMENTAL FRIENDLY SYNTHESIS OF CARBON NANOTUBES: INTRODUCTION

Synthesis of new generation nanomaterials including carbon-based materials is a topic of discussion nowadays. Several precursors like methane, benzene, toluene, etc. can be used for the synthesis of carbon-based materials like CNTs and graphene. These carbon feedstock materials are derived from fossil fuels which are nonrenewable resources. This is an emerging challenge since this will form an energy crisis in the present scenario. In order to overcome this challenge and to develop more competitive carbon-based materials, several methods are being introduced to synthesize such materials from natural resources. The main advantage of such methods is that they offer green alternatives and the raw materials are cheap also. Availability of raw materials is extended to the future since they are renewable. This will strongly support the principles of green chemistry. However, the process should possess maximum atom economy, that is, the process should include maximum incorporation of constituent atoms into the final product. Natural substances and waste materials which are hydrocarbon sources are available on earth in much abundance. Plant-derived precursors are also a big source of carbon. They are having some basic hydrocarbon skeletal structures. When compared with fossil fuels, these natural resources are quite inexpensive too. These materials can be transformed effectively into 1D and 2D nanomaterials. These materials pave a way to environmentally friendly large-scale production of CNTs and graphene.[14]

11.2.1 NATURAL HYDROCARBON PRECURSORS

Single-walled and multiwalled carbon nanotubes and large area, layered graphene has been widely produced from natural hydrocarbon resources by chemical vapor deposition and spray pyrolysis methods. Natural hydrocarbon precursors include various plant-based oils like turpentine oil, eucalyptus oil, palm oil, neem oil, sunflower oil, etc., biodiesel, tea tree extract, honey, dairy products like milk and butter, sugar, egg, etc. These raw materials are environment friendly, cost-effective and are having high atom economy too. The prepared materials possess novel electronic, optical, and energy applications. Figure 11.5 represents different natural sources and waste materials that are widely used for the synthesis of graphene and CNTs.

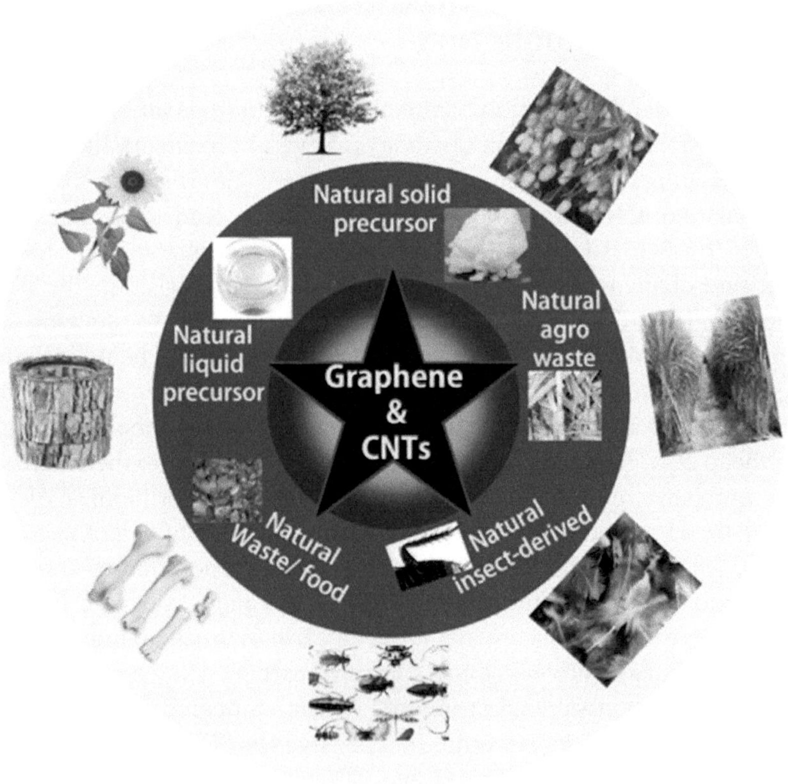

FIGURE 11.5 Natural and waste liquid and solid hydrocarbon precursors used for the synthesis of graphene and CNTs.

Source: Reprinted with permission from Ref. 17. © 2015 Elsevier.

Turpentine oil is considered as a good natural carbon source. It is derived from *Pinus palustris* of Pinaceae family. It is a very common plant species found in all over the parts of the world. One of the most common applications of turpentine oil is its use as a thinner in paints and varnishes. Turpentine oil is a mixture of essential oils and terpenes. By the spray pyrolysis process, CNTs can be formulated from turpentine oil. In this method, a quartz boat with a suitable catalyst is introduced into a quartz tube. Generally, a bimetallic catalyst of cobalt and iron supported on silica gel is employed as the catalyst. Catalyst has a key role in the nucleation of carbon nanotube. Size of the catalyst determines the formation of SWNTs and MWNTs and their diameter. Quartz tube is connected to a spray nozzle and water bubble. An inert atmosphere is created by purging N_2 gas to it. The system is then kept in a suitable temperature and carbon source is allowed to flow through this. Deposition of CNTs is recorded in different time intervals. This method is suitable for single-walled and multiwalled carbon nanotubes. Usually, low temperature is suitable for the synthesis of MWCNTs. A temperature range of 500°C–900°C is considered as favorable range. Studies have shown that at higher temperatures, catalyst gets agglomerated and this will result in the formation of thicker nanotubes.[15] When ferrocene is used as a catalyst, it is found that bundles of aligned CNTs (ACNTs) can effectively be synthesized.[16] Figure 11.6 shows the schematic representation of spray pyrolysis.

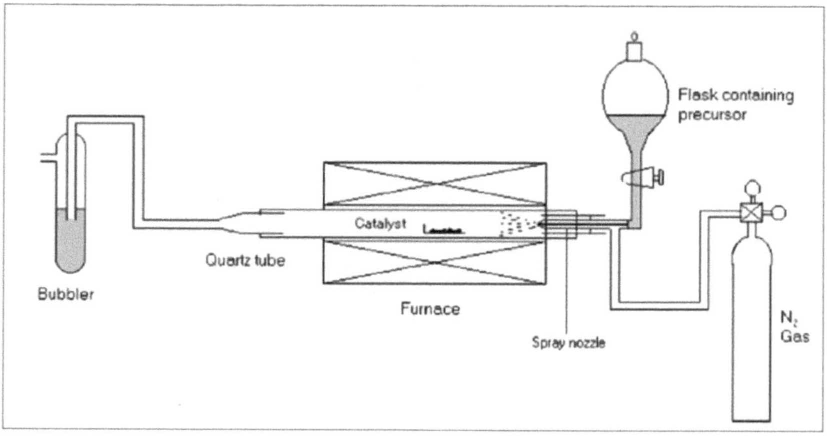

FIGURE 11.6 Schematic representation of spray pyrolysis setup used for synthesis of carbon nanotubes at different reaction temperatures.

Source: Reprinted with permission from Ref. 18. © 2006 Elsevier.

SEM and TEM images of CNTs show that they are well grown at all temperatures. Temperature below 500°C is insufficient for the pyrolysis and therefore CNTs are not well formed over this range. On the other hand, a profound formation of CNTs with uniform diameter is found at 700°C. Figure 11.7 represents the TEM images of as-grown CNTs at different temperatures.

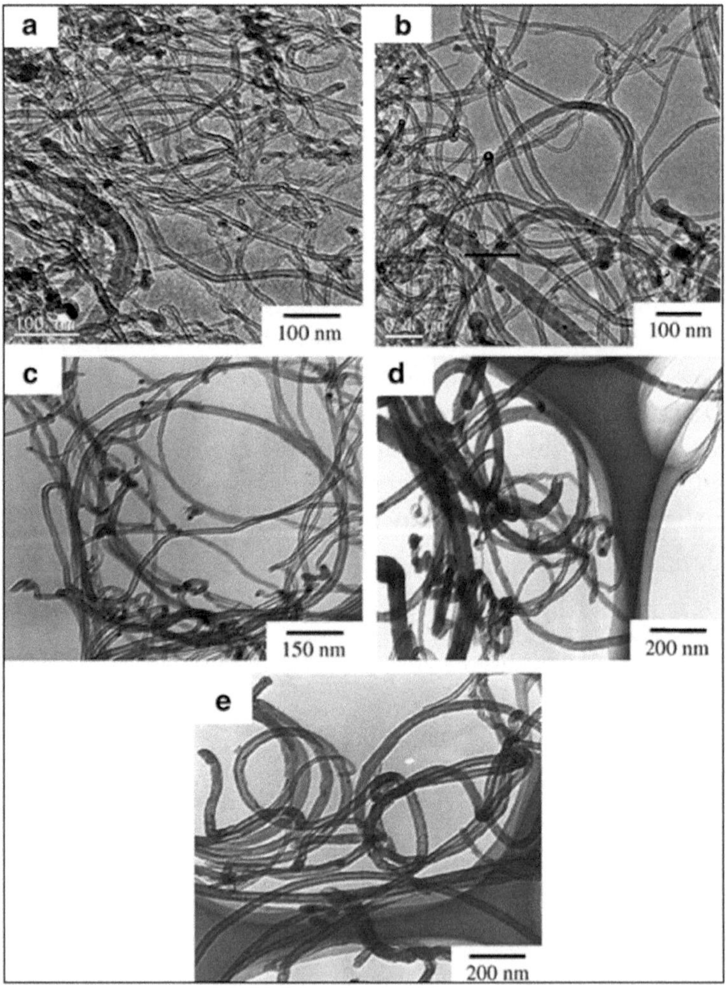

FIGURE 11.7 TEM images of as-grown CNTs at different temperatures: (a) 500°C, (b) 600°C, (c) 700°C, (d) 800°C, and (e) 900°C.

Source: Reprinted with permission from Ref. 18. © 2006 Elsevier.

Another important source of CNTs is the eucalyptus oil. Using eucalyptus oil, single-walled carbon nanotubes can be efficiently synthesized using spray pyrolysis method. In this method, Y-zeolite is employed as a catalyst. Eucalyptus oil consists of many bicyclic chemical compounds, chances of contamination in low traces is pretty much high. However, studies have shown that the CNTs prepared have high purity and uniform diameter too. Most probably, zeolite-assisted synthesis produces SWNTs in larger proportions.[17] SEM micrographs clearly confirm the deposition of CNTs on the surface of zeolite particles. Raman spectroscopy also helps to identify the CNTs thus formed. Figure 11.4 represents the Raman spectra of CNTs synthesized at 850°C. Spectrum is characterized by the strong G band at 1587 $cm^{-1}$and the weak D band at 1335 cm^{-1}. Also, sharp radial breathing modes are also observed at 100–400 cm^{-1}. Splitting of G band confirms the presence of SWNTs (in Figure 11.8b).[18]

Another important plant-based carbon feed stocks are palm oil, neem oil, sunflower oil, and castor oil. CNTs can be synthesized from palm oil by aerosol–assisted catalytic CVD method. Suriani and coworkers have successfully synthesized CNTs from palm oil for the first time. Pure, multiwalled CNTs were synthesized at that time. Here, ferrocene is generally used as a catalyst. This method is cost-effective and also it provides a green alternative for the industrial-scale production of CNTs. SEM analysis of the materials shows the uniform distribution of CNTs and they are arranged in strands. Diameter exactly lies in the nano range of ~ 20–30 nm. EDX analysis of the sample prepared shows ~ 13.57 At% C. Traces of elements like Si, Fe, and O were also found. Presence of these elements is due to the catalyst and impurities also.[19] Thermogravimetric studies show the weight loss at lower temperatures which arises due to the decomposition of impurities. Significant weight loss is observed in the temperature range of 400°C–700°C is due to the decomposition of CNTs. Remaining loss is due to the presence of catalyst and other elements.[6]

Similar results has also obtained when using neem oil as a natural precursor using the same spray pyrolysis method. It is inexpensive too. SEM images are in good agreement with the formation of ACNTs as before. XRD data also support the formation of CNTs. Peak at [002] indicates the cylindrical nature of graphene sheet and also the formation of multiwalled tubes. Peak at [101] also represents the hexagonal graphite. Interplanar spacing is also in good agreement with the TEM results. (JCPDS # 13-0148 Graphite, carbon). Raman spectroscopy can be used as another tool to confirm the formation of CNTs. On a range of 200–2000 cm^{-1}, two peaks are observed at 1348 and 1585 cm^{-1} which represent the D and G bands, respectively. G band corresponds to E_{2g} mode of pyrolytic graphite which confirms the presence of crystalline graphite in CNTs. D band arises due to the distortions in the tubes. Generally, G bands are much more intense than D band and indicate the high crystallinity.[20] Figure 11.9 (a) and (b) represents the XRD and Raman Spectra of CNTs prepared from neem oil.

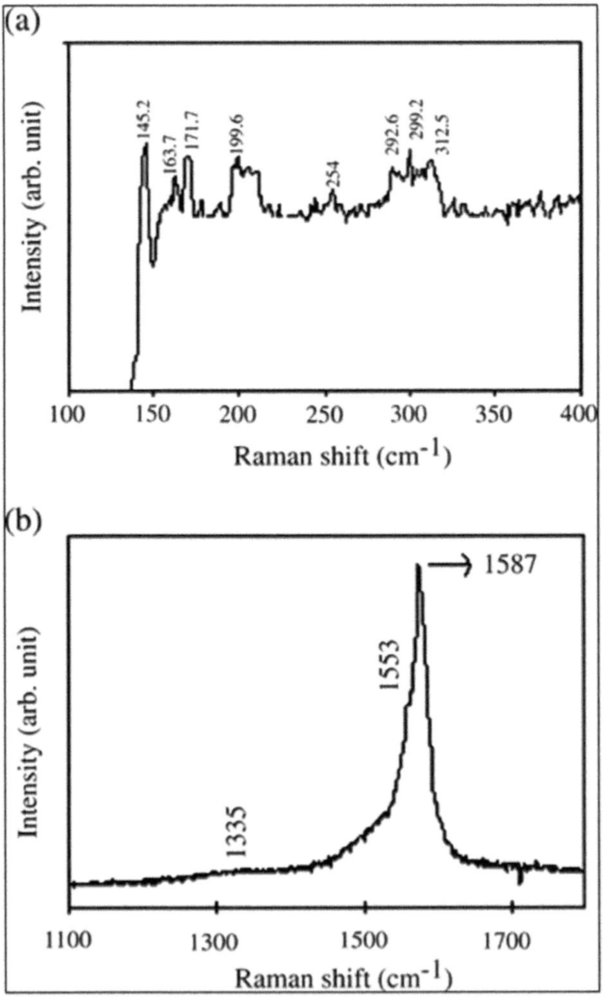

FIGURE 11.8 Typical Raman spectrum (green laser, wavelength = 532 nm) of the SWNTs grown at 850°C Ref. [21].

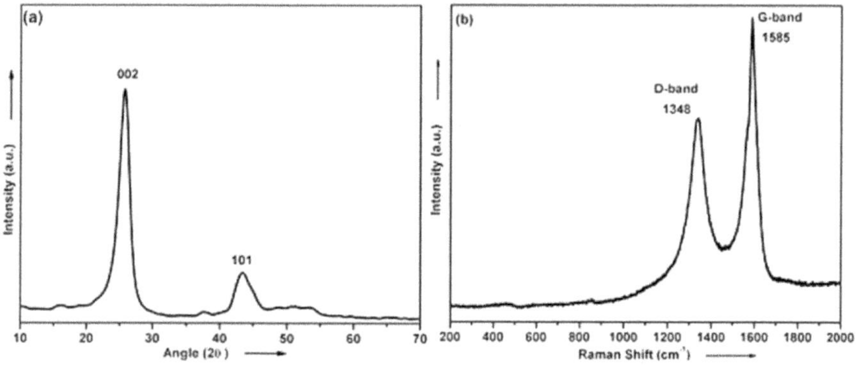

FIGURE 11.9 (a) XRD and (b) Raman spectra of the as-grown ACNTs at 825°C.
Source: Reprinted with permission from Ref. 23. © 2020 Elsevier.

Studies are also carried out using sunflower oil as feedstock. It shows that insertion of ammonia into sunflower oil followed by the spray pyrolysis gives bamboo-shaped CNTs. But quality is inferior to those produced from palm oil and neem oil. Another important triglyceride used is castor oil derived from castor beans. MWNTs with diameters 20–60 nm can effectively produce using this precursor. Here also, the insertion of ammonia will give bamboo-shaped CNTs. Studies show that microwave-assisted process will facilitate the transformation.

CNTs can also be synthesized from biodiesel. Biodiesel can be produced from plants. Biodiesel consists of monoester of fatty acids. Here also, spray pyrolysis method can be used for the synthesis of CNTs. Afre et al. reported that for better nucleation of CNTs, Co impregnated with silica gel can be used as the catalyst. But for better results, a combination of Fe and Co is used. Results have shown that MWNTs are produced when using biodiesel as feedstock.[21] Plasma-assisted CVD process can be used to prepare CNTs from tea tree extract which is a carbon-rich and abundant natural resource. Produced vertically aligned CNTs are having large surface area and possess unique functionalities.

Also, there are some eatable products from which CNTs can be synthesized. Plasma technique can be effectively used to synthesize CNTs from eatables like sugar, butter, cheese, and milk. The process does not require any catalyst at all. The same method can be adapted for the synthesis of CNTs from chicken eggs.

11.2.2 CNT SYNTHESIS FROM WASTE MATERIALS

Waste materials can be exploited for producing carbonized materials such as CNTs. The waste that can be used is divided into two categories. The first category includes low-valued waste materials mainly composed of carbohydrates which are naturally degraded. This includes food wastes, agricultural by-products, and animal/bird/insects wastes. These materials are having less impact on environment. Second category consists of nondegradable wastes which possess a large negative impact on the environment. They are not undergoing a small extent of degradation or sometimes no degradation at all. This category includes man-made materials like plastics and other polymeric products, industrial effluents, and municipal wastes. Conversion of these waste materials into carbon-based materials reduces the cost of production and minimizes environmental pollution too. Preparation of carbon-based materials from waste mainly consists of the following steps, Activation method, pyrolysis method and hydrothermal carbonization (HTC) method. The method will give carbon-based materials with high carbon content.[22]

11.2.2.1 CNT SYNTHESIS FROM AGRO-WASTES

CNTs can also be synthesized from agricultural wastes. Agricultural wastes consisting of biomolecules like cellulose, hemicellulose, and other complex organic polymers like lignin. These organic biomolecules are rich sources of carbon and used for the synthesis of CNTs. Rice by-products, sugarcane bagasse, cotton stalks can be used as the carbon precursors for the synthesis of carbon nanomaterials like CNTs, graphene, etc. Goodell et al. reported that CNTs with 10–20 nm can be effectively synthesized from wood fiber through oxidation process carried out at 240°C followed by cyclic oxidations at much more elevated temperatures under controlled oxygen supply. They also reported that the synthesis of CNTs from pure cellulose and lignin is limited. CNTs produced from wood fiber possess hexagonal arrangement of carbon atoms in the graphite sheets. Pyrolysis of sugarcane bagasse and corn grains also produces CNTs by the catalytic conversion of bioethanol by-products at temperature in the range of 600°C–1000°C. Sugarcane bagasse will produce longer MWNTs with length 50 μm and diameters 20–50 nm (Alvesa et al.). Corn grains will give rope-like structures of carbon nanotubes with diameters in the range 100–300 nm. Bamboo charcoal can also be used as a carbon feedstock to prepare CNTs through CVD method (Zhu et al.). Here, a catalyst like calcium silicate is employed to enhance the nucleation process. Recent studies also show the synthesis of CNTs from rice straw by hydrothermal treatment of rice straw with a transition metal catalyst like Fe or Ni to get a hydrochar which is used as the carbon source. Graphenated CNTs can be produced from rice husk waste by microwave plasma irradiation (MPI). Produced structures consist of graphene standing on the walls of CNTs.[23] Figure 11.10 represents the SEM images of CNTs produced from sugarcane bagasse. The SEM images show cylindrical nanomaterials on the surface of the catalyst. It is evident that the CNTs produced are long, well-separated and straight and have diameter in the range of 20–80 nm.[24]

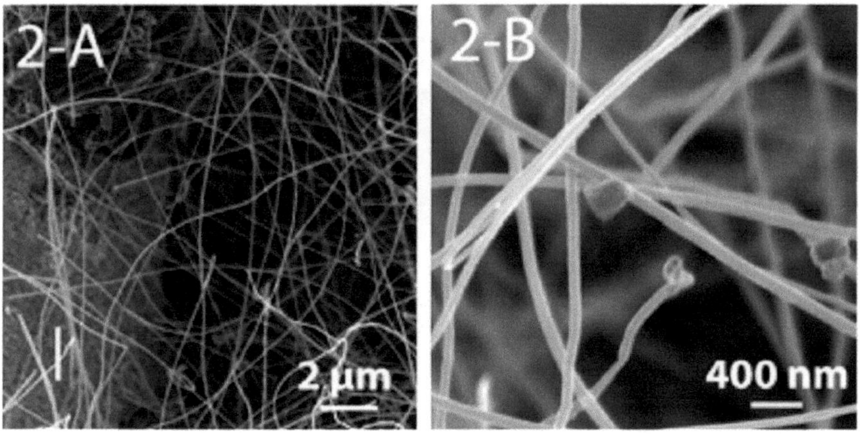

FIGURE 11.10 SEM images taken at low magnification (A) and high magnification (B) of typical CNMs synthesized at optimum operation conditions using carbon feedstock from waste sugar cane bagasse.

Source: Reprinted with permission from Ref. 27. © 2012 American Chemical Society.

11.2.2.2 CNT SYNTHESIS FROM AGRO-WASTES: CHICKEN FEATHER

Poultry feather is an abundant and inexpensive source of carbon and can be used for the synthesis of doped CNTs. Nowadays, poultry feathers are deposited in landfills without proper waste management. This will lead to environment problems. Efficient utilization of these waste materials can give astounding results. Chicken feathers can be effectively used for the synthesis of N-doped CNTs. Chicken feathers are good source of nitrogen too. In super critical carbon dioxide systems, chicken feathers are dissociated to simple molecules and the process resembles the CVD method. Here, when we use a transition metal catalyst like Ni, N-doped CNTs are formed. Usually, nickel acetate tetrahydrate is used as the precursor for Ni and when it enters the system, this will decompose to form Ni nanoparticles.[25] Figure 11.11 represents the whole process by which N-doped CNTs are prepared from chicken feathers.

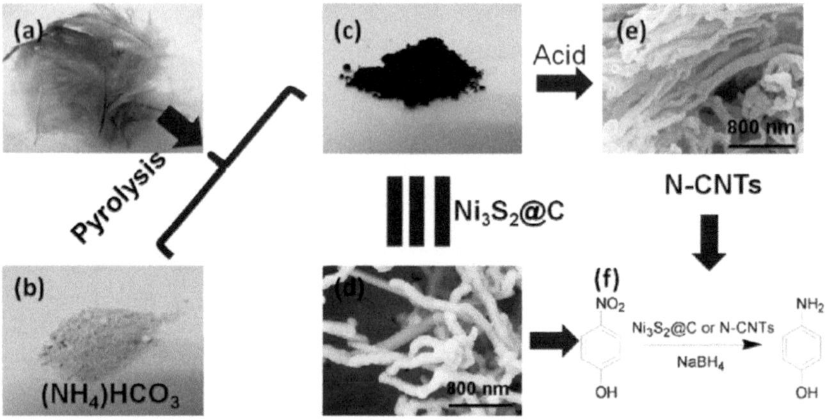

FIGURE 11.11 Schematic diagram of the total experimental process. Optical photos of the raw materials and products, (a) chicken feathers, (b) the white product formed at the top of the autoclave, (c) the black product formed at the bottom of the autoclave, and FESEM images of (d) the sample Ni3S2@C, (e) the sample N-CNTs, in which the Ni3S2 was removed, (f) the catalytic reaction of the 4-NP to 4-AP by $NaBH_4$ by the sample Ni3S2@ C or N-CNTs.

Source: Reprinted with permission from Ref. 28. © 2014 American Chemical Society.

11.2.3 CNTs FROM NATURAL RESOURCES

The low-cost bulk production of CNT has become an increasing demand in the field of nanotechnology. The preparation methods and the cost of the catalyst make the synthesis industrially less viable. The CNT production from natural resources is considered as cost-effective and greener. The natural source utilized for the synthesis of carbon nanotubes is coal and biomass. A novel thought in this category can be from the city gas with the help of a catalyst like garnet powder. Iron particles grown on the surface of sand enhances the growth of the crystalline CNTs. The most important advantage of this method is that the CNTs can be easily separated and there is very less or no usage of strong acids. The garnet powder requires an air oxidation at very high temperature and the city gas must be diluted with argon or hydrogen. Both these are mixed in a reaction tube and a heating of about 800°C[26] is allowed. After about an hour, crystalline CNTs were separated on the top layer.

Multiwall carbon nanotubes can be obtained from natural resources using extremely greener routes. The lava grown MWCNTs is an alternate thought in this regard. Lava can be used as a catalyst for deriving carbon nanotubes and fibers. The volcanic eruption of the famous Mount Etna provided lava in abundance for the process and the lava was usually rich in metallic ores. The Fe_2O_3 composition was about 11%. For the synthesis, the lava stone was crushed and powdered and put in the reactor in H_2 environment at high temperature of about 700°C. After the chemical vapor deposition process, carbon nanotubes with an inner diameter of about 10 nm were obtained.[27]

Coal is a worthy source for many carbon nanomaterials. Different varieties of carbon nanomaterials, such as nanotubes, nanoparticles, nanofibers, graphene and it derivatives can be obtained from coal. Similarly, coal serves as the most widely used natural source for obtaining single-walled and multiwalled nanotubes. Different techniques are utilized with coal and it can be with and without a catalyst. Arc-discharge method is one such and it reduces the usage of acids and other reagents.[28,29] High purity CNT samples can be obtained using the arc-discharge method and the SWCNTs exist as bundles or entanglements. The average diameter was between 10 and 20 nm. The Raman spectra confirmed the characteristic peaks of CNTs as G band at 1500 cm^{-1} and the D band at1349 cm^{-1}.[30] Instead of direct arc process, a coal/arc jet process where the coal is subjected to plasma arc can be employed for CNT production.[31] Qiu and coworkers examined and compared the CNTs from 11 caking coals using arc plasma technique and CNTs with similar properties as that of coal are obtained. The advantage of the method was that fullerene was obtained as by-product.[32] MWCNTs can be synthesized without using any catalysts by electric arc-discharge method with annealed bituminous coal electrode in the presence of iron and Ni–Y as catalysts.[33] The purity of the carbon nanotubes obtained from arc-discharge method depends on the concentration of the catalyst.[34] Carbon nanotubes with interesting surface morphology can be derived from the coal.[35] Y and T joined carbon nanotubes present interest in terms of morphology and applications. The arc discharge with copper as a catalyst can be utilized for the production of aforementioned branched morphology. The inner and the outer diameter of the branched CNTs were estimated to be nearly 40–50 and 50–60 nm, respectively.[36] On account of branch A that it would appear as a semi (or quasi) T-type cylinder somewhat, all projective angles were found to be obtuse. For branch B that is a normal Y-intersection tube, in which the projective angle between two upper branches is smaller than 90°, while the projective points between the lower pair of the branches are equivalent and assessed to be 137°. It ought to be noticed that by and large, the focal pits in all parts of BCNTs are empty what is more, ceaseless, however, at times, some BCNTs with bamboo-like branches can be seen.

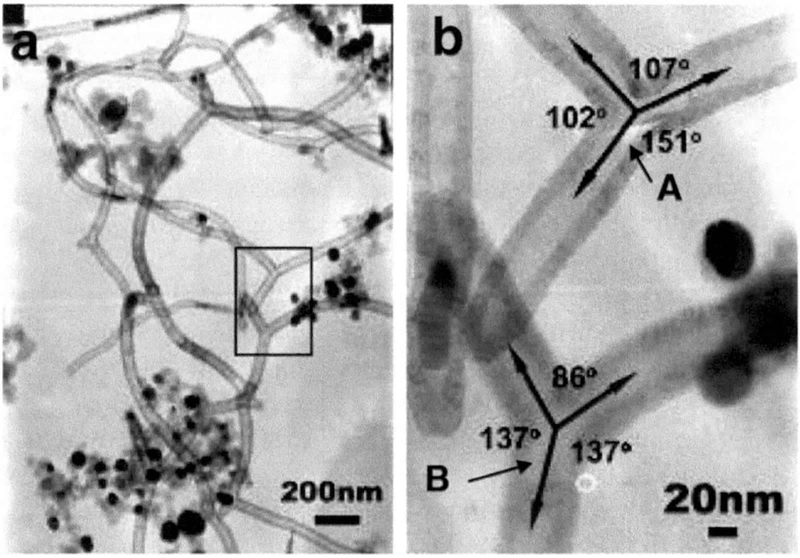

FIGURE 11.12 TEM images of branched single-walled carbon nanotubes.

Source: Reprinted with permission from Ref. 36. © 2006 Elsevier.

Other than the regular tubular shape of ordinary carbon nanotubes, they exist in some special morphology like helix, bamboo-shaped fish bone and chain. Among these, bamboo-shaped CNTs caught much attention. The carbon nanotubes that are synthesized from coal using arc-discharge method exhibited bamboo-shaped CNTs. In the bamboo CNTs, hollow compartments are seen that are separated by conical graphene layers. The hollow compartments are somewhat regular. The bamboo-shaped CNTs are important in the energy applications, especially in the batteries.[37,38]

Charcoal is another source of CNTs. The inorganic materials, especially the metals present in it are made use as catalyst for the nanotube production and radiofrequency plasma-assisted chemical vapor deposition process (RFCVD) can be used. Coconut shell charcoal was used to synthesize CNTs of diameter 25–30 nm using the RFCVD process. The coconut shell-derived charcoal avoided the metal impregnation process as it contained calcium, potassium, magnesium, and sodium.[39,40] Zhu and researchers reported a similar method of CNT synthesis and they have utilized bamboo charcoal. In this method,first, fresh bamboo was heated at above 1500°C and then the charcoal was produced. Bamboo charcoal was rich in magnesium silicate and calcium silicate, and these will enhance the growth of CNTs. During the heating or pyrolysis to obtain charcoal, the minerals present in the charcoal will form droplets. The carbon decomposes on the surface of the minerals becomes dissolved inside them and develop nanotubes (Table 11.2).[41]

TABLE 11.2 Summary of the Different CNTs Obtained From Coal As a Source.

Type of CNT	Type of coal	Catalyst used	Specifications of CNT	Yield	Refs.
SWCNT	Bituminous	Ni-Y	Diameter ~1 nm	45%	[32]
SWCNT	Bituminous	Zr/Ni	Diameter 1–2 nm	46 mg	[37]
SWCNT	Bituminous	Fe, Ni-Y	Diameter 1.2–1.7 nm		[36]
Bamboo-shaped CNTs	Anthracite	Fe	Diameter 40–60 nm		[41]
Bamboo-shaped CNTs	Bitumite	Ni-Sm_2O_3	Length several micrometers		[40]
MWCNT	Anthracite		Diameter 2–15 nm	8.17%	[42]
SWCNT	Anthracite	Fe	Diameter 1.24–2.19		[33]
Branched CNTs	Anthracite	CuO	Diameter 40–60 nm	<1%	[39]

11.2.4 GREEN CATALYTIC METHODS OF SYNTHESIS (PLANT EXTRACTS, FRUIT JUICES, SEED EXTRACTS ETC.)

Apart from the biomass, the carbon nanotubes can be synthesized using plant extract-mediated methods. Plant extracts have been utilized in the synthesis of a large number of nanoparticles and hybrid nanomaterials. In the case of carbon nanotubes also, they will act as catalyst for the growth. Even though the reports are very limited with carbon nanotubes, some noted literatures are available. Generally, the CNT synthesis from plants uses the oils or hydrocarbon present in it and was explained well in the previous section. In addition, plant extracts can be used as precursors for CNTs. *Moringa oleifera* esters can generate nanotubes at 600°C with an I_G/I_D ratio of about 0.84. The yield was about 61%.[43] Carbon nanotubes can be grown on different plant extracts and at lower temperatures. Garden grass, rose, neem and wallnut plant extracts can be utilized. The chemical vapor deposition at 575°C leads to the production of both aligned and horizontal CNTs. The yield was 0.0113 g with a G to D band ratio 0.59. As reported by Tripati et al., the walnut extract produced single-walled CNTs, whereas rose, neem, and garden grass gave MWCNTs.[44]

11.3 PROPERTY ANALYSIS OF CNTs FROM GREEN ROUTES

Carbon nanotubes are well-known for their superior properties. They typically have outstanding conducting and magnetic properties. They have high aspect ratio and so possess high mechanical strength. The synthetic route normally affects the morphology and properties of CNTs. It is stated by Yosof and coworkers that the carbon precursors play a key role in the growth characteristics and properties of CNTs because of their binding energy, type, and the role of reactive groups as well as thermodynamic properties.[45] But a detailed property analysis of virgin CNTs derived from greener routes has not been explored yet. The properties of results under consideration for CNTs from the aforementioned routes are the ratio of G band to D band represented as ($I_G/I_{D)}$ and the dimensions like length, inner and outer diameter for MWCNT (diameter for SWCNT), and aspect ratio.[46] The CNTs were added to different materials like transition metals, other carbon nanomaterials, such as graphene and added to polymers for hybrid and composite preparation, respectively. When carbon nanotubes are selected for specific applications like energy storage, the specific conductance and related properties will be studied. When hierarchical porous carbon nanotubes where analyzed for supercapacitor applications, it was found that the bulk resistance (R_s) of electrode, the charge-transfer resistance (R_{ct}) at the electrode/electrolyte interface were much smaller like 0.31 and 0.24 Ω, respectively. The reduced internal resistance of the

CNT electrode was responsible for the high-rate power delivery. Another important property required for conductive applications is the cyclic stability.[47]

11.4 CHARACTERIZATION TECHNIQUES USED FOR CNT FROM GREENER ROUTES

The carbon nanotubes from greener synthetic methods are characterized as normal CNTs. The main techniques practiced for property analysis are spectral studies like Raman spectra and FTIR, X-ray diffraction studies, microscopic analysis like scanning electron microscopy, transmission electron microscopy, etc. and mechanical property analysis. Spectral studies are mainly done to identify the structure of carbon nanotubes. FTIR spectra are often considered for the functionalization process to confirm the introduction of groups on the surface of the nanotubes. The Raman spectroscopy is an unavoidable tool for CNT characterization. It will decide the CNT formation and its structure. In the carbon nanotubes, the G band will be in the range of 1540 cm^{-1} and the deformation band or D band will be in the 1330 cm^{-1} range. The I_G/I_D which is the ratio of intensity of the G and D bands occurs in the Raman spectra. For high purity and defect-free carbon nanotubes, the value of the ratio should be less than 2%.[48] Morphology is another important aspect in the case of nanotubes. Transmission electron microscopy is more dedicated to explain this. The parameters, such as length and diameter are measured from TEM images.[49,50] The SEM and FESEM are used for the structural analysis (Figure 11.13).[51]

In the X-ray diffraction analysis, the graphitic material mainly displays two characteristic peaks at diffraction angles (2θ) of nearly 25 and 43 degrees. In this, the peak around 2θ of 25 degree correspond to the (002) plane and at around 43 stands for (100) plane. The development or growth, purity, and crystallinity or amorphous nature will be clear from the XRD images. The very small peaks or humps in the XRD pattern usually represent the impurities.[52] Electrochemical characterization is necessary as mentioned earlier for electronic and electrical applications. As a well-known fact, CNTs are used in capacitors, in conducting materials as fillers, in sensing, in energy storage devices, such as solar cells, lithium batteries and so on. The current voltage relationship (or cyclic response) is the tool to understand about the electrical behavior of CNTs.[53]

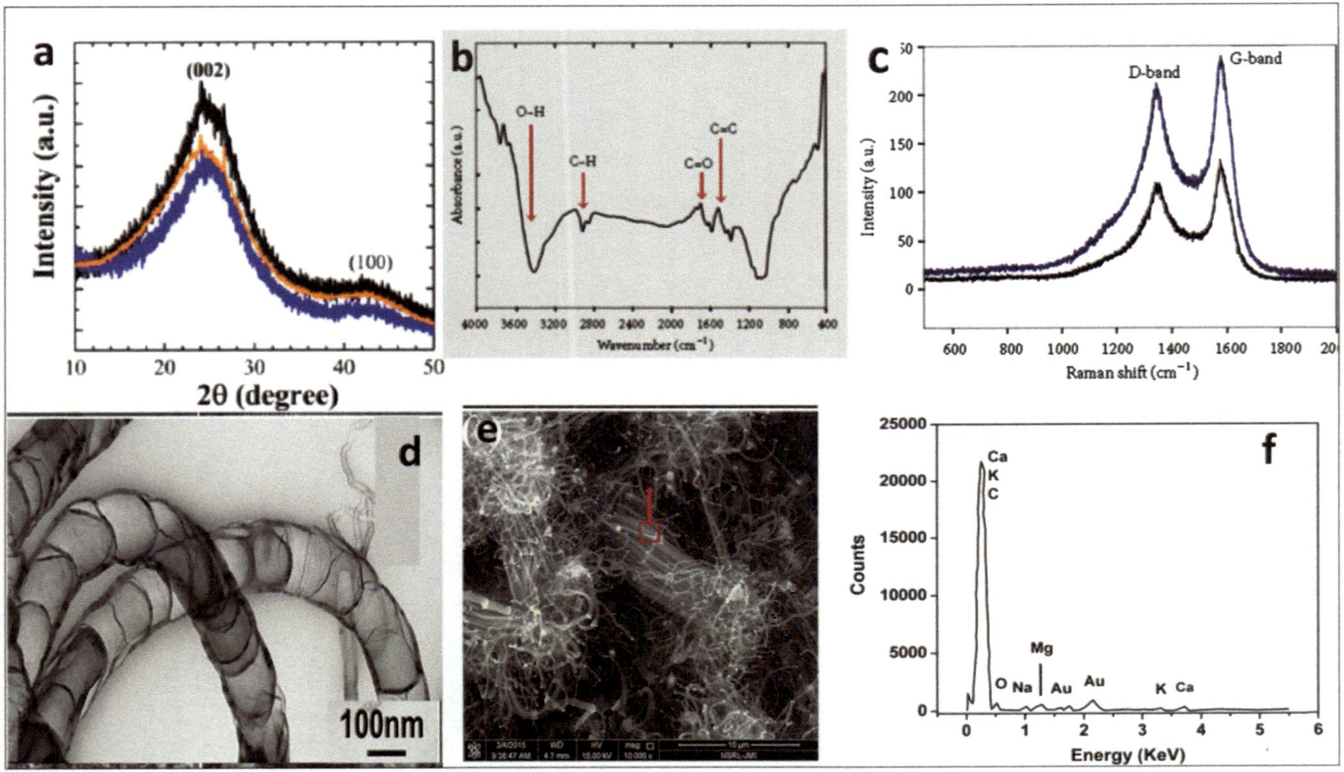

FIGURE 11.13 The results obtained for carbon nanotubes from different green synthetic routes. (a) XRD curves. (b) FTIR spectra. (c) Raman spectra of SWCNT and MWCNT from coconut shell charcoal. (d) HRTEM image of bamboo-shaped CNT. (e) FESEM image of CNT bundles using plant extract. (f) Elemental analysis of CNTs obtained from camphor oil.

11.5 CONCLUSION

This chapter reviewed the various methods for the environmental friendly synthesis of CNT. Compared with conventional chemical synthesis, green synthesis has got so many advantages which include eliminating the use of expensive chemicals, it produces less hazardous products exempting the requirement of high temperature, pressure, energy, and toxic chemicals, safer solvents and auxiliary substances, and economical and eco-friendly approach. Many plant-based oils were found to be a good source of hydrocarbon which is the precursor for the synthesis of CNT. Waste materials were exploited for the synthesis of CNT. Both agro and animal waste were utilized for the synthesis of CNT. Some natural resources like coal and biomass were found to be the effective starting material for the production of CNT. Apart from biomass, the carbon nanotubes can be synthesized using plant extract-mediated methods. CNTs prepared via these techniques have got excellent properties. So, green synthesis of CNT is an effective strategy to minimize waste and it thus saves our environment.

KEYWORDS

- **CNT**
- **green synthesis**
- **natural hydrocarbon resources**
- **agro-waste**

REFERENCES

1. Dai, H. Carbon Nanotubes: Synthesis, Integration, and Properties. *Acc. Chem. Res.* **2002,** *35* (12), 1035–1044.
2. Peigney, A.; Coquay, P.; Flahaut, E.; Vandenberghe, R. E.; De Grave, E.; Laurent, C. A Study of the Formation of Single-And Double-Walled Carbon Nanotubes by a CVD Method. *J. Phys. Chem. B* **2001,** *105* (40), 9699–9710.
3. Shah, K. A.; Tali, B. A. Synthesis of Carbon Nanotubes by Catalytic Chemical Vapour Deposition: A Review on Carbon Sources, Catalysts and Substrates. *Mater. Sci. Semiconduct. Process.* **2016,** *41*, 67–82.

4. Arora, N.; Sharma, N. N. Arc Discharge Synthesis of Carbon Nanotubes: Comprehensive Review. *Diamond Relat. Mater.* **2014,** *50*, 135–150.
5. Koziol, K.; Boskovic, B. O.; Yahya, N. Synthesis of Carbon Nanostructures by CVD Method. In *Carbon and Oxide Nanostructures*; Springer: Berlin, Heidelberg, 2010; pp 23–49.
6. Wang, Q.; Wang, X.; Chai, Z.; Hu, W. Low-Temperature Plasma Synthesis of Carbon Nanotubes and Graphene Based Materials and Their Fuel Cell Applications. *Chem. Soc. Rev.* **2013,** *42* (23), 8821–8834.
7. Tsai, S. H.; Chao, C. W.; Lee, C. L.; Shih, H. C. Bias-Enhanced Nucleation and Growth of the Aligned Carbon Nanotubes with Open Ends Under Microwave Plasma Synthesis. *Appl. Phys. Lett.* **1999,** *74* (23), 3462–3464.
8. Chrzanowska, J.; Hoffman, J.; Małolepszy, A.; Mazurkiewicz, M.; Kowalewski, T. A.; Szymanski, Z.; Stobinski, L. Synthesis of Carbon Nanotubes by the Laser Ablation Method: Effect of Laser Wavelength. *Phys. Status Solidi (B)* **2015,** *252* (8), 1860–1867.
9. Zahid, M. U.; Pervaiz, E.; Hussain, A.; Shahzad, M. I.; Niazi, M. B. K. Synthesis of Carbon Nanomaterials from Different Pyrolysis Techniques: A Review. *Mater. Res. Express* **2018,** *5* (5), 052002.
10. Hsu, W. K.; Hare, J. P.; Terrones, M.; Kroto, H. W.; Walton, D. R. M.; Harris, P. J. F. Condensed-Phase Nanotubes. *Nature* **1995,** *377* (6551), 687–687.
11. Schwandt, C.; Dimitrov, A. T.; Fray, D. J. High-Yield Synthesis of Multi-Walled Carbon Nanotubes from Graphite by Molten Salt Electrolysis. *Carbon, 50* (3), 1311–1315.
12. Novoselova, I. A.; Oliinyk, N. F.; Volkov, S. V.; Konchits, A. A.; Yanchuk, I. B.; Yefanov, V. S.; Karpets, M. V. Electrolytic Synthesis of Carbon Nanotubes from Carbon Dioxide in Molten Salts and Their Characterization. *Phys. E: Low-Dimensional Syst. Nanostruct.* **2008,** *40* (7), 2231–2237.
13. Tratnyek, P. G.; Johnson, R. L. Nanotechnologies for Environmental Cleanup. *Nano Today* **2006,** *1* (2), 44–48.
14. Kumar, R.; Singh, R. K.; Singh, D. P. Natural and Waste Hydrocarbon Precursors for the Synthesis of Carbon Based Nanomaterials: Graphene and CNTs. *Renew. Sustain. Energ. Rev.* **2016,** *58*, 976–1006.
15. Afre, R. A.; Soga, T.; Jimbo, T.; Kumar, M.; Ando, Y.; Sharon, M.; Umeno, M. Carbon Nanotubes by Spray Pyrolysis of Turpentine Oil at Different Temperatures and Their Studies. *Micropor. Mesopor. Mater.* **2006,** *96* (1–3), 184–190.
16. Awasthi, K.; Kumar, R.; Tiwari, R. S.; Srivastava, O. N. Large Scale Synthesis of Bundles of Aligned Carbon Nanotubes Using a Natural Precursor: Turpentine Oil. *J. Exp. Nanosci.* **2010,** *5* (6), 498–508.
17. Janas, D. From Bio to Nano: A Review of Sustainable Methods of Synthesis of Carbon Nanotubes. *Sustainability* **2020,** *12* (10), 4115.
18. Ghosh, P.; Afre, R. A.; Soga, T.; Jimbo, T. A Simple Method of Producing Single-Walled Carbon Nanotubes from a Natural Precursor: Eucalyptus Oil. *Mater. Lett.* **2007,** *61*(17), 3768–3770.
19. Maryam, M.; Suriani, A. B.; Shamsudin, M. S.; Rusop Mahmood, M. Synthesis of Carbon Nanotubes from Palm Oil Precursor by Aerosol-Assisted Catalytic CVD method. In *Applied Mechanics and Materials*, Vol. 229; Trans Tech Publications Ltd, 2012; pp 247–251.

20. Kumar, R.; Tiwari, R. S.; Srivastava, O. N. Scalable Synthesis of Aligned Carbon Nanotubes Bundles Using Green Natural Precursor: Neem Oil. *Nanoscale Res. Lett.* **2011,** *6* (1), 92.
21. Karthikeyan, S.; Mahalingam, P. Synthesis and Characterization of Multi-Walled Carbon Nanotubes from Biodiesel Oil: Green Nanotechnology Route. *Int. J. Green Nanotechnol. Phys. Chem.* **2010,** *2* (2), P39–P46.
22. Ma, Q.; Yu, Y.; Sindoro, M.; Fane, A. G.; Wang, R.; Zhang, H. Carbon-Based Functional Materials Derived from Waste for Water Remediation and Energy Storage. *Adv. Mater.* **2017,** *29* (13), 1605361.
23. Fathy, N. A.; Basta, A. H.; Lotfy, V. F. Novel Trends for Synthesis of Carbon Nanostructures from Agricultural Wastes. In *Carbon Nanomaterials for Agri-Food and Environmental Applications*; Elsevier, 2020; pp 59–74.
24. Zhuo, C.; Alves, J. O.; Tenorio, J. A.; Levendis, Y. A. Synthesis of Carbon Nanomaterials Through Up-Cycling Agricultural and Municipal Solid Wastes. *Indust. Eng. Chem. Res.* **2012,** *51* (7), 2922–2930.
25. Gao, L.; Li, R.; Sui, X.; Li, R.; Chen, C.; Chen, Q. Conversion of Chicken Feather Waste to N-doped Carbon Nanotubes for the Catalytic Reduction of 4-Nitrophenol. *Environ. Sci. Technol.* **2014,** *48* (17), 10191–10197.
26. Endo, M.; Takeuchi, K.; Kim, Y. A.; Park, K. C.; Ichiki, T.; Hayashi, T.; Dresselhaus, M. S. Simple Synthesis of Multiwalled Carbon Nanotubes from Natural Resources. *ChemSusChem* **2008,** *1* (10), 820–822.
27. Su, D. S.; Chen, X. W. Natural Lavas as Catalysts for Efficient Production of Carbon Nanotubes and Nanofibers. *Angewandte Chemie* **2007,** *119* (11), 1855–1856.
28. Mathur, R. B.; Lal, C.; Sharma, D. K. Catalyst-Free Carbon Nanotubes from Coal-Based Material. *Energ. Sources Part A* **2007,** *29* (1), 21–27.
29. Williams, K. A.; Tachibana, M.; Allen, J. L.; Grigorian, L.; Cheng, S. C.; Fang, S. L.; ... & Eklund, P. C. Single-Wall Carbon Nanotubes from Coal. *Chem. Phys. Lett.* **1999,** *310* (1–2), 31–37.
30. Qiu, J.; Li, Y.; Wang, Y.; Wang, T.; Zhao, Z. High-Purity Single-Wall Carbon Nanotubes Synthesized from Coal by Arc Discharge. *Carbon (New York, NY)* **2003,** *41* (11), 2170–2173.
31. Tian, Y.; Zhang, Y.; Wang, B.; Ji, W.; Zhang, Y.; Xie, K. Coal-Derived Carbon Nanotubes by Thermal Plasma Jet. *Carbon* **2004,** *42* (12–13), 2597–2601.
32. Qiu, J. S.; Zhang, F.; Zhou, Y.; Han, H. M.; Hu, D. S.; Tsang, S. C.; Harris, P. J. F. Carbon Nanomaterials from Eleven Caking Coals. *Fuel* **2002,** *81* (11–12), 1509–1514.
33. Awasthi, S.; Awasthi, K.; Ghosh, A. K.; Srivastava, S. K.; Srivastava, O. N. Formation of Single and Multi-Walled Carbon Nanotubes and Graphene from Indian Bituminous Coal. *Fuel* **2015,** *147*, 35–42.
34. Kumar, R.; Singh, R. K.; Ghosh, A. K.; Sen, R.; Srivastava, S. K.; Tiwari, R. S.; Srivastava, O. N. Synthesis of Coal-Derived Single-Walled Carbon Nanotube from Coal by Varying the Ratio of Zr/Ni as Bimetallic Catalyst. *J. Nanopart. Res.* **2013,** *15* (1), 1406.
35. Moothi, K.; Iyuke, S. E.; Meyyappan, M.; Falcon, R. Coal as a Carbon Source for Carbon Nanotube Synthesis. *Carbon* **2012,** *50* (8), 2679–2690.
36. Wang, Z.; Zhao, Z.; Qiu, J. Synthesis of Branched Carbon Nanotubes from Coal. *Carbon N Y* **2006,** *44*, 1321–1324.

37. Li, Y. F.; Qiu, J. S.; Zhao, Z. B.; Wang, T. H.; Wang, Y. P.; Li, W. Bamboo-Shaped Carbon Tubes from Coal. *Chem. Phys. Lett.* **2002,** *366*, 544–550.
38. Li, J.; Cao, Y.; Wang, L.; Jia, D. Cost-Effective Synthesis of Bamboo-Structure Carbon Nanotubes from Coal for Reversible Lithium Storage. *RSC Adv.* **2017,** *7*, 34770–34775.
39. Araga, R.; Sharma, C. S. One Step Direct Synthesis of Multiwalled Carbon Nanotubes from Coconut Shell Derived Charcoal. *Mater. Lett.* **2017,** *188*, 205–207.
40. Melati, A.; Hidayati, E. Synthesis and Characterization of Carbon Nanotube from Coconut Shells Activated Carbon. *J. Phys. Conf. Ser.***2016,** *694* (1), 012073).
41. Zhu, J.; Jia, J.; Kwong, F. L.; Ng, D. H. L.; Tjong, S. C. Synthesis of Multiwalled Carbon Nanotubes from Bamboo Charcoal and the Roles of Minerals on Their Growth. *Biomass Bioenerg.* **2012,** *36*, 12–19.
42. Qiu, J.; Li, Y.; Wang, Y.; Li, W. Production of Carbon Nanotubes from Coal. *Fuel Process Technol.* **2004,** *85*, 1663–1670.
43. Carbon Nanotubes from Plant Derived Hydrocarbon—An Efficient Renewable Precursor. *J. Environ. Nanotechnol.* **2018.**
44. Tripathi, N.; Pavelyev, V.; Islam S. S. Synthesis of Carbon Nanotubes Using Green Plant Extract as Catalyst: Unconventional Concept and Its Realization. *Appl. Nanosci.* **2017,** *7*, 557–566.
45. Abdullah, H. B.; Ramli, I.; Ismail, I.; Yusof, N. A. Hydrocarbon Sources for the Carbon Nanotubes Production by Chemical Vapour Deposition: A Review. *Pertanika J. Sci. Technol.* **2017,** *25*, 379–396.
46. Paul, S.; Samdarshi, S. K. A Green Precursor for Carbon Nanotube Synthesis. *New Carbon Mater.* **2011,** *26*, 85–88.
47. Storage, E.; Paul, S.; Samdarshi, S. K.; Tripathi, N.; Pavelyev, V.; Islam, S. S. Synthesis of Carbon Nanotubes Using Green Plant Extract as Catalyst: Unconventional Concept and Its Realization. *Appl. Nanosci.* **2017,** *7*, 557–566.
48. Ryzhkov, V. A. Carbon Nanotube Production by a Cracking of Liquid Hydrocarbons. *Phys. B Condens. .Matter* **2002,** *323*, 324–326.
49. Fathy, N. A.; Lotfy, V. F.; Basta, A. H. Comparative Study on the Performance of Carbon Nanotubes Prepared from Agro- and Xerogels as Carbon Supports. *J. Anal. Appl. Pyrolysis.* **2017,** *128*, 114–120.
50. Sarno, M.; Senatore, A.; Scarpa, D.; Cirillo, C. "Green" Synthesis of Nanocarbons for Reduced Friction and Wear. *Lubricants* **2020,** *8*.
51. Mehare, M. D.; Deshmukh, A.D.; Dhoble, S. J. Preparation of Porous Agro-Waste-Derived Carbon from Onion Peel for Supercapacitor Application. *J. Mater. Sci.* **2020,** *55*, 4213–4224.
52. Adhikari, M. P.; Adhikari, R.; Shrestha, R. G.; Rajendran, R.; Adhikari, L.; Bairi, P., et al. Nanoporous Activated Carbons Derived from Agro-Waste Corncob for Enhanced Electrochemical and Sensing Performance. *Bull. Chem. Soc. Jpn.* **2015,** *88*, 1108–1115.
53. Osman, A. I.; Farrell, C.; Al-Muhtaseb, A. H.; Harrison, J.; Rooney, D. W. The Production and Application of Carbon Nanomaterials from High Alkali Silicate Herbaceous Biomass. *Sci. Rep.* **2020,** *10*, 1–13.

Index